An Introduction to
Engineering Fluid Mechanics

Other Macmillan titles of related interest

An Introduction to Engineering Fluid Mechanics

J. A. FOX

Department of Civil Engineering
University of Leeds

MACMILLAN

First published 1974 by
THE MACMILLAN PRESS LTD
London and Basingstoke
Associated companies in New York Dublin
Melbourne Johannesburg and Madras

SBN 333 14587 9

Set in IBM Press Roman and printed by
William Clowes & Sons Ltd., London, Colchester & Beccles

Contents

Contents

Contents

Contents

Preface

This book is for undergraduates and HNC/HND students in both civil and mechanical engineering. The accent throughout has been placed upon the engineering aspects of the subject but it is hoped that the more mathematically minded reader will find sufficient to interest him.

Assumptions upon which analyses are based have been carefully specified. Any analysis is only as accurate as its underlying assumptions and so the reader should develop the habit of assessing the value of a piece of theory by considering the applicability of its assumptions in the context of the problem under examination.

Both engineers and mathematicians have contributed to the study of fluid mechanics and of recent years there has been a marked tendency to use mathematical methods in place of the empiricism that was used in the past. I believe that this trend will continue and academic courses will become progressively more mathematical in their approach.

The systems of units that have been used are the British system and the SI system. Even after the SI system has been completely introduced in the UK and Europe, the British system will still be used in many areas of the world. It will therefore be necessary for British engineers designing projects in these areas to know both systems.

At the end of each chapter questions have been included which it is hoped will be of assistance in understanding the chapter. They are set in both systems of units, the SI values being enclosed in square brackets. Some questions come from examination papers of the University of London, the University of Leeds and the Part II hydraulics examinations of the Institution of Civil Engineers, and I gratefully acknowledge permission to use them; others have been evolved for this book. The answers supplied are of course my own.

The subject is very large and it is not possible to cover every topic in detail. The student will need to read further and a list of suggested reading is included.

I would like to thank Mr J. Higgins, of the Faculty of Applied Science, the University of Leeds, who prepared the drawings.

J. A. Fox

Department of Civil Engineering,
University of Leeds

List of Principal Symbols

As far as possible all symbols used have been defined in the text as they occur, so any ambiguities arising out of the use of the same symbol to denote different variables can be easily resolved by reference to the text. The dimensions are given in parentheses.

a and A area of flow (L^2)

A constant in the pump characteristic equation (Chapter 10) (LT^2)

α angle (dimensionless)

constant in the equation that describes the variation of μ with temperature $(\theta^{-1},$ where θ is the dimension of temperature)

energy coefficient (dimensionless)

b and B breadth of a channel (L)

surface breadth of a channel (L)

b breadth of a lamina (L)

B constant in the pump characteristic equation (Chapter 10)

breadth of a runner or impeller in the axial direction (L)

β the momentum coefficient (dimensionless)

constant in the equation that describes how μ varies with temperature (θ^{-2})

exit angle of a moving blade in a rotadynamic machine (dimensionless)

c or C general constant

c wave velocity (Chapters 6, 8 and 9) (LT^{-1})

C_d coefficient of discharge of an orifice, notch or weir (dimensionless)

C_D coefficient of drag (dimensionless)

C_v coefficient of velocity of an orifice (dimensionless)

C_c coefficient of contraction of an orifice (dimensionless)

C_l coefficient of lift (dimensionless)

C Chezy constant in free-surface flows $(L^{1/2}T^{-1})$

constant in the pump equation (Chapter 10) (T^2L^{-5})

d depth of a flow (L)

pipe diameter (L)

d_s height of opening below a sluice gate (Chapter 7) (L)

d_c critical depth of a free-surface flow (L)

δ boundary layer thickness (L)

δ' laminar sublayer thickness (L)

δ^* boundary layer displacement thickness (L)

δ mean depth of a flow (Chapter 7) (L)

δ small increment or decrement of a variable

Δ a finite difference

E specific energy (energy of flow referred to bed of flow) (L)

Young's modulus of the pipe wall material (Chapter 8) $(ML^{-1}T^{-2})$

E_h ideal or hydraulic efficiency (dimensionless)

E_0 overall efficiency (dimensionless)

E_m mechanical efficiency (dimensionless)

e exponential constant (2.71826) (dimensionless)

η fractional valve opening (dimensionless)

Fn Froude number (dimensionless)

f frequency of a vortex trail (T^{-1})

Darcy friction coefficient (T^{-1})

F force (MLT^{-2})

strength of a doublet (L^2T^{-1})

ϕ velocity potential (L^2T^{-1})

List of Principal Symbols

g	acceleration due to gravity (LT^{-2})
Γ	circulation around a fluid boundary (L^2T^{-1})
γ	angle of shear deformation in a solid (dimensionless)
h	height of a rectangular lamina (L)
	potential head (pressure head plus height above datum) (L)
	height of a bed hump (L)
h_f	energy loss/unit weight due to friction (L)
h_i	inertia head (L)
h_m	level difference between the two menisci of a manometer (L)
H	total energy/unit of fluid referred to a horizontal datum (L)
	horizontal component of force acting on a lamina (MLT^{-2})
H_s	static head (L)
I	second moment of area (L^4)
i	bed slope of a channel (dimensionless)
	hydraulic gradient (dimensionless)
j	energy loss/unit weight/unit length of a channel flow (dimensionless)
k	constant
	roughness number, Manning's, Kutta's or Strickler's (dimensionless)
k_g	radius of gyration (L)
K	bulk modulus of a fluid ($ML^{-1}T^{-2}$)
l	length of pipe (L)
	distance along a channel (L)
L	lift force (MLT^{-2})
λ	kineticity of a flow (Chapter 7) (dimensionless)
m	hydraulic mean radius (in pipe flow) (L)
	hydraulic mean depth (in free-surface flows) (L)
M	total momentum flow rate (MLT^{-1})
μ	coefficient of dynamic viscosity ($ML^{-1}T^{-1}$)
n	index (dimension variable)
	roughness number
N	rotational speed of a pump or turbine (T^{-1})
N_s	specific speed of a pump or turbine (dimensionless)
ν	coefficient of kinematic viscosity (L^2T^{-1})
Ω	angular velocity of a forced vortex, a rotodynamic machine or a fluid element (T^{-1})
p	pressure ($ML^{-1}T^{-2}$)
P	force (MLT^{-2})
	wetted perimeter of a flow channel (L)
	total force acting on a curved lamina (MLT^{-2})
π	constant (3.14159) (dimensionless)
π	dimensionless groups
ψ	stream function (L^2T^{-1})
q	flow per unit width of a rectangular channel (L^2T^{-1})
	flow leaving per unit length of a pipe (L^2T^{-1})
	flow entering or leaving a channel per unit length (Chapter 7) (L^2T^{-1})
Q	flow (L^3T^{-1})
r	radius (L)
R	radius (L)
	area ratio (Chapter 9) (dimensionless)
Re	Reynolds number (dimensionless)
ρ	mass density (ML^{-3})
	pipe characteristic (Chapter 8) (ML^{-3})
s	specific gravity of a fluid (dimensionless)
	distance (L)
	slope of sides of a channel (dimensionless)
S	specific force of a free surface flow (L^2)
σ	coefficient of surface tension (MT^{-2})
	Poisson's ratio (Chapter 8) (dimensionless)

t	time (T)
	thickness (L)
T	pipe period (T)
τ	viscous shear stress ($ML^{-1}T^{-2}$)
τ_0	viscous shear at a boundary ($ML^{-1}T^{-2}$)
θ	angle (dimensionless)
u	local velocity (LT^{-1})
	velocity component in the x direction (LT^{-1})
	peripheral velocity of a rotating blade (LT^{-1})
u'	time varying component of velocity in x direction in turbulent flow (LT^{-1})
$\bar{u}$	steady component of velocity in x direction in turbulent flow (LT^{-1})
U	undisturbed stream velocity (LT^{-1})
v	stream velocity (LT^{-1})
	velocity component in the y direction (LT^{-1})
v'	time varying velocity component in the y direction in a turbulent flow (LT^{-1})
$\bar{v}$	mean velocity of a flow (LT^{-1})
V	velocity of a fluid particle on the centreline of a pipe velocity (LT^{-1})
	vertical component of a force acting upon a lamina (MLT^{-2})
V^*	shear velocity (LT^{-1})
V_w	velocity of a surge wave in a free-surface flow (LT^{-1})
	velocity of whirl (Chapter 10) (LT^{-1})
V_f	velocity of flow (Chapter 10) (LT^{-1})
V_r	relative velocity (Chapter 10) (LT^{-1})
w	specific weight of fluid ($ML^{-2}T^{-2}$)
	component of velocity in the z direction (MLT^{-1})
w'	time varying velocity component in z direction in turbulent flow (LT^{-1})
$\bar{w}$	steady velocity component in z direction in turbulent flow (LT^{-1})
W	weight of a fluid body (MLT^{-2})
	weight flow (MLT^{-3})
x	distance (L)
X	body force acting in the x direction (MLT^{-2})
$\bar{x}$	distance along a lamina from its point of intersection with the free surface to the centroid (L)
y	distance, usually in the direction of the earth's gravity field (L)
	distance from a pipe wall (L)
Y	body force acting in the y direction (MLT^{-2})
z	elevation of a point above a datum (L)
	distance down from the surface of an area element (L)
$\bar{z}$	depth below the centroid of the cross sectional area of a flow (L)
ξ	vorticity (T^{-1})

1 Definitions and Hydrostatics

From the viewpoint of a man living on the surface of the earth, the world appears to be a ball of rock covered with an envelope of air, with water forming the oceans. This superficial view might lead him to believe that, similarly, the universe consists of solid bodies. But this is of course not so. The earth has a solid crust but its core probably consists of molten iron, while the stars are balls of very hot plasma and the galaxies appear to rotate as vortices. Where galaxies collide the pattern of star streams observed closely resembles turbulent fluid flow. Solid bodies are found only on the surface of planets, in asteroids and in the small-scale debris of planetary space.

The description of fluid motion is thus clearly a matter of great interest and importance. Much of physics and mathematics is concerned with the study of fluid properties and motion, and in engineering where these disciplines are applied the study of fluid mechanics is vital.

1.1 Basic definitions

Fluid mechanics may be defined as the study of the stresses and velocities that occur in a fluid in motion; a fluid at rest is regarded as a special case of this general study. This definition leaves open the question of what precisely is meant by a fluid. The statement that a fluid is a substance that is not solid, although it includes both gases and liquids, is not adequate since fluids exist which under the action of small shear stresses behave as solids but under large stresses behave as liquids. Such substances are usually complex organic compounds with large molecules and are frequently colloidal.

Liquids and gases

A liquid is a substance that under suitable conditions of temperature will in time deform to take up the shape of any container into which it is placed. A free surface will develop if the container is located in a gravity field.

1

A gas is a substance that will deform and expand to occupy the entire volume of any container in which it is placed without developing a free surface.

An ideal or perfect fluid is one in which the viscosity (defined below) is zero. Needless to say, no such fluids exist.

1.2 Viscosity

Real fluids develop shear stresses within themselves when they move, these stresses being dependent on the rate of shear deformation. In this respect they differ from solids, in which shear stresses depend on the shear deformation itself. The relationship between the shear stress and the rate of shear can be used to classify various types of fluid. The shear stress τ can be related to the shear rate $\dot{\gamma}$ by the equation

$$\tau = f(\dot{\gamma})$$

where $\dot{\gamma}$ denotes the rate at which a fluid element is being deformed and f means 'function of'. The simplest form of this equation is

$$\tau = \mu \dot{\gamma}$$

where μ is a constant. This defines a linear relationship between shear stress and shear rate. This particular relationship was first suggested by Newton and it has been found to describe very accurately the behaviour of many fluids including most gases, water, many lubricating oils and the lower hydrocarbons. For each fluid there is a particular value for the constant μ at a given temperature. μ is called the coefficient of dynamic viscosity and is a true constant for fluids that obey this linear relationship. Such fluids are called newtonian. $\dot{\gamma}$, the rate of shear deformation, will later be shown to be numerically equal to $\partial u/\partial y$ for a uni-dimensional flow parallel to the x-axis in a cartesian coordinate system (Section 2.3). The parameter μ is not dependent upon $\dot{\gamma}$ for newtonian fluids but it does vary with temperature. In liquids its value decreases with increasing temperature and in gases it increases with increasing temperature. The difference in the behaviour of liquids and gases can be explained as follows. A gas is a collection of molecules that are so far apart that molecular forces of attraction are negligibly small. Their motion is determined by their momenta and to a first approximation they may be imagined as infinitely small spheres in random motion. If the gas is in motion a translational velocity is imposed upon these random motions; that is the random movements result in a net movement in space of the entire collection of molecules in the direction of the velocity. In other words the translational velocity is the mean velocity of the collection. If adjacent layers of gas move at different velocities, the random motions of individual molecules will cause transfer of molecules between the layers. The higher the temperature the greater the randomly directed velocities will be and thus the rate of transfer of molecules from one layer to the other will increase. Molecules moving between the two layers will carry momentum with them, the net effect being to transfer momentum from the fast layer to the slow layer. The rate of transfer of momentum is equivalent to a shear stress so that as the rate of transfer increases with temperature the equivalent shear stress will also increase with temperature, resulting in an increase in the coefficient of viscosity.

In a liquid the molecules are much more closely packed and any given molecule is within the zone of attraction of the surrounding molecules. When adjacent layers of a liquid move relative to one another the attractive bonds between the molecules in the layers must be broken and reformed continuously. This explains the relation between the shear stress and the shear rate. If the temperature is increased the spacing between the molecules increases as the liquid expands and consequently the attractive forces between them weaken. Hence the shear force and the coefficient of dynamic viscosity decrease with increasing temperature.

Generally the behaviour of the coefficient of viscosity of a liquid is described by the equation

$$\mu = \mu_0/(1 + \alpha t + \beta t^2)$$

and for a gas by

$$\mu = \mu_0 + \alpha t - \beta t^2$$

For water; in the cgs system μ is in poise and

$$\mu_0 = 17 \cdot 90 \times 10^{-3}$$
$$\alpha = 33 \cdot 68 \times 10^{-3}$$
$$\beta = 22 \cdot 1 \times 10^{-3}$$

In the Imperial system, μ is in slugs/foot second and

$$\mu_0 = 37 \cdot 5 \times 10^{-6}$$
$$\alpha = 33 \cdot 68 \times 10^{-3}$$
$$\beta = 2 \cdot 21 \times 10^{-3}$$

In the SI system, μ is in kilograms/metre second and

$$\mu_0 = 1 \cdot 79 \times 10^{-3}$$
$$\alpha = 33 \cdot 68 \times 10^{-3}$$
$$\beta = 2 \cdot 21 \times 10^{-3}$$

For air, if t is in degrees centigrade

$$\mu = 355\,300 + 1680t - 2 \cdot 48t^2 \text{ slugs/ft s}$$

or

$$\mu = (17\,600 + 83 \cdot 1t - 0 \cdot 123t^2)10^{-3} \text{ kg/m s}$$

1.3 Non-newtonian fluids

As mentioned earlier, other fluids exist for which the simple newtonian equation relating the shear stress τ to the shear rate $\dot{\gamma}$ does not apply.

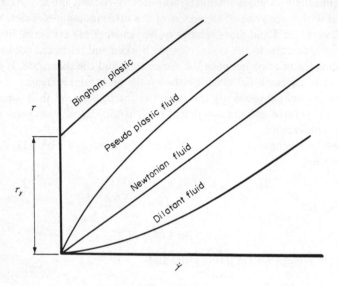

Fig. 1.1

The main classes of non-newtonian fluids as illustrated in Fig. 1.1 are

(i)   Plastics, in which Bingham plastics are included as a special case
(ii) Pseudoplastics
(iii) Dilatants

Plastics are different from the other two classes in that they possess a yield stress which must be exceeded before flow can commence. Substances such as tomato ketchup, butter and many other organic colloids are such fluids. Blood has a very low yield stress and so must be classed as a plastic. For all these fluids the shear stress is uniquely defined by the shear rate but other fluids exist in which the shear stress depends not only upon the shear rate but also upon the shear history of the fluid. Such fluids are called thixotropes. Fluid mechanics is the study of newtonian fluids and rheology the study of all fluids. On this definition fluid mechanics is a subdivision of rheology. Non-newtonian fluids are very interesting and exhibit many unusual properties but their behaviour cannot be dealt with in this book for reasons of space.

1.4 Specific mass, weight and gravity

The specific mass of a fluid is the mass of fluid contained within a unit volume. It is usually denoted by the symbol ρ (rho). The value of the specific mass of a gas depends of course on its pressure but that of a liquid can be considered constant for most purposes. When pressure transients are caused by sudden or rapid velocity changes in a flow the assumption of constant density cannot be made but in steady liquid flows it is justified in almost all circumstances. The specific mass units in common use are

SI	kilograms/metre^3
cgs	grams/centimetre^3
British physicists	pounds/foot^3
British and American engineers	slugs/foot^3

The specific weight of a fluid in the weight of the fluid per unit volume and in this book it will be denoted by w. The units in common use are

SI	newtons/metre^3
cgs	dynes/centimetre^3
British physicists	poundals/foot^3
British and American engineers	pounds force/foot^3

The word slug was used above to define the mass unit in the British engineers' system. The reason for the introduction of this unit is as follows. It is essential for engineers to develop an intuitive appreciation of the magnitude of the forces, and velocities that arise in any problem they are attempting to solve. Every person has an intuitive awareness of the magnitude of the force unit to which they are accustomed, the pound weight, the kilogram weight or whatever it might be. In most of the English speaking world the unit of force in everyday use is the pound weight and most people including engineers are instinctively aware of the magnitude of this force. Very few people have any feeling for the magnitude of forces measured in poundals (one poundal is approximately one half of an ounce weight). Engineers must be able to judge the accuracy of an answer to a problem and this can be difficult if it is expressed in strange units. Consider Newton's second law which is the basis of all calculations concerning dynamic phenomena: $F = ma$. F is the force, m is the mass of the body and a is the acceleration. If mass is in pounds mass and acceleration in ft/s^2 the force will be in poundals. If it is required that forces are to be in pounds force (lbf) and accelerations in ft/s^2, then because the force unit is the numerical value of g times as large as the poundal, the mass unit must be as many times as large as the pound mass. Denote the numerical value of g by $|g|$ then 1 slug = $|g|$ lb mass.

For example, in the case of water $\rho = 62 \cdot 4$ lb/ft^3 and this is the correct unit to use in the physicists' system. In the engineers' system $\rho = 62 \cdot 4/32 \cdot 2$ slug/ft^3 = $1 \cdot 94$ slug/ft^3 and the specific weight $w = \rho g = 1 \cdot 94 \times 32 \cdot 2 = 62 \cdot 4$ lbf/ft^3.

As stated above, in the SI system the unit of mass density is the kilogram per cubic metre. In the cgs system water weighs 1 gram weight per cubic centimetre so the weight of 1 cubic metre of water is 1 000 000 grams weight and the specific mass of 1 cubic metre of water is therefore 1000 kilograms per cubic metre. The corresponding specific weight of water is ρg, that is, 9810 newtons/metre³.

The specific gravity (s) of a fluid is the ratio of the weight of a volume of the fluid to the weight of an equal volume of water.

$$s = w_{\text{fluid}}/w_{\text{water}}$$

Water, by definition, has a specific gravity of unity so that in the cgs system the specific gravity of water equals the numerical value of its specific weight.

1.5 Pressure at a point in a fluid

In a stationary newtonian fluid no shear stresses can exist as the fluid is not undergoing a shearing process. Pressures or direct stresses do occur however. Consider a small element of area δa in the fluid. Let the force acting across it be δp. Then the pressure on the area is $\delta p/\delta a$. If the element is made extremely small and taken to the limit the pressure at the point is given by $\lim_{\delta a \to 0} \delta p/\delta a$. The units of pressure are dynes/cm² in the cgs system, N/m² in the SI system, pdl/ft² in the British physicists' system and lbf/ft² in the British and American engineers' system. These are all basic fundamental units but in practice are not very well suited to everyday use as they are all too small. The units commonly used are 1bf/in² and kilograms wt/cm².

Proof that pressure acts equally in all directions

At a point in a fluid the pressure acts equally in all directions. This statement can be proved as follows. Consider the forces acting upon a triangular prism of unit

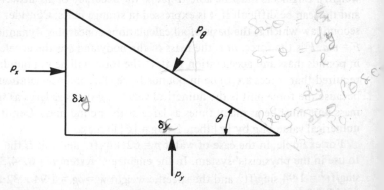

Fig. 1.2

length (see Fig. 1.2). Resolving forces in the x direction

$$p_x \, \delta y = \left(p_\theta \, \frac{\delta y}{\sin \theta} \right) \sin \theta$$

$$\therefore \qquad p_x = p_\theta$$

Resolving vertically (remembering to include the weight of the fluid)

$$p_y \, \delta x = \left(p_\theta \, \frac{\delta x}{\cos \theta} \right) \cos \theta + \tfrac{1}{2} \, \delta x \, \delta y \, w$$

When δx and δy become very small $p_y = p_\theta$.

Thus at a point in a fluid $p_x = p_y = p_\theta$ and as θ can take any value this proves that pressures act equally in all directions in a stationary fluid.

1.6 Pressure distribution in the atmosphere

Consider a small cylindrical element of gas with its axis vertical at a height h above a given level. Let its cross sectional area be A and its thickness δh (see Fig. 1.3).

Let the pressure acting on its lower horizontal surface be p, then the pressure on its upper horizontal surface will be $p + (\mathrm{d}p/\mathrm{d}h) \, \delta h$. The net upward vertical force acting on it will be $-(\mathrm{d}p/\mathrm{d}h) \, \delta h \, A$ and this will have to balance its weight which will be $A\rho g \, \delta h$.

$$\therefore \qquad -(\mathrm{d}p/\mathrm{d}h) \, \delta h \, A = A\rho g \, \delta h$$

so

$$\mathrm{d}p/\mathrm{d}h = -\rho g$$

Using the universal gas law, $p/\rho = gRT$ (R is a constant and T the absolute temperature) and assuming that the gas temperature is everywhere constant (that is, the process is isothermal) then by substituting

$$\rho = p/gRT$$

$$\mathrm{d}p/\mathrm{d}h = -(p/gRT)g = -p/RT$$

$$\therefore \qquad \mathrm{d}p/p = \mathrm{d}h/RT$$

Integrating

$$\log_e p = -h/RT + B$$

where B is a constant of integration.

When $h = 0$, let the pressure p be p_0, so that $\log_e p_0 = B$, then

$$\log_e p = -h/RT + \log_e p_0$$

$$\log_e (p/p_0) = -h/RT$$

If the gas process is not isothermal, but adiabatic, then the above result does not apply. The equation $dp/dh = -\rho g$ is still valid but the adiabatic equation $p/\rho^\gamma = k$ (where k is a constant) will apply instead. By substituting for ρ from $p/\rho^\gamma = k$ into the first equation

$$dp = -(p/k)^{1/\gamma} g \, dh$$

$$\therefore \qquad dp/p^{1/\gamma} = -(1/k)^{1/\gamma} g \, dh$$

Integrating

$$\frac{\gamma}{\gamma - 1} p^{(\gamma-1)/\gamma} = -(1/k)^{1/\gamma} gh + C$$

where C is a constant of integration. When $h = 0$, $p = p_0$

$$\therefore \qquad \frac{\gamma}{\gamma - 1} p_0^{(\gamma-1)/\gamma} = C$$

$$\therefore \qquad \frac{\gamma}{\gamma - 1} (p^{(\gamma-1)/\gamma} - p_0^{(\gamma-1)/\gamma}) = -(1/k)^{1/\gamma} gh$$

but

$$(1/k)^{1/\gamma} = \rho_0/p_0^{1/\gamma} = \rho/p^{1/\gamma}$$

$$\therefore \qquad \frac{\gamma}{\gamma - 1} \left(\frac{p_0}{w_0} - \frac{p}{w} \right) = h$$

Alternatively

$$\frac{\gamma}{\gamma - 1} \frac{p_0}{w_0} \left[1 - \left(\frac{p}{p_0} \right)^{(\gamma-1)/\gamma} \right] = h$$

In the atmosphere, approximations to both isothermal and adiabatic processes occur. In the troposphere, the process most closely approximates to an adiabatic process. Such a case is said to be one of convective equilibrium. In the stratosphere, conditions approximate to an isothermal equilibrium and such a case is said to be one of conductive equilibrium.

Convective equilibrium is said to occur when two equal masses of gas can be exchanged between two different levels and they exchange pressures and temperatures without the addition or subtraction of heat.

1.7 Hydrostatic pressures in incompressible fluids

At the base of a column of fluid of cross sectional area A (see Fig. 1.3) the pressure force pA must act upwards to balance the weight of the column. The weight of the column is $\int_0^h A\rho_x g \, dx$ where ρ_x denotes the mass per unit volume (or specific mass) of the fluid in the elemental volume at height x above the base

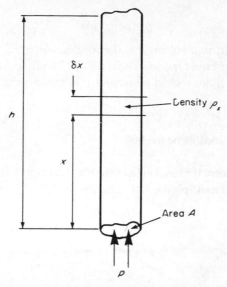

Fig. 1.3

of the column. If the fluid is incompressible the mass density does not change with the pressure, so $\rho_x = \rho$. The column weight = ρgAh where h is the total column height up to the free liquid surface.

Therefore

$$pA = \rho gAh$$

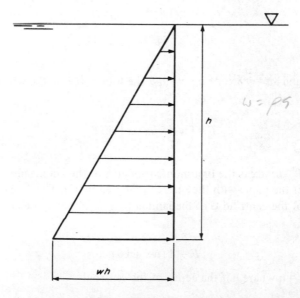

Fig. 1.4

so

$$p = \rho g h = wh$$

where w is the weight/unit volume, i.e. the specific weight.

This means that pressure increases linearly with depth in a liquid (see Fig. 1.4), but in compressible fluids such as gases, pressures vary logarithmically with depth.

1.8 Force on an inclined plane lamina

The force on an elemental plane lamina (see Fig. 1.5) equals the pressure at the depth of the element multiplied by its area, that is

$$\delta F = p \delta a$$

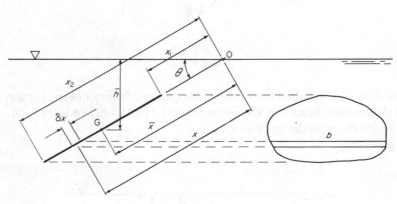

Fig. 1.5

Now $p = wh$ and $\delta a = b \, \delta x$. As $h = x \sin \theta$ the total force on the whole lamina is given by

$$F = \left(w \int_{x_1}^{x_2} bx \, \mathrm{d}x \right) \sin \theta \tag{1.1}$$

The integral $\int_{x_1}^{x_2} bx \, \mathrm{d}x$ is the first moment of area of the lamina about the line of intersection of the plane with the surface (denoted by O). This is $A\bar{x}$ where $\bar{x}$ is the distance of the centroid G of the lamina from O, A being the area of the lamina.

Therefore

$$F = A (w\bar{x} \sin \theta)$$

and as $\bar{x} \sin \theta = \bar{h}$ where $\bar{h}$ is the depth of the centroid

$$F = \text{area of lamina} \times \text{the pressure at the centroid.} \tag{1.2}$$

It is often easy to specify the position of the centroid and so to be able to calculate the pressure at the centroid. This approach avoids the difficulties that may be encountered when the direct integration of the basic equation is attempted.

Location of the centre of pressure of a plane lamina

The effect of the pressure system acting upon a plane inclined lamina can be represented by a force of magnitude given by equation (1.1) located at a position

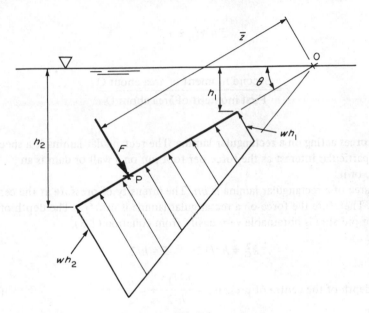

Fig. 1.6

on the lamina such that the moment of the force about O is the same as that of the pressure distribution about O. The point at which the equivalent force must be located is termed the *centre of pressure* of the lamina. Its position can be calculated as follows.

$F \times \bar{z}$ = moment of the pressure system about O as $\bar{z}$ is the distance from O measured along the plane to the centre of pressure P.

Therefore $F \times \bar{z} = \int_{x_1}^{x_2} b \, dx \, wx \sin \theta \, x$, that is the area of element x pressure x moment arm.

Therefore

$$F \times \bar{z} = w \sin \theta \int_{x_1}^{x_2} bx^2 \, dx$$

Now $\int_{x_1}^{x_2} bx^2 \, dx$ is the second moment of area of the lamina taken about O. This is usually denoted by I. So

$$F \times \bar{z} = w \sin \theta I.$$

But $I = I_G + Ax^2 = Ak_G^2 + Ax^2$ (from the parallel axis theorem) where k_G is the radius of gyration of the lamina about the centroid G. As the force F is given by $wAx \sin \theta$ (see equation (1.2))

$$wA\bar{x} \sin \theta \bar{z} = w \sin \theta A (k_G^2 + \bar{x}^2)$$

so

$$\bar{z} = k_G^2 / \bar{x} + \bar{x} \tag{1.3}$$

or

$$\bar{z} = \frac{\text{Second moment of area about O}}{\text{First moment of area about O}} \tag{1.4}$$

1.8.1 Forces acting on a rectangular lamina

The rectangular lamina is a special case of particular interest as the force per foot run on a wall or dam is an example of it.

The area of a rectangular lamina is bh. The intensity of pressure at the centroid is $wh/2$. Therefore the force on a rectangular lamina is $wbh^2/2$. The depth of the centre of pressure is obtainable very easily from equation (1.3).

$$k_G^2 = h^2/12; \qquad \bar{x} = h/2$$

so the depth of the centre of pressure $= \dfrac{h^2/12}{h/2} + \dfrac{h}{2} = \tfrac{2}{3}h \tag{1.5}$

In the case of a rectangular lamina, the breadth being constant, the centroid of the *pressure diagram* is coincident with the centre of pressure. The pressure diagram is triangular and its centroid is at a depth of $\tfrac{2}{3}h$ which agrees with the result obtained above in equation (1.5).

1.9 Forces acting on curved surfaces

The force on an element of a curved surface per unit width $= \delta P = wh \, \delta s$ (see Fig. 1.7). Resolving this force into its horizontal and vertical components, H and V

$$\delta H = \delta P \sin \theta = wh \, \delta s \sin \theta$$

but

$$\delta s \sin \theta = \delta h$$

so

$$\delta H = wh \, \delta h$$

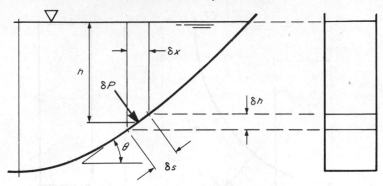

Fig. 1.7

Now this is the force on the projection of the area of the element onto the vertical plane. Therefore

$$H = wh^2/2 \tag{1.6}$$

$$\delta V = \delta P \cos \theta = wh \, \delta s \cos \theta$$

but

$$\delta s \cos \theta = \delta x$$

so

$$\delta V = wh \, \delta x$$

and therefore δV is the weight of the volume of fluid above the element δs.

The total vertical force must therefore be the weight of the fluid enclosed by the surface.

The two components of the force can now be calculated and the force P itself found

$$P = (H^2 + V^2)^{1/2} \tag{1.7}$$

The angle it makes to the horizontal is

$$\alpha = \tan^{-1}(V/H) \tag{1.8}$$

The location of the line of the resultant force

The horizontal component of the total force acts through the centre of pressure of the projection of the surface upon the vertical plane. If this projected area is a rectangle it is at a depth of $\frac{2}{3}h$. The vertical component of the force must act through the centroid G of the enclosed fluid mass.

The value of $\bar{x}$ in Fig. 1.8 can either be calculated or obtained by graphical methods. The point of intersection of the horizontal and vertical components locates a point on the line of action of the resultant force—the resultant and its components must pass through the same point. The direction and magnitude of the resultant are specified by equations (1.7) and (1.8), thus it is totally defined.

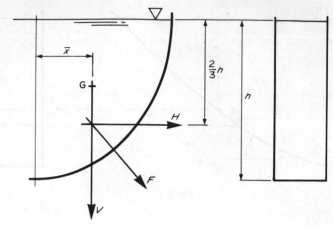

Fig. 1.8

If the curved surface is curved in three dimensions the vertical projection of the surface may not be rectangular but the technique described is still applicable when appropriately modified.

The special case of a curved surface which is part of a cylinder or a sphere is somewhat easier to deal with than the general case. All pressure forces acting upon such a surface must act through its centre of curvature. Thus the resultant of all such pressure forces must also act through the centre as shown in Fig. 1.9.

The vertical and horizontal component of the forces can be very easily calculated, and thus the magnitude of the resultant found. The angle of the resultant to the horizontal is then found from equation (1.8) and the resultant must pass through the centre of curvature O. The distance z is then equal to $R \sin \theta$. The resultant is thus completely defined in magnitude, position and direction but it has not been necessary to calculate the position of the centre of gravity of the enclosed fluid mass.

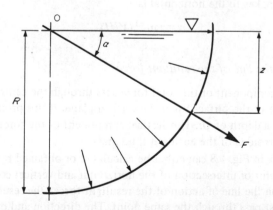

Fig. 1.9

1.10 Surface tension

The molecules comprising any substance attract one another and it is these attractive forces which make the material cohesive. A steel bar breaks when placed under sufficiently large tension because the tension is larger than the cohesive stress generated by the steel. In a fluid, cohesive forces are present and account for the development of surface tension. Surrounding any molecule within

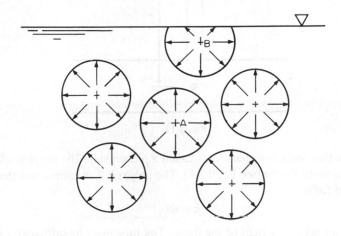

Fig. 1.10

the fluid is a zone within which another molecule will attract it. Such a molecule is shown at A in Fig. 1.10. All molecules within the fluid are acted upon by other molecules in this way so that a cohesive stress is set up. There is no resultant force acting upon a molecule such as the one located at A because all the forces acting upon it are balanced by others in each direction. However a molecule located at the fluid surface such as the one at B does experience a net resultant force acting in a direction perpendicular to the surface. It is this force that creates the defined surface exhibited by liquids. (It must be remembered that the zone around any molecule within which it is subjected to attractive forces from other molecules is extremely small, its radius being of the order of one millionth of a centimetre.) The surface layer of a liquid is thus in a different state from that prevailing in the rest of the fluid and it can be seen from Fig. 1.8 that it is in a state of uniform tension. This tension is called the surface tension of the liquid and it is usually denoted by σ. It is measured in units of force per unit length. Its dimensions are M/T^2. Water at ordinary temperatures has a value of σ of 75 dynes/cm. The value of σ decreases with increasing temperature and becomes zero at the critical point—in this state a liquid and its vapour become indistinguishable.

1.11 Manometry

The body of techniques that exists for measuring pressure and differential pressures is called manometry.

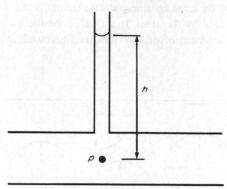

Fig. 1.11

A device that measures pressures is called a piezometer. The simplest of these consists of a vertical tube (see Fig. 1.11). The height h is measured and the pressure p calculated from

$$p = wh_1$$

where w is the specific weight of the fluid . The tube must be sufficiently large for surface tension effects to be negligible. Obviously such a device can only be used to measure the pressure of a liquid.

Pressure gauges

A mechanical device, a Bourdon gauge, is available for measuring pressures. It consists of a curved tube which tends to straighten when placed under pressure (see Fig. 1.12). The movement of the free end of this tube is used to drive a multiplying mechanism which rotates a pointer over a dial. These gauges should be regularly calibrated on a dead-weight tester.

Most pressure measuring devices measure the pressure relative to atmospheric pressure which can conveniently taken as zero. Pressures measured in this way are called gauge pressures. Such pressures are sometimes written as p lbf/in^2 g the g denoting gauge. If absolute pressures are required, the magnitude of the atmospheric pressure should be added to the gauge pressures. Atmospheric pressure can be measured by an Aneroid or mercury barometer.

Differential manometers

A U tube manometer is shown in Fig. 1.13. Let the specific weight of the working fluid be w_f and the specific weight of the manometer fluid be w_m. The pressure

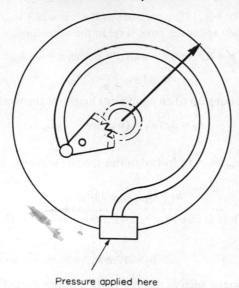

Pressure applied here

Fig. 1.12

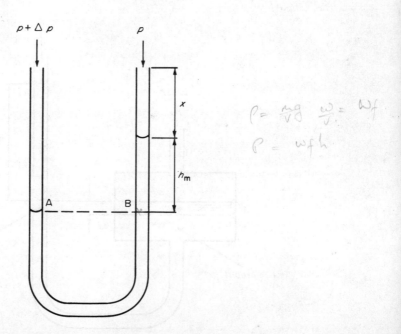

Fig. 1.13

at $A = p + \Delta p + w_f(x + h_m)$. The pressure at $B = p + w_f x + w_m h_m$. These pressures must be equal as they are at the same level in the same fluid.

$$\therefore \qquad p + \Delta p + w_f x + w_f h_m = p + w_f x + w_m h_m$$

$$\therefore \qquad \Delta p = (w_m - w_f)h_m$$

Converting this pressure Δp to an equivalent height of the working fluid

$$h_f = \Delta p/w_f = (w_m/w_f - 1)h_m$$

but

$$w_m/w_f = s_m/s_f \text{ (s denotes specific gravity)}$$

so

$$h_f = (s_m/s_f - 1)h_m \qquad (1.15)$$

If the working fluid is water $s_f = 1$ and if the manometer fluid is mercury $s_m = 13\cdot6$

then

$$h_f = 12\cdot6\, h_m$$

A U tube manometer such as this is a fairly unsophisticated instrument. To measure h_m two readings are necessary; namely the heights of the left hand and right hand menisci. By enlarging the left hand limb so that its cross sectional area is very much larger than that of the right hand limb the surface movement in it can be made almost negligible (see Fig. 1.14).

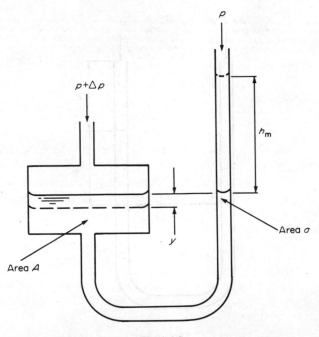

Fig. 1.14

The meniscus in the right hand limb rises by an amount h_m while that in the left hand limb falls only by y.

$$Ay = ah_m$$

As for the U tube

$$h_f = \Delta p/w = (s_m/s_f - 1)(h_m + y)$$
$$= (s_m/s_f - 1)(h_m + h_m a/A)$$
$$\therefore \quad h_f = (s_m/s_f - 1)(1 + a/A)h_m$$

If $a \ll A$

$$h_f \approx (s_m/s_f - 1)h_m$$

but if the calibration of the scale for measuring h_m is made such that

$$1 \text{ scale unit} = \frac{1}{1 + a/A} \text{ true units}$$

then the value measured on this modified scale will be larger than the true value, so that

$$h_{m \text{ measured}} = (1 + a/A)h_{m \text{ true}}$$

Then

$$h_f = (s_m/s_f - 1)h_{m \text{ measured}}$$

and the value of h_f will be accurately measured by the manometer

The inverted U tube

If the value of p is large, it will not be possible to use two separate piezometers. A mercury-water differential manometer could be considered for this application but if Δp is small this would give too small a reading. By using an air–liquid inverted differential manometer and by pressurising the air a quite satisfactory reading may be obtained (see Fig. 1.15).

As before

$$h_f = \Delta p/w = (1 - w_{air}/w_f)h_m$$

As w_{air} is very much less than w_f the value of w_{air}/w_f is very much less than 1. For instance, if water is the working fluid $w_{air}/w_f = 1/780$. The buoyant effect of the air can therefore be neglected and

$$h_f = h_m$$

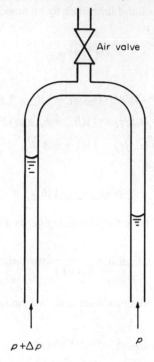

Fig. 1.15

Micromanometers

A number of devices are available for the measurement of extremely small pressures.

In the range of small, but significant pressures, the slant tube manometer can be used (see Fig. 1.16).

The value of l_m can be converted to the equivalent head reading by multiplying by $\sin \theta$.

For small pressures multi-fluid manometers can be used (see Fig. 1.17).

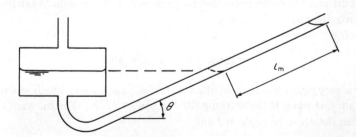

Fig. 1.16

By suitable choice of the two fluids in this manometer the sensitivity can be made high. The two fluids must be immiscible and have specific gravities of nearly the same value, for example, water and paraffin. If the specific weights of the two fluids are w_1 and w_2, the equation is

$$\Delta p = h_m \left[w_2(1 + a/A) - w_1(1 - a/A) \right]$$

This assumes that the working fluid is a gas for which the specific weight is negligible.

A Casella micromanometer is shown in Fig. 1.18. The inverted reflection of the needle tip in the fluid surface and the image of the needle tip itself can be made

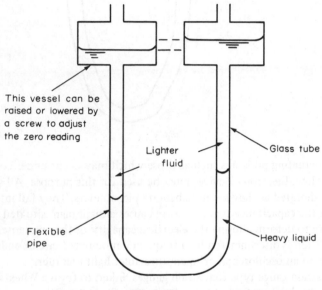

This vessel can be raised or lowered by a screw to adjust the zero reading

Lighter fluid

Glass tube

Flexible pipe

Heavy liquid

Fig. 1.17

coincident by raising or lowering the tank in the right hand limb using a micrometer screw. The amount through which it is raised or lowered gives the pressure difference in terms of head of manometer fluid.

The Chattock micromanometer is an old but extremely sensitive device for measuring very small pressure differences. It is awkward to use but is probably the cheapest of the very sensitive micromanometers.

Electronic pressure transducers

The manometers mentioned are all well suited to measuring steady pressure and have the advantage of being absolute measuring devices; the conversion of the reading into pressure depending solely upon the geometry of the instrument and the density of the manometer fluid. They are quite unsuitable for measuring

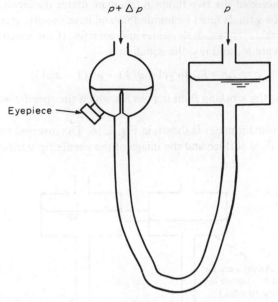

Eyepiece

Fig. 1.18

rapidly fluctuating pressures such as those which may occur under conditions of unsteady flow. Electronic devices must be used for this purpose. All such devices must be calibrated as they are not absolute piezometers. They fall into two classes. In the capacitance type, pressure causes a diaphragm situated close to a metal plate to deform altering the electrical capacity of the condenser so formed. This alteration is discriminated by a frequency modulated circuit and the output can be fed to an oscilloscope or to an ultraviolet light recorder.

In the strain gauge type four strain gauges linked to form a Wheatstone net are fixed to a diaphragm of suitable thickness (See Fig. 1.19). These strain gauges are usually silicon semiconductors. A voltage is applied across the net and the output fed to the galvanometer of an ultraviolet recorder.

The frequency response of such circuits must be carefully examined before use as it is very easy—especially in the case of output to an ultraviolet light

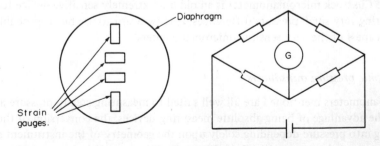

Diaphragm

Strain
gauges.

Fig. 1.19

recorder—to think that high frequency components of a pressure transient are being recorded when they are not. This is due to the low-frequency—high sensitivity response characteristic of the galvanometer. The frequency response not only of the transducer but of the entire recording system should be considered. Such transducers as these can be used to investigate pressure ranges from as low as 0–0·001 psi up to 0–1000 psi depending upon the diaphragm diameter and thickness employed.

Table 1.1
Specific weights of some common fluids

Fluid	Specific weight lbf/ft^3	kN/m^3	Specific gravity
Water (fresh)	62·4	9·81	1
Water (salt)	64·0	10·7	1·03
Benzene	54·9	8·6	0·88
Carbon tetrachloride	99·9	15·3	1·59
Petrol	42·0	6·61	0·675
Paraffin	50·0	7·89	0·803
Ethyl alcohol	49·3	7·74	0·79
Mercury	847	134	13·6
Oil	54	8·51	0·869
	to	to	to
	60	9·48	0·965

Worked examples

(1) A dam has an upstream profile as illustrated in Fig. 1.20. Calculate the force per metre run acting upon it, specifying its magnitude, direction and location.

Horizontal component of the force

$$H = \tfrac{1}{2} \times 9810 \times 1 \times (10 + 1 + 5)^2$$
$$= 1\cdot26 \text{ meganewtons}$$

Vertical component of the force

$$V = \text{weight of contained fluid}$$
$$= \text{area} \times w$$
$$= (6 \times 10 + 1 \times 5 + \tfrac{1}{2} \times 10 \times 10 + \tfrac{\pi}{4} \times 1^2) \times 9810$$
$$= 1\cdot14 \text{ meganewtons}$$

Resultant force

$$R = \sqrt{(H^2 + V^2)}$$
$$= \sqrt{(1\cdot26^2 + 1\cdot14^2)} \text{ meganewtons}$$
$$= 1\cdot69 \text{ meganewtons}$$

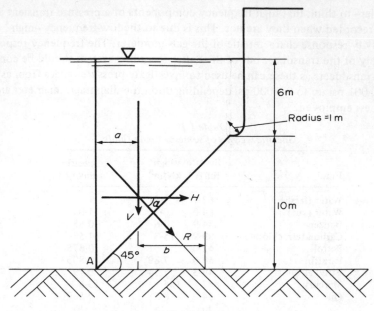

Fig. 1.20

Direction of resultant:
Let α be the angle the resultant makes to the horizontal.
Then

$$\alpha = \arctan (H/V)$$
$$= \arctan (1\cdot26/1\cdot14)$$
$$= 47° \ 53'$$

Position of resultant:
It is first necessary to locate the centroid of the mass of water contained between a vertical through A and the dam profile. Let the horizontal distance from A of the centroid be a.

$$a = \frac{6 \times 10 \times 5 + \frac{1}{2} \times 10^2 \times \frac{10}{3} + 5 \times 1 \times 10\cdot5 + (10 + 0\cdot4244) \times \frac{\pi}{4}}{60 + 50 + 5 + \frac{\pi}{4}}$$

(Note that the distance of the centroid of a quadrant of a circle is $0\cdot4244 \times R$ from either straight edge).

∴
$$a = 4\cdot55$$
$$b = 5\cdot33/\tan \alpha = 5\cdot33/1\cdot11$$
$$= 4\cdot82 \text{ m}$$

The resultant force of 1·69 meganewtons intersects the base at an angle of 47° 53' to the horizontal at a distance of 9·37 metres from the upstream edge of the base of the dam.

(2) A flat rectangular plate 1 metre wide by 2 metres long is immersed in water of density 1000 kg/m³. The shorter side is parallel to the water surface and 0·5 metres below it. The longer side is inclined at 60° to the vertical. (See Fig. 1.21). Calculate the magnitude, direction and location of the force acting upon one side of the plate due to water pressure.

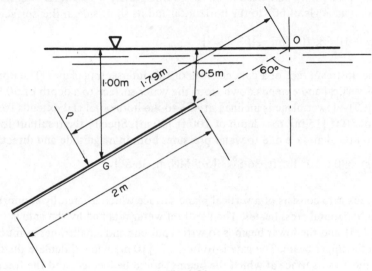

Fig. 1.21

$$\text{Area of plate} = 2 \text{ m}^2$$

$$\text{Depth of centroid} = 0·5 + 1 \cos 60°$$

$$= 1·00$$

$$\text{Force} = \text{area} \times \text{pressure at centroid}$$

$$= 2 \times 9810 \times 1·00$$

$$= 19·6 \text{ kilonewtons}$$

This force acts perpendicularly to the plate so it is acting in a direction which is 30° to the vertical.

Location of centre of pressure:

$$X = \bar{x} + k_G^2/\bar{x} \qquad k_G^2 = d^2/12 = 2^2/12 = 0·333$$

$$\bar{x} = 0·5/\cos 60° + 1 = 2·0$$

∴

$$X = 2·00 + \frac{0·333}{2·00} = 2·17 \text{ m}$$

Questions

Questions throughout this book are set in both the British and the SI systems of units. The values quoted in square brackets are in the SI system. Note they do not exactly correspond to the British values; this is to avoid unnecessarily awkward arithmetic. The answers are treated similarly.

(1) Calculate the force acting on one face of a rectangular lamina 3 ft wide [1 m] and 12 ft long [4 m] immersed in oil of specific gravity 0·9 with its long side inclined at an angle of 60° to the horizontal and its short side in the surface.

Answer: 10·5 x 10³ lbf   [61·2 kN].

(2) The upstream face of a dam consists of two intersecting planes. The upper plane is vertical and extends down from the water surface to a depth of 50 ft [15 m]. The lower plane is inclined at 45° to the horizontal and extends from the depth of 50 ft [15 m] to a depth of 100 ft [30 m]. Specify the resultant force acting on the dam face due to water pressures both in magnitude and direction.

Answer: 390 x 10³ lbf/ft, 36·8°   [5·5 MN/m, 36·8°].

(3) A lock gate consists of a vertical plane surface which is laterally supported by two horizontal cross beams. The depth of water retained by the gate is 20 ft [7 m] and the lower beam is to withstand one and a half times as much load as the upper beam. The gate is to be 30 ft [10 m] wide. Calculate the depths below the water surface at which the beams should be located, and the forces acting on the beams.

Answer: 8·43 ft, 16·6 ft, 150 x 10³ lbf, 225 x 10³ lbf   [2·93 m, 5·84 m, 960 kN, 1440 kN].

(4) The upstream profile of a dam is defined by the equation $y = 0·25x^2$. Calculate the magnitude and direction of the resultant thrust per unit width of dam when the upstream depth of fresh water is 100 ft [30 m].

Answer: 323 x 10³ lbf/ft at 15° to horizontal   [4·91 MN/m at 26·0° to horizontal].

(5) A bridge across a river has an elevation as shown in Fig. 1.22. When the river is in flood the upstream water surface is level with the road surface and the downstream level is 15 ft [5 m] below it. Assuming the pressure distribution to be hydrostatic calculate (a) the horizontal thrust upon the bridge and (b) the overturning moment acting upon it.

Answer: 1·14 x 10⁶ lbf, 36·9 x 10⁶ ft lbf [6·17 MN, 57·5 MN m].

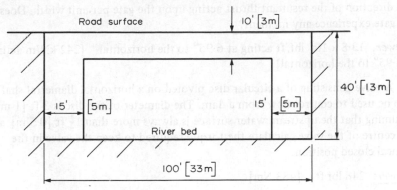

Fig. 1.22

(6) A rectangular leaf gate is fitted into a side wall of a reservoir mounted in a vertical position. Its width is 4 ft [1·3 m] and its height is 10 ft [3 m]. Its upper edge is hinged to the reservoir wall and its lower edge is equipped with a retaining device that holds the gate vertical against the forces applied to its upstream face by the water. Its downstream face is at atmospheric pressure. Calculate the force on the hinge and the force on the retaining device when the water surface is 20 ft [7 m] above the level of the gate hinge.

Answer: 29·2 x 10³ lbf, 33·2 x 10³ lbf   [153kN, 172kN].

(7) A spherical vessel of radius 12 ft [4 m] is filled with water. The pressure at the highest point is atmospheric. Calculate the magnitude and direction of the resultant force acting on the half of the sphere created by the intersection of a vertical plane through the centre of the sphere and the sphere itself. Also locate the point of intersection of the resultant with the surface of the sphere.

Answer: Force: 407 x 10³ lbf, angle: 33·7° to horizontal, point of intersection: 6·6 ft below horizontal diametral plane [Force: 2·39 MN, angle: 33·7°, point of intersection: 2·23 m below horizontal diametral plane].

(8) A radial gate has the dimensions given in Fig. 1.23. Calculate the magnitude

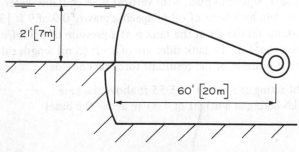

Fig. 1.23

and direction of the resultant thrust acting upon the gate per unit width. Does the gate experience any moment?

Answer: 13·8 x 10^3 lbf/ft acting at 6·95° to the horizontal   [242 kN/m acting at 6·95° to the horizontal].

(9) A valve consisting of a circular disc pivoted on a horizontal diametral shaft is to be used to control flow from a dam. The diameter of the disc is 3 ft [1 m]. Assuming that the upstream water surface is always more than 1½ ft [0·5 m] above the centre of the valve calculate the torque required to keep the valve in the vertical closed position.

Answer: 248 lbf ft   [483 Nm].

(10) A bridge across a river has the profile illustrated in Fig. 1.24. The width of the bridge is 30 ft [10 m]. Calculate the upthrust on the bridge when the river

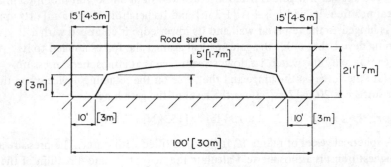

Fig. 1.24

level on both sides of the bridge is level with the road surface. Also calculate the resultant force acting on one half of the underside of the bridge from springing to crown in magnitude and direction.

Answer: 2·21 x 10^6 lbf, 1·16 x 10^6 lbf at 18·2° to the horizontal
    [11·5 MN, 6·11 MN at 20·3° to the horizontal].

(11) A cubical tank, square in plan, with vertical sides contains water to a depth of 3 ft [1 m] overlain by a layer of oil of specific gravity 0·9 of 9 ft [3 m] thickness. Above the oil the gas in the tank is at a pressure of 1·5 lbf/in^2 [10 x 10^3 newtons/m^2]. If the tank sides are of 15 ft [5 m] length calculate the magnitude and the position of the resultant force on one side.

Answer: 110 lbf acting at a height of 5·55 ft above the base
    [605 kN acting at a height of 1·81 m above the base].

2 Hydrodynamics

The study of hydrodynamics involves a mathematically rigorous approach to the analysis of fluid motions. The concepts upon which it is based are the same as those employed in engineering fluid mechanics but the results obtained are usually expressed in the form of differential equations and are usually of more general application.

2.1 The continuity equation

The law of conservation of mass leads to the differential form of the continuity equation. Consider the flow into and out of a cubical element ABCDEFGH, as shown in Fig. 2.1.

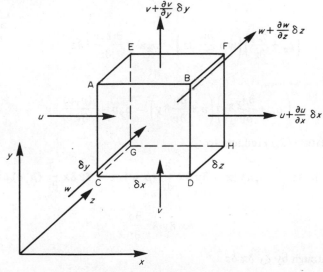

Fig. 2.1

Fluid entering face AEGC will have a mass density ρ_x and when it leaves across face BFHD it will have a mass density of $\rho_x + (\partial\rho_x/\partial x)\,\delta x$. Similarly for faces ABDC and EFHG the densities will be ρ_z and $\rho_z + (\partial\rho_z/\partial z)\,\delta z$ and for faces CGHD and AEFB ρ_y and $\rho_y + (\partial\rho_y/\partial y)\,\delta y$.

The difference between the amount of fluid entering and leaving the element per second must be stored within the element and this is possible if the fluid within the element is compressible. The rate of storage within the element is $\delta x\,\delta y\,\delta z\,(\partial\rho/\partial t)$ where ρ is the mean value of the density, that is $(\rho_x + \rho_y + \rho_z)/3$.

$$\therefore \qquad \delta y\,\delta z\left[\rho_x u - \left(\rho_x + \frac{\partial\rho_x}{\partial x}\delta x\right)\left(u + \frac{\partial u}{\partial x}\delta x\right)\right]$$

$$+ \delta y\,\delta x\left[\rho_z w - \left(\rho_z + \frac{\partial\rho_z}{\partial z}\delta z\right)\left(w + \frac{\partial w}{\partial z}\delta z\right)\right]$$

$$+ \delta z\,\delta x\left[\rho_y v - \left(\rho_y + \frac{\partial\rho_y}{\partial y}\delta y\right)\left(v + \frac{\partial v}{\partial y}\delta y\right)\right]$$

$$= \delta x\,\delta y\,\delta z\,\frac{\partial\rho}{\partial t}$$

Now

$$\left(\rho_x + \frac{\partial\rho_x}{\partial x}\delta x\right)\left(u + \frac{\partial u}{\partial x}\delta x\right) - \rho_x u = \frac{\partial(\rho_x u)}{\partial x}\delta x$$

similarly

$$\left(\rho_z + \frac{\partial\rho_z}{\partial z}\delta z\right)\left(w + \frac{\partial w}{\partial z}\delta z\right) - \rho_z w = \frac{\partial(\rho_z w)}{\partial z}\delta z$$

and

$$\left(\rho_y + \frac{\partial\rho_y}{\partial y}\delta y\right)\left(v + \frac{\partial v}{\partial y}\delta y\right) - \rho_y v = \frac{\partial(\rho_y v)}{\partial y}\delta y$$

so that equation (2.1) reduces to

$$- \delta y\,\delta z\,\frac{\partial}{\partial x}(\rho_x u)\,\delta x - \delta y\,\delta x\,\frac{\partial}{\partial z}(\rho_z w)\,\delta z - \delta z\,\delta x\,\frac{\partial}{\partial y}(\rho_y v)\,\delta y$$

$$= \delta x\,\delta y\,\delta z\,\frac{\partial\rho}{\partial t}$$

Dividing through by $\delta y\,\delta x\,\delta z$

$$\frac{\partial}{\partial x}(\rho_x u) + \frac{\partial}{\partial y}(\rho_y v) + \frac{\partial}{\partial z}(\rho_z w) + \frac{\partial\rho}{\partial t} = 0$$

At a point in the fluid when $\delta x \to \delta y \to \delta z \to 0$

$$p_x \to p_y \to p_z \to p$$

so

$$\frac{\partial}{\partial x}(\rho u) + \frac{\partial}{\partial y}(\rho v) + \frac{\partial}{\partial z}(\rho w) + \frac{\partial \rho}{\partial t} = 0$$

If the fluid is incompressible ρ must be constant so

$$\frac{\partial u}{\partial x} + \frac{\partial v}{\partial y} + \frac{\partial w}{\partial z} = 0$$

This is equivalent to saying that the *divergence* of the velocity is zero or that the volume of a fluid element is constant.

If the fluid is compressible, but the flow is steady

$$\frac{\partial \rho}{\partial t} = 0$$

$$\therefore \qquad \frac{\partial(\rho u)}{\partial x} + \frac{\partial(\rho v)}{\partial y} + \frac{\partial(\rho w)}{\partial z} = 0$$

2.2 The Euler equation

This equation is a differential form of Newton's second law of motion.

Consider an element, as before, in a fluid flow in which pressure fields exist, but in which friction is absent (see Fig. 2.2).

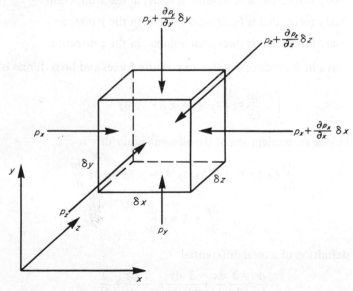

Fig. 2.2

There will be a force acting on the element in the x direction, another in the y direction and a third in the z direction. There may be body forces acting upon the fluid. Such forces are forces which are proportional to the volume of the fluid. An example of such a body force is weight, the body force being the specific weight w. If the x and z axes are located in a horizontal plane the only body force that exists is the weight acting in the y direction but if the coordinate system has no particular orientation relative to the earth's gravitational field— the vertical—then weight body-forces can exist in all three directions (see Fig. 2.3).

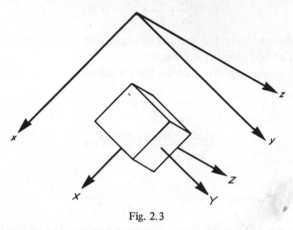

Fig. 2.3

These body forces are usually denoted by

X for the body force, that is force/unit volume, in the x direction

Y for the body force, that is force/unit volume, in the y direction

Z for the body force, that is force/unit volume, in the z direction

Then the force in the x direction due to pressure forces and body forces is

$$\left(-\frac{\partial p}{\partial x}\,\delta x\right)\delta y\,\delta z + X\,\delta x\,\delta y\,\delta z$$

This must cause an acceleration of the element of du/dt.

$$\therefore \qquad -\frac{\partial p}{\partial x}\,\delta x\,\delta y\,\delta z + X\,\delta x\,\delta y\,\delta z = \rho\,\delta x\,\delta y\,\delta z\,\frac{du}{dt}$$

$$\therefore \qquad -\frac{\partial p}{\partial x}+X=\rho\,\frac{du}{dt}$$

Now, by definition of a total differential

$$\frac{d}{dt}=\frac{\partial}{\partial x}\frac{dx}{dt}+\frac{\partial}{\partial y}\frac{dy}{dt}+\frac{\partial}{\partial z}\frac{dz}{dt}+\frac{\partial}{\partial t}$$

so

$$\frac{du}{dt} = u\frac{\partial u}{\partial x} + v\frac{\partial u}{\partial y} + w\frac{\partial u}{\partial z} + \frac{\partial u}{\partial t}$$

$\therefore$

$$-\frac{\partial p}{\partial x} + X = \rho\left(u\frac{\partial u}{\partial x} + v\frac{\partial u}{\partial y} + w\frac{\partial u}{\partial z} + \frac{\partial u}{\partial t}\right)$$

By an exactly similar process equations can be developed that describe the force—acceleration relationships in the y and z directions, that is

$$-\frac{\partial p}{\partial y} + Y = \rho\left(u\frac{\partial v}{\partial x} + v\frac{\partial v}{\partial y} + w\frac{\partial v}{\partial z} + \frac{\partial v}{\partial t}\right)$$

$$-\frac{\partial p}{\partial z} + Z = \rho\left(u\frac{\partial w}{\partial x} + v\frac{\partial w}{\partial y} + w\frac{\partial w}{\partial z} + \frac{\partial w}{\partial t}\right)$$

These three equations are known as the Euler equations. The groups of derivatives

$$u\frac{\partial u}{\partial x} + v\frac{\partial u}{\partial y} + w\frac{\partial u}{\partial z} \tag{2.1}$$

$$u\frac{\partial v}{\partial x} + v\frac{\partial v}{\partial y} + w\frac{\partial v}{\partial z} \tag{2.2}$$

$$u\frac{\partial w}{\partial x} + v\frac{\partial w}{\partial y} + w\frac{\partial w}{\partial z} \tag{2.3}$$

are called convective accelerations because they describe velocity changes caused by movements in space.

The Euler equations describe relationships between forces and accelerations and are therefore differential forms of momentum equations. If they are integrated with respect to distance (in the direction for which they apply) energy relationships are obtained.

If the coordinate system is taken so that x and z are located in an horizontal plane then the X and Z body forces are zero and the Y body force is $-w_f$, the specific weight of the fluid.

If the flow is steady, that is the derivatives $\partial u/\partial t$, $\partial v/\partial t$ and $\partial w/\partial t$ are all zero, the equations become

$$\frac{\partial p}{\partial x} = -\rho\left(u\frac{\partial u}{\partial x} + v\frac{\partial u}{\partial y} + w\frac{\partial u}{\partial z}\right)$$

$$\frac{\partial p}{\partial y} = -\rho\left(u\frac{\partial v}{\partial x} + v\frac{\partial v}{\partial y} + w\frac{\partial v}{\partial z}\right) - w_f$$

$$\frac{\partial p}{\partial z} = -\rho\left(u\frac{\partial w}{\partial x} + v\frac{\partial w}{\partial y} + w\frac{\partial w}{\partial z}\right)$$

Now

$$dp = \frac{\partial p}{\partial x}\,\delta x + \frac{\partial p}{\partial y}\,\delta y + \frac{\partial p}{\partial z}\,\delta z$$

$$\therefore \quad p = -\rho\left[\int\left(u\frac{\partial u}{\partial x} + v\frac{\partial u}{\partial y} + w\frac{\partial u}{\partial z}\right)dx + \int\left(u\frac{\partial v}{\partial x} + v\frac{\partial v}{\partial y} + w\frac{\partial v}{\partial z}\right)dy\right.$$

$$\left. + \int\left(u\frac{\partial w}{\partial x} + v\frac{\partial w}{\partial y} + w\frac{\partial w}{\partial z}\right)dz\right] - \rho g y \qquad (2.4)$$

Also

$$\delta u = \frac{\partial u}{\partial x}\,\delta x + \frac{\partial u}{\partial y}\,\delta y + \frac{\partial u}{\partial z}\,\delta z$$

when the flow is steady.

$$\therefore \quad u\frac{du}{dx} = u\left(\frac{\partial u}{\partial x} + \frac{\partial u}{\partial y}\frac{dy}{dx} + \frac{\partial u}{\partial z}\frac{dz}{dx}\right)\text{ and } u = \frac{dx}{dt}$$

so

$$u\frac{dy}{dx} = \frac{dy}{dt} = v \text{ and } u\frac{dz}{dx} = \frac{dz}{dt} = w$$

$$\therefore \quad u\frac{du}{dx} = u\frac{\partial u}{\partial x} + v\frac{\partial u}{\partial y} + w\frac{\partial u}{\partial z}$$

so the convective accelerations in the equations (2.1), (2.2), (2.3), can be written

$$u\frac{du}{dx}, v\frac{dv}{dy} \text{ and } w\frac{dw}{dz}$$

Substituting into equation (2.4) gives

$$p = -\rho\left[\int u\frac{du}{dx}dx + gy + \int v\frac{dv}{dy}dy + \int w\frac{dw}{dz}dz\right]$$

$$\therefore \quad p + w_f y + \rho\left(\frac{u^2 + v^2 + w^2}{2}\right) = \text{constant}$$

$$\therefore \quad \frac{p}{w_f} + \frac{V^2}{2g} + y = \text{constant} \qquad (2.5)$$

as $u^2 + v^2 + w^2 = V^2$ where V is the local velocity. Equation (2.5) is Bernoulli's equation.

2.3 Normal strain and deformation of a fluid element

It is necessary to describe, mathematically, the displacements, distortions and rotations of any fluid element. Consider an element that is a right parallelepiped. It can deform in two ways as illustrated in Figs. 2.4 and 2.5 and can also rotate without necessarily deforming. These two types of deformation can cause internal stresses within the fluid.

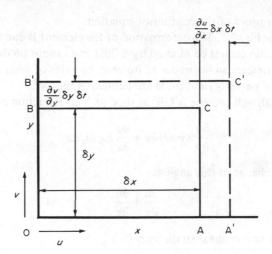

Fig. 2.4

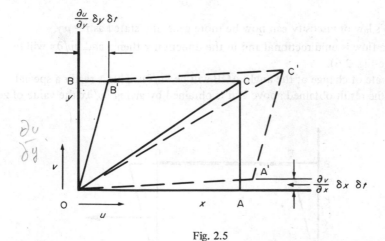

Fig. 2.5

In Fig. 2.4 normal strain is occurring: point B moves relative to point O (also a moving point) by an amount of $(\partial v/\partial y)\, \delta y\, \delta t$ in time δt. Similarly point A moves relative to point O an amount $(\partial u/\partial x)\, \delta x\, \delta t$.

If the fluid is incompressible, area $\delta y \, \delta x$ must be constant (considering a two dimensional case)

$\therefore$

$$\frac{\partial v}{\partial y} \delta y \, \delta t \, \delta x = -\frac{\partial u}{\partial x} \delta x \, \delta t \, \delta y$$

$\therefore$

$$\frac{\partial u}{\partial x} + \frac{\partial v}{\partial y} = 0$$

which is another proof of the continuity equation.

Next consider Fig. 2.5. The deformation of the element is due to shearing effects. Shear strain cannot be resisted by a fluid and cannot produce shear stresses. Shear stresses can be produced however by rates of shear straining (Newton's law of viscosity) and these are defined as follows.

The angle AOB will become A'OB' in time δt. This reduction in the angle is

$$\frac{\partial u}{\partial y} \delta y \, \delta t / \delta y + \frac{\partial v}{\partial x} \delta x \, \delta t / \delta x$$

and the rate of change of this angle is

$$\frac{\partial u}{\partial y} + \frac{\partial v}{\partial x}$$

Denoting the rate of shear strain by $\dot{\gamma}$ then

$$\dot{\gamma} = \frac{\partial u}{\partial y} + \frac{\partial v}{\partial x}$$

Newton's law of viscosity can now be more generally stated as $\tau = \mu \dot{\gamma}$.

If the flow is unidirectional and in the direction x then v and $\partial v/\partial x$ will be zero (see Fig. 2.6).

The rate of change of the angle AOB will be $\partial u/\partial y$. This is really a special case of the result obtained above and is obtained by giving $\partial v/\partial x$ the value of zero.

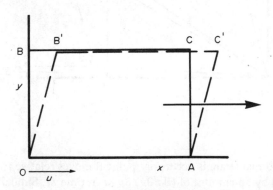

Fig. 2.6

2.4 Rotation of a fluid element

The rotation of a fluid element is illustrated in Fig. 2.7. In time δt OA rotates to position OA', that is through an angle

$$\frac{\partial v}{\partial x} \delta x \, \delta t / \delta x \quad \text{that is } \frac{\partial v}{\partial x} \delta t$$

and line OB rotates to position OB' i.e. through an angle

$$-\frac{\partial u}{\partial y} \delta y \, \delta t / \delta y \quad \text{that is } -\frac{\partial u}{\partial y} \delta t$$

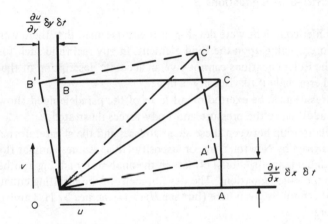

Fig. 2.7

Note that for line OB to rotate as shown in Fig. 2.7 the value of $\partial u/\partial y$ must be negative (point B moving in the x direction at a smaller velocity than point O).

The diagonal OC will rotate through an angle of

$$\frac{1}{2}\left(\frac{\partial v}{\partial x} - \frac{\partial u}{\partial y}\right) \delta t$$

and its rate of rotation will be

$$\frac{1}{2}\left(\frac{\partial v}{\partial x} - \frac{\partial u}{\partial y}\right)$$

Thus Ω, the mean angular velocity is

$$\Omega = \frac{1}{2}\left(\frac{\partial v}{\partial x} - \frac{\partial u}{\partial y}\right)$$

Vorticity (which will be dealt with later) is usually denoted by the symbol ζ (zeta) and has a value of twice the angular velocity so

$$\zeta = \frac{\partial v}{\partial x} - \frac{\partial u}{\partial y}$$

Thus a fluid for which the vorticity is zero is one in which all the fluid elements have zero angular velocity, that is they are not rotating. The case illustrated in Fig. 2.7 is an example of rotation without shearing deformation. In Fig. 2.5 a case is illustrated in which both shearing deformation and rotation is occurring.

2.5 The Navier-Stokes equations

When the Euler equations were developed it was assumed that there were no frictional forces acting upon the fluid element. In any real fluid such forces must exist and the Euler equations cannot give an accurate description of the behaviour of the fluid especially if its viscosity is large.

Shear stresses must be exerted on all faces of the parallelepiped shown in Fig. 2.2 in addition to the pressure and body forces illustrated.

If the relationship between these shear stresses and the shear deformations occurring is given by Newton's law of viscosity (this assumes laminar flow) it is possible to include the frictional effects in the analysis by adding another set of terms to the Euler equations. The development of the resulting equations is lengthy and cannot be given here (but see *Hydrodynamics* by H. Lamb). These equations are

$$-\frac{\partial p}{\partial x} + X = \rho \left(u \frac{\partial u}{\partial x} + v \frac{\partial u}{\partial y} + w \frac{\partial u}{\partial z} + \frac{\partial u}{\partial t} \right) - \mu \left(\frac{\partial^2 u}{\partial x^2} + \frac{\partial^2 u}{\partial y^2} + \frac{\partial^2 u}{\partial z^2} \right)$$

$$-\frac{\partial p}{\partial y} + Y = \rho \left(u \frac{\partial v}{\partial x} + v \frac{\partial v}{\partial y} + w \frac{\partial v}{\partial z} + \frac{\partial v}{\partial t} \right) - \mu \left(\frac{\partial^2 v}{\partial x^2} + \frac{\partial^2 v}{\partial y^2} + \frac{\partial^2 v}{\partial z^2} \right)$$

$$-\frac{\partial p}{\partial z} + Z = \rho \left(u \frac{\partial w}{\partial x} + v \frac{\partial w}{\partial y} + w \frac{\partial w}{\partial z} + \frac{\partial w}{\partial t} \right) - \mu \left(\frac{\partial^2 w}{\partial x^2} + \frac{\partial^2 w}{\partial y^2} + \frac{\partial^2 w}{\partial z^2} \right)$$

These are the Navier-Stokes equations and describe accurately and completely the dynamics of an *incompressible* viscous fluid. A general solution of these equations has not yet been achieved and there seems little likelihood of one ever being found. In certain special circumstances simplifying assumptions make their integration possible.

2.6 The velocity potential

In the preceding paragraphs expressions have been obtained which describe the rate of rotation of infinitesimal elements of the fluid. If the rate of rotation is everywhere zero then the motion is called *irrotational* and the vorticity ζ is zero, that is

$$\zeta = 2\Omega = \frac{\partial v}{\partial x} - \frac{\partial u}{\partial y} = 0$$

so for such a flow

$$\frac{\partial v}{\partial x} = \frac{\partial u}{\partial y} \tag{2.6}$$

Similarly for three dimensional flow

$$\frac{\partial u}{\partial z} = \frac{\partial w}{\partial x} \tag{2.7}$$

and

$$\frac{\partial w}{\partial y} = \frac{\partial v}{\partial z} \tag{2.8}$$

Imagine that a function φ (phi) exists such that

$$u = \partial\varphi/\partial x$$
$$v = \partial\varphi/\partial y$$
$$w = \partial\varphi/\partial z$$

Then by substituting into equations (2.6), (2.7), (2.8)

$$\frac{\partial^2\varphi}{\partial z\,\partial y} = \frac{\partial^2\varphi}{\partial y\,\partial z}; \frac{\partial^2\varphi}{\partial x\,\partial z} = \frac{\partial^2\varphi}{\partial z\,\partial x}; \frac{\partial^2\varphi}{\partial y\,\partial x} = \frac{\partial^2\varphi}{\partial x\,\partial y}$$

It is clear that these results could only be obtained if the vorticity in the three directions is zero and no function such as φ could exist if these results could not be shown to hold. Thus, if the vorticity of a flow is zero the velocity potential φ can exist. The flow is irrotational and because the velocity potential φ exists such a flow is often called *potential* flow.

The continuity equation for an incompressible fluid is

$$\frac{\partial u}{\partial x} + \frac{\partial v}{\partial y} + \frac{\partial w}{\partial z} = 0$$

Substituting for u, v and w, we get the *Laplace* equation

$$\frac{\partial^2\varphi}{\partial x^2} + \frac{\partial^2\varphi}{\partial y^2} + \frac{\partial^2\varphi}{\partial z^2} = 0 \tag{2.9}$$

or

$$\nabla^2\varphi = 0$$

The operator ∇ denotes

$$\left(\frac{\partial}{\partial x} + \frac{\partial}{\partial y} + \frac{\partial}{\partial z}\right)$$

and ∇^2 denotes

$$\left(\frac{\partial^2}{\partial x^2} + \frac{\partial^2}{\partial y^2} + \frac{\partial^2}{\partial z^2}\right)$$

It is used in vector algebra which permits the generalisation of results obtained in one system of coordinates to any other.

If the motion is two dimensional i.e. there is no flow velocity in, say, the z direction the process of obtaining velocity potentials is greatly simplified.

The vorticity has only one component so

$$\frac{\partial v}{\partial x} = \frac{\partial u}{\partial y}$$

$$u = \frac{\partial \varphi}{\partial x} \quad \text{and} \quad v = \frac{\partial \varphi}{\partial y}$$

so

$$\frac{\partial^2 \varphi}{\partial x\, \partial y} = \frac{\partial^2 \varphi}{\partial y\, \partial x}$$

also the Laplace equation becomes

$$\frac{\partial^2 \varphi}{\partial x^2} + \frac{\partial^2 \varphi}{\partial y^2} = 0$$

2.7 The stream function

If a function ψ (psi) is postulated such that $u = (\partial\psi/\partial y)$ and $v = -(\partial\psi/\partial x)$ the continuity equation will demonstrate its nature

$$\frac{\partial u}{\partial x} + \frac{\partial v}{\partial y} = 0$$

so

$$\frac{\partial^2 \psi}{\partial x\, \partial y} = \frac{\partial^2 \psi}{\partial y\, \partial x}$$

A streamline is a line in the flow field to which the adjacent flow is parallel. By this definition the flow velocity is tangential to the streamline (see Fig. 2.8) so that

$$\frac{v}{u} = \frac{dy}{dx}$$

$$\therefore \qquad u\,\delta y - v\,\delta x = 0$$

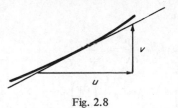

Fig. 2.8

Substituting for u and v gives

$$\frac{\partial \psi}{\partial y} \, \delta y - \left(-\frac{\partial \psi}{\partial x}\right) \delta x = 0$$

so

$$\frac{\partial \psi}{\partial y} \, \delta y + \frac{\partial \psi}{\partial x} \, \delta x = 0$$

This, from the definition of a total differential, gives $\partial \psi = 0$. So along a streamline $\partial \psi = 0$ and therefore $\psi = $ constant. Thus the function ψ defines the streamlines of a fluid motion. For an irrotational flow from equation (2.6)

$$\frac{\partial v}{\partial x} = \frac{\partial u}{\partial y}$$

substituting for v and u gives

$$\frac{\partial^2 \psi}{\partial x^2} + \frac{\partial^2 \psi}{\partial y^2} = 0$$

so the stream function must also satisfy the Laplace equation.

From the definitions of the velocity potential and the stream function it follows that

$$u = \frac{\partial \psi}{\partial y} = \frac{\partial \varphi}{\partial x}$$

and

$$v = -\frac{\partial \psi}{\partial x} = \frac{\partial \varphi}{\partial y}$$

For a streamline

$$\frac{dy}{dx} = \frac{v}{u}$$

Along a line of constant velocity potential, that is an equipotential line

$$\partial \varphi = \frac{\partial \varphi}{\partial x} \, \delta x + \frac{\partial \varphi}{\partial y} \, \delta y = 0$$

$$\therefore \qquad 0 = u \, \delta x + v \, \delta y$$

so for an equipotential line

$$\frac{dy}{dx} = -\frac{u}{v}$$

∴
$$\left(\frac{dy}{dx}\right)_{streamline} = -\left(\frac{dy}{dx}\right)_{equipotential\ line}^{-1}$$

This effectively means that the stream lines intersect equipotential lines orthogonally, that is at right angles.

2.7.1 The meaning of the stream function Consider two stream lines in an
x-y plane as shown in Fig. 2.9.

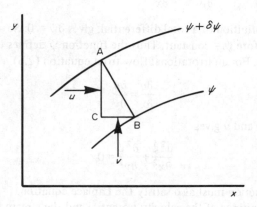

Fig. 2.9

Between these streamlines a fluid flux occurs and as no flow can cross a
streamline (by definition) this flux δQ must be made up of flow across AB
which in turn must be the same as the sum of the flow across CB and that
across AC. The flow across CB is $-v\,\delta x$ (δx is negative). Thus

$$\delta Q = -v\,\delta x + u\,\delta y$$

$$\delta\psi = \frac{\partial\psi}{\partial x}\delta x + \frac{\partial\psi}{\partial y}\delta y$$

But

$$\frac{\partial\psi}{\partial x} = -v$$

and

$$\frac{\partial\psi}{\partial y} = u$$

so $\qquad\qquad \delta\psi = -v\,\delta x + u\,\delta y$

∴ $\qquad\qquad \delta Q \equiv \delta\psi$

Thus streamlines, which have been defined as lines of constant value of the stream function, must also be lines of constant flux; the flux between a streamline of stream function ψ and one having a value for its stream function of zero is numerically equal to ψ.

2.8 Circulation

The circulation of a fluid element is defined as the line integral of the velocity taken once around the element's boundary (see Fig. 2.10), that is the circulation Γ (gamma) is given by

$$\Gamma = \oint v \cos \alpha \, ds$$

Consider motion in a circular path, that is a vortex (Fig. 2.11). The value of Γ

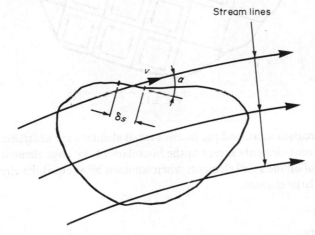

Fig. 2.10

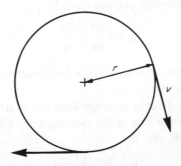

Fig. 2.11

for the element is simply $2\pi r v$. If the fluid motion is like a solid-body rotation $v = r\,\Omega$ (forced vortex) and then

$$\Gamma = 2\pi r^2 \Omega$$

and if the motion is that of a free vortex and given by the equation

$$v = C/r$$

then

$$\Gamma = 2\pi C$$

The concept of circulation is not limited to small elements. If a large element is drawn and this is split up into small elements as shown in Fig. 2.12 it will be

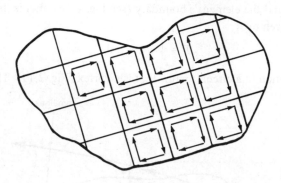

Fig. 2.12

seen that circulations around the interior small elements are neutralised by those around adjacent elements except at the boundary of the large element. Thus the circulation of all the small elements when summed must equal the circulation around the large element.

2.9 Vorticity

The vorticity defined previously as ζ was given by

$$\zeta = \frac{\partial v}{\partial x} - \frac{\partial u}{\partial y}$$

and was shown to describe the rotation of a very small element and to be equal to twice the angular velocity.

Vorticity may also be thought of as the limit of the circulation around a very small element divided by the area of the element when the area of the element is made infinitesimally small, that is

$$\zeta = \lim_{\delta a \to 0} \delta\Gamma/\delta a$$

If the element is circular then as discussed above $\delta\Gamma = 2\pi r^2 \Omega$

so

$$\zeta = \lim_{r \to 0} \frac{2\pi r^2 \Omega}{\pi r^2} = 2\Omega$$

If a rectangular element is considered as shown in Fig. 2.13

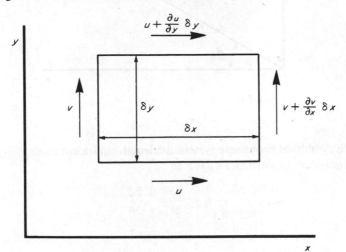

Fig. 2.13

$$\zeta = \lim_{\delta x \to \delta y \to 0} \left[\left(v + \frac{\partial v}{\partial x}\,\delta x \right)\delta y + u\,\delta x - v\,\delta y - \left(u + \frac{\partial u}{\partial y}\,\delta y \right)\delta x \right] \Big/ \delta x\,\delta y$$

$$\therefore \quad \zeta = \frac{\partial v}{\partial x} - \frac{\partial u}{\partial y}$$

using the convention that circulation anticlockwise around the element is positive. This result is the same as that obtained previously.

Before continuing to describe how these concepts of stream function, velocity potential, circulation and vorticity can be used to analyse some flow situations it is necessary to give the expressions for velocities in terms of the derivatives of the velocity potential and stream function expressed in cylindrical coordinates.

The velocity V can either be resolved into two velocity components u and v in the x and y directions or into two velocities in a cylindrical coordinate system (see Fig. 2.14), one in the direction of r, u_r, and the other in the direction of θ, u_θ,

Now

$$u = \frac{\partial \varphi}{\partial x} = \frac{\partial \psi}{\partial y} \quad \text{and} \quad v = \frac{\partial \varphi}{\partial y} = -\frac{\partial \psi}{\partial x}$$

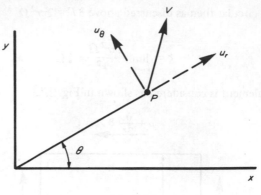

Fig. 2.14

In the cylindrical coordinate system the length equivalent to δx is δr and the length equivalent to a distance δy is $r\,\delta\theta$

$\therefore$

$$u_r = \frac{\partial \varphi}{\partial r} = \frac{1}{r}\frac{\partial \psi}{\partial \theta}$$

and

$$u_\theta = \frac{1}{r}\frac{\partial \varphi}{\partial \theta} = -\frac{\partial \psi}{\partial r}$$

2.10 The source

Certain idealised fluid motions can be considered which can in fact not occur in the real universe but which when compounded with other motions give resultant fluid situations that are of very great significance. These are the source, the sink, the doublet and the vortex, and uniform flow.

A source is a point in space from which fluid is emitted at constant rate. This concept cannot of course represent a real situation. In this book we will consider only two dimensional flows. A source in this case will be a line from which fluid is emitted uniformly in all directions in the plane normal to the line. This is equivalent to saying that in the plane perpendicular to the line source the two dimensional flow created by the line source is radiating from a point (see Fig. 2.15).

At a distance r from such a source the fluid velocity will be directed away from the source in a radial direction so

$$u_r = \frac{Q}{2\pi r} \quad \text{and} \quad u_\theta = 0$$

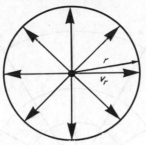

Fig. 2.15

where Q = flow emitted from the line source per unit length.

Now

$$\frac{1}{r}\frac{\partial\psi}{\partial\theta} = u_r = \frac{Q}{2\pi r} \quad \text{and} \quad \frac{\partial\psi}{\partial r} = -u_\theta = 0$$

$$\therefore \quad \delta\psi = \frac{\partial\psi}{\partial r}\delta r + \frac{\partial\psi}{\partial\theta}\delta\theta = 0 \times \delta r + r \times \frac{Q}{2\pi r}\delta\theta$$

$$\therefore \quad \delta\psi = \frac{Q}{2\pi}\delta\theta$$

$$\therefore \quad \psi = \frac{Q}{2\pi}\theta + \text{constant} = \frac{Q}{2\pi}\tan^{-1}\frac{y}{x} + \text{const}$$

The expression for the stream function of a source given by this equation agrees with the idea that the stream function is the flux. If Q units of fluid are being emitted in all directions every second, then $(Q/2\pi)\theta$ units will be emitted in an angle θ.

From the expressions for u_r and u_θ in terms of the partial derivatives of the velocity potential, the equation for the velocity potential can be obtained

$$u_r = \frac{\partial\varphi}{\partial r} = \frac{Q}{2\pi r}$$

$$u_\theta = \frac{1}{r}\frac{\partial\varphi}{\partial\theta} = 0$$

Now

$$\delta\varphi = \frac{\partial\varphi}{\partial r}\delta r + \frac{\partial\varphi}{\partial\theta}\delta\theta = \frac{Q}{2\pi}\frac{\delta r}{r} + 0$$

$$\therefore \quad \varphi = \frac{Q}{2\pi}\log_e r + \text{constant}$$

or

$$\varphi = \frac{Q}{4\pi}\log_e(x^2 + y^2) + \text{constant}$$

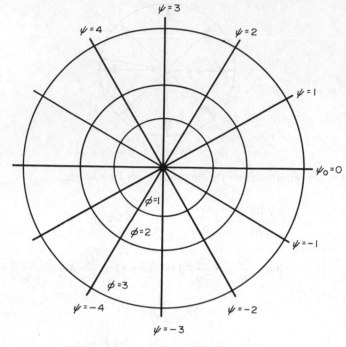

Fig. 2.16

Summarising

$$\psi_{\text{source}} = \frac{Q}{2\pi}\theta = \frac{Q}{2\pi}\tan^{-1}\left(\frac{y}{x}\right)$$

$$\varphi_{\text{source}} = \frac{Q}{2\pi}\log_e r = \frac{Q}{4\pi}\log_e(x^2 + y^2)$$

omitting constants. These functions are illustrated in Fig. 2.16.

2.11 The Sink

A sink is the opposite of a source, that is it is a point in space at which fluid disappears. Its stream function and velocity potential is simply obtained by using a negative value for the total flux Q.

2.11.1 Combination of source and sink

A combination of a source and an equal sink a distance s apart produces streamlines which are circular arcs. See Fig. 2.17.

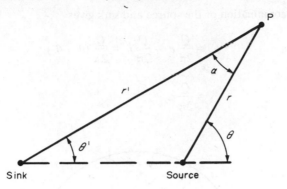

Fig. 2.17

The stream function at the point P of the combined flow ψ_c due to the source and the sink can be obtained by adding the stream function of the source and sink—this is equivalent to vector addition. This can be justified as follows

$$u_c = u_{source} + u_{sink}$$

$$v_c = v_{source} + v_{sink}$$

These two statements are equivalent to the following two statements

$$\therefore \quad \left(\frac{\partial \psi}{\partial y}\right)_c = \left(\frac{\partial \psi}{\partial y}\right)_{source} + \left(\frac{\partial \psi}{\partial y}\right)_{sink}$$

$$-\left(\frac{\partial \psi}{\partial x}\right)_c = -\left(\frac{\partial \psi}{\partial x}\right)_{source} - \left(\frac{\partial \psi}{\partial x}\right)_{sink}$$

And

$$\delta\psi_c = \left(\frac{\partial \psi}{\partial x}\right)_c \delta x + \left(\frac{\partial \psi}{\partial y}\right)_c \delta y$$

$$= \left(\frac{\partial \psi}{\partial x}\right)_{source} \delta x + \left(\frac{\partial \psi}{\partial x}\right)_{sink} \delta x + \left(\frac{\partial \psi}{\partial y}\right)_{source} \delta y + \left(\frac{\partial \psi}{\partial y}\right)_{sink} \delta y$$

$$\therefore \quad \psi_c = \psi_{source} + \psi_{sink}$$

The foregoing argument applies to the process of combining any set of stream functions and an exactly similar argument can be used to show that velocity potentials can be added and also result in vector addition. This ability to perform a vector addition by the addition of velocity potentials or stream functions is the reason for their use.

Thus the combination of the source and sink gives

$$\psi_c = \frac{Q}{2\pi}\theta - \frac{Q}{2\pi}\theta' = \frac{Q}{2\pi}(\theta - \theta')$$

$$= \frac{Q}{2\pi}\alpha$$

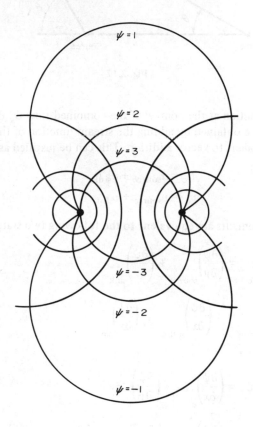

Fig. 2.18

For a particular value of ψ_c the value of α must be constant. The streamline equivalent to ψ_c must be a circular arc (angles subtended by the same chord are equal) as shown in Fig. 2.18. Similarly $\varphi_c = \varphi_{\text{source}} + \varphi_{\text{sink}}$

$$\therefore \qquad \varphi_c = \frac{Q}{2\pi}\log_e\left(\frac{r}{r'}\right)$$

2.12 The doublet

From Fig. 2.19 the velocity potential and stream function are as follows:

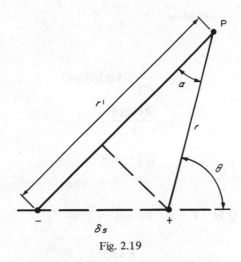

Fig. 2.19

Velocity potential

$$\varphi_c = \frac{Q}{2\pi} \log_e \frac{r}{r'} = -\frac{Q}{2\pi} \log_e \frac{r'}{r}$$

but

$$r' \approx r + \delta s \cos \theta$$

∴

$$\frac{r'}{r} \approx 1 + \frac{\delta s}{r} \cos \theta$$

Now

$$\log_e \left(\frac{r'}{r} \right) = \frac{\delta s \cos \theta}{r} - \frac{1}{2} \left(\frac{\delta s \cos \theta}{r} \right)^2 + \ldots$$

∴ $$\varphi_c = -\frac{Q}{2\pi} \frac{\delta s}{r} \cos \theta \text{ to the first order of small quantities.}$$

If Q is increased as δs is decreased in such a manner that $Q \, \delta s$ remains constant a doublet results.

Thus

$$\varphi_{\text{doublet}} = -\frac{F \cos \theta}{r}$$

where

$$F = +\frac{Q\,\delta s}{2\pi}$$

$\therefore$

$$\varphi_{\text{doublet}} = -\frac{F\cos\theta}{r} = -\frac{Fx}{x^2 + y^2}$$

Stream function

$$\psi_c = \frac{Q}{2\pi}\alpha \quad \text{(as before)}$$

$$\alpha \approx \frac{\delta s\,\sin\theta}{r}$$

$\therefore$

$$\psi_c = \frac{Q\,\delta s}{2\pi}\frac{\sin\theta}{r}$$

as before

$$Q\,\delta s \to F \quad \text{as} \quad \delta s \to 0$$

so

$$\psi_{\text{doublet}} = \frac{F\sin\theta}{r} = \frac{Fy}{x^2 + y^2}$$

2.13 The vortex

The stream function and velocity potential for a vortex are as follows (Fig. 2.20).

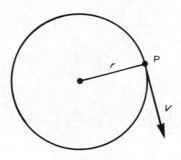

Fig. 2.20

Stream function

$$u_\theta = V = -\frac{\partial\psi}{\partial r} = \frac{1}{r}\frac{\partial\varphi}{\partial\theta}$$

$$u_r = 0$$

$\therefore$

$$\delta\psi = -V\,\delta r$$

Using the concept of circulation

$$\Gamma = 2\pi r V$$

$$\therefore \quad V = \frac{\Gamma}{2\pi r}$$

$$\therefore \quad \psi = -\frac{\Gamma}{2\pi} \int \frac{\delta r}{r}$$

$$\psi = -\frac{\Gamma}{2\pi} \log_e r + \text{constant}$$

Thus for a vortex the stream lines consist of concentric circles.

Velocity potential

$$\varphi = \int V r \, d\theta = \int \frac{\Gamma}{2\pi r} r \, d\theta$$

$$\therefore \quad \varphi = \frac{\Gamma}{2\pi} \theta + \text{constant.}$$

Thus the stream function of a vortex can be obtained from the expression for the velocity potential for a source by replacing Q by Γ; and the velocity potential of a vortex from the expression for the stream function of a source by replacing Q by Γ. So a vortex is the inverse of a source.

2.14 The uniform wind

Consider a uniform flow at velocity V inclined at an angle α to the x axis (see Fig. 2.21).

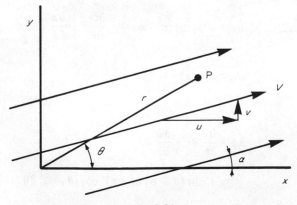

Fig. 2.21

Stream function

$$\delta\psi = \frac{\partial\psi}{\partial x}\delta x + \frac{\partial\psi}{\partial y}\delta y$$

$$u = \frac{\partial\psi}{\partial y} \text{ and } v = -\frac{\partial\psi}{\partial x}$$

so

$$\delta\psi = -v\,\delta x + u\,\delta y$$

$$\therefore \qquad \psi = uy - vx + \text{constant}$$

Velocity potential

Similarly $\varphi = ux + vy + \text{constant}$.

Stream function (in polar coordinates)

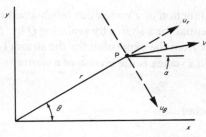

Fig. 2.22

From Fig. 2.22

$$u_\theta = -V\sin(\theta - \alpha) = -\frac{\partial\psi}{\partial r} = \frac{1}{r}\frac{\partial\varphi}{\partial\theta}$$

$$u_r = V\cos(\theta - \alpha) = \frac{1}{r}\frac{\partial\psi}{\partial\theta} = \frac{\partial\varphi}{\partial r}$$

but

$$\delta\psi = \frac{\partial\psi}{\partial r}\delta r + \frac{\partial\psi}{\partial\theta}\delta\theta$$

$$= V\sin(\theta - \alpha)\,\delta r + rV\cos(\theta - \alpha)\,\delta\theta$$

$$\psi = Vr\sin(\theta - \alpha) + \text{constant}$$

Velocity potential (in polar coordinates)

$$\delta\varphi = \frac{\partial\varphi}{\partial r}\,\delta r + \frac{\partial\varphi}{\partial\theta}\,\delta\theta$$

$$= V\cos(\theta - \alpha)\,\delta r - rV\sin(\theta - \alpha)\,\delta\theta$$

$$\therefore \qquad \varphi = rV\cos(\theta - \alpha) + \text{constant}$$

These results can be checked by back differentiation. The equation above giving the expression $u_\theta = -V\sin(\theta - \alpha)$ may need a little explanation. If θ is greater than α, $V\sin(\theta - \alpha)$ is positive yet u_θ must be negative as it is directed in the direction of θ becoming negative. Summarising

$$\psi = uy - vx = Vr\sin(\theta - \alpha)$$
$$\varphi = ux + vy = Vr\cos(\theta - \alpha)$$

In the special case when the wind is horizontal

$$\psi = uy = ur\sin\theta$$
$$\varphi = ux = ur\cos\theta$$

Up to this point the flows that have been analysed may seem to have little application to any real situations. Combinations of various of these flows will now be investigated and it will be seen that in this way reasonably accurate descriptions of real flows can be obtained.

2.15 Combinations of flow patterns

2.15.1 The uniform wind plus a source From Fig. 2.23

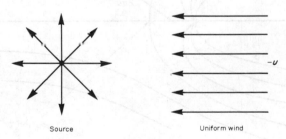

Source   Uniform wind

Fig. 2.23

$$\psi_c = \frac{Q}{2\pi}\theta - Ur\sin\theta$$

$$\text{(source) (uniform wind)}$$

$$\varphi_c = \frac{Q}{2\pi}\log_e r - Ur\cos\theta$$

It is interesting to examine the $\psi_c = 0$ streamline.
If $\psi_c = 0$

$$\frac{Q}{2\pi} \theta = Ur \sin \theta$$

When $\theta = 0$ the equation is correct irrespective of the value of r so one solution of the $\psi_c = 0$ streamline must be a horizontal line through the origin. For very small values of θ the small-angle relationship holds so $\theta \to \sin \theta$

$$\therefore \qquad\qquad\qquad r_{\theta=0} = \frac{Q}{2\pi U}$$

When $\theta = \pi/2$,

$$r = \frac{Q}{4U}$$

When $\theta \to \pi$,

$r \sin \theta$ (the semi thickness of the body, that is $y_{x \to -\infty}$) tends to $Q/2U$.

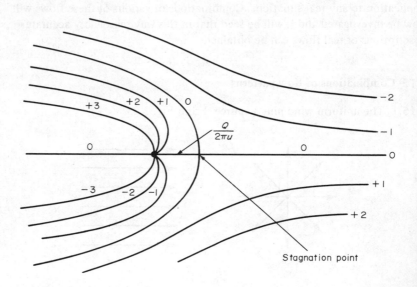

Fig. 2.24

The zero streamline can be regarded as a boundary (as no flow crosses it) so it will represent the nose of a streamlined body. It is called a semi-streamlined body (see Fig. 2.24).

2.15.2 The streamlined body The combination of a source, sink and a uniform flow gives the profile of a fully streamlined body (see Fig. 2.25).

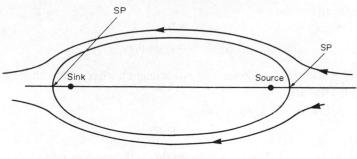

Fig. 2.25

In a real fluid the viscosity causes the fluid motion to be rotational and gives rise to velocity gradients near the boundary. As a consequence, in areas of divergent flow, fluid motions do not conform to the patterns predicted by stream function theory even approximately. The results obtained are good for the upstream end of the streamlined body but due to the formation of vortex motions or boundary layer separation effects (see later) are not at all applicable except at very low velocities (strictly, at low Reynolds' numbers).

2.15.3 Flow around a cylinder By combining the stream functions and velocity potentials of a doublet and a uniform flow the appropriate functions for flow around a cylinder can be obtained (see Fig. 2.26).

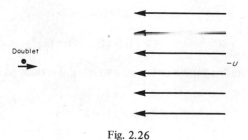

Fig. 2.26

(The arrow beneath the doublet indicates the direction of the infinitely short line from the sink to the source.)

The stream function is

$$\psi = -Ur \sin \theta + F \sin \theta / r$$

and the velocity potential

$$\varphi = -Ur\cos\theta - F\cos\theta/r$$

On the zero streamline

$$\psi = 0$$

$$Ur\sin\theta = F\sin\theta/r$$

$\sin\theta = 0$ is a solution for any value of r—a straight horizontal line through the origin of the doublet.

Also $r^2 = F/U$ when $\sin\theta \to \theta \to 0$ so

$$r = \sqrt{(F/U)}$$

Label this value of r, a. This gives a circle (that is the cross section of a cylinder). Thus the zero stream function is a straight line intersecting a circle as shown in Fig. 2.27.

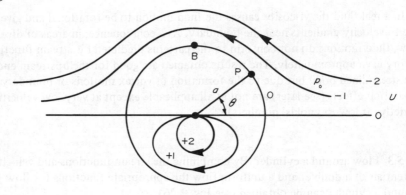

Fig. 2.27

In general

$$\psi = -Ur\sin\theta \left(1 - a^2/r^2\right)$$

Any other streamline can be obtained by fixing the value of ψ; for example, if $\psi = 1$

$$1 = -Ur\sin\theta \left(1 - a^2/r^2\right)$$

and

$$\sin\theta = \frac{1}{Ur(a^2/r^2 - 1)} \, .$$

For any value of r, θ can then be calculated and the coordinates of the streamline obtained. Two solutions can be obtained for θ; one describes the flow external to the cylinder and the other the flow inside the cylinder. This internal flow, being of little interest, is usually ignored.

2.16 Pressure distribution around a cylinder in a uniform flow

The velocity V at a point P on the cylinder $\psi = 0$ can be obtained from

$$V = (u_\theta^2 + u_r^2)^{1/2}$$

Differentiating ψ partially with respect to r and θ, we get

$$\psi = - Ur \sin \theta \, (1 - a^2/r^2)$$

$$\therefore \quad u_\theta = - \frac{\partial \psi}{\partial r} = U \sin \theta + \frac{Ua^2}{r^2} \sin \theta$$

but $r = a$ on the cylinder so

$$u_{\theta(r=a)} = 2U \sin \theta$$

u_r must be zero at the boundary but the theory will in any case show this to be true:

$$u_r = \frac{1}{r} \frac{\partial \psi}{\partial \theta} = (- Ur \cos \theta \, [1 - a^2/r^2])/r$$

but at the boundary, $r = a$ so $u_{r=a} = 0$ as stated above.

$$\therefore \quad V = 2U \sin \theta$$

Applying Bernoulli's equation (2.5) to the upstream condition where the flow is uniform and also to the point P on the cylinder gives

$$\frac{p_0}{w} + \frac{U^2}{2g} = \frac{p_P}{w} + \left(\frac{2U \sin \theta}{2g} \right)^2$$

$$\therefore \quad \frac{p_P - p_0}{w} = \frac{U^2}{2g} (1 - 4 \sin^2 \theta) \qquad (2.10)$$

When

$$\theta = \frac{\pi}{6}, \quad \frac{p_P - p_0}{w} = 0$$

$$\theta = 0°, \quad \frac{p_P - p_0}{w} = \frac{U^2}{2g}, \text{ the stagnation pressure}$$

$$\theta = \frac{\pi}{2}, \quad \frac{p_P - p_0}{w} = - \frac{3U^2}{2g}$$

The curve of $(p_P - p_0)/w$ plotted against θ is shown in Fig. 2.28.

This theory, as has been emphasised several times before, is developed on the assumption that the fluid is 'ideal' or 'perfect' that is that it has zero viscosity.

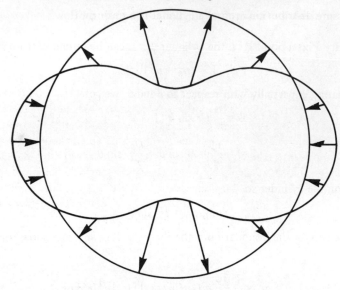

Fig. 2.28

The behaviour of a real fluid is quite accurately described by this theory for the values of θ lying between $\pm(\pi/2)$. Unfortunately, real fluids do not, even approximately, conform to the theory over the downstream face because of boundary layer separation effects. This phenomenon is described in detail in the chapter on boundary layer flows (Chapter 5) but here only a brief explanation will be given.

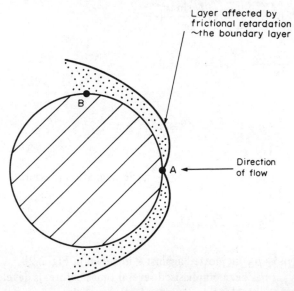

Fig. 2.29

The fluid moving over the surface of the cylinder experiences frictional forces which retard it. As the fluid moves around the cylinder it will be gradually retarded and the layer of fluid affected by these forces will become thicker (see Fig. 2.29).

At point B the fluid will have a low pressure because its velocity is high: $V = 2U$. Downstream of the cylinder the fluid velocity will be low again, that is $\approx U$ so the pressure will increase in the downstream direction. The fluid outside the boundary layer will obey Bernoulli's equation but the fluid within the boundary layer will not, having lost much of its kinetic energy and hence momentum. Consequently the fluid in the boundary layer will not be able to move forward against the positive pressure gradient and will have its slow, forward directed velocity reversed. It will consequently be compelled to leave the boundary and will become involved in complex vortex motions (see Fig. 2.30). The pressure over the downstream face of the cylinder will not obey the equations derived and the pressure distribution prevailing will be of the form shown in Fig. 2.31.

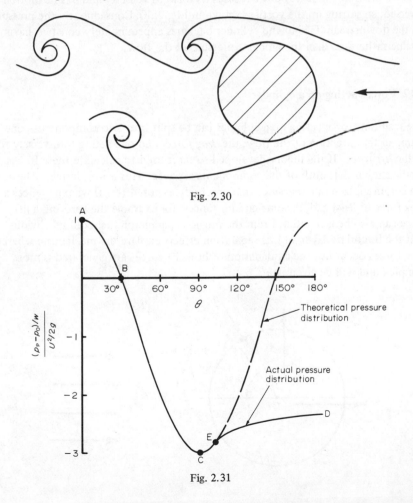

Fig. 2.30

Fig. 2.31

The portion of the curve ABC is accurately described by equation (2.10) but the shape of the portion CD is not so easily defined. The point E at which it deviates from the theoretical pressure distribution line depends upon the nature of the flow in the boundary layer. If this is laminar E is near to C—at approximately $\theta = 100°$ but if the boundary layer flow is turbulent the point at which the flow deviates from the theoretical equation moves further downstream, that is θ increases. This is explained by the fact that in a boundary layer in which flow is turbulent the velocities of fluid near the boundary are higher than they are in a laminar boundary layer because momentum is being transferred to fluid layers near the boundary by a mass transfer mechanism in turbulent flow, whereas in a laminar boundary layer momentum transfer is by viscous shears and this is a much weaker mechanism. Thus in turbulent boundary layers the fluid near the boundary has relatively large momentum and can proceed further against the adverse positive pressure gradient before it loses its forward velocity and has its motion reversed, so setting up the vortices shown in Fig. 2.30. Consequently the pressure on the downstream face of the cylinder becomes approximately constant having a value rather less than the ambient pressure of the flow.

2.17 Forces acting on a cylinder

The resultant force acting on a cylinder can be split into two components—one acting in the direction of the flow, the *drag force*; and one acting transversely to it, the *lift force*. If the theory developed so far is used to calculate these forces it can be shown that both of the components are zero. This is not, in fact, true as it is a commonplace of experience that any body exposed to a flow experiences a drag force at least and, in some circumstances, for example the aerofoil, a lift force can also be generated. From the previous paragraph it should be obvious that the boundary layer and its separation effects explain the mechanism whereby drag forces occur but the explanation of how lift forces are generated is more complex and will be given later.

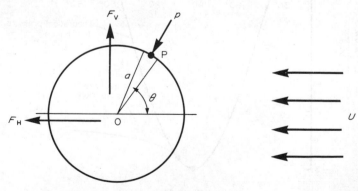

Fig. 2.32

From Fig. 2.32 the pressure acting at point P is p. There is thus a force $\delta F = pa\ \delta\theta$ acting on the element of the circumference of the cylinder directed towards O. Resolving this force into the horizontal and transverse directions produces two components δF_H and δF_V.

$$\delta F_H = - pa\ \delta\theta\ \cos\theta$$

and

$$\delta F_V = - pa\ \delta\theta\ \sin\theta$$

$$F_H = - \int_0^{2\pi} pa\ \cos\theta\ d\theta \qquad (2.11)$$

$$F_V = - \int_0^{2\pi} pa\ \sin\theta\ d\theta \qquad (2.12)$$

Substituting for p into these two expressions gives

$$F_H = - a \int_0^{2\pi} (p_0 + \tfrac{1}{2}\rho U^2(1 - 4\sin^2\theta))\cos\theta\ d\theta \qquad (2.13)$$

and

$$F_V = - a \int_0^{2\pi} (p_0 + \tfrac{1}{2}\rho U^2(1 - 4\sin^2\theta))\sin\theta\ d\theta \qquad (2.14)$$

Evaluating these integrals gives zero results for both F_H and F_V. Thus it is predicted from potential theory that a body moving in a fluid experiences no force at all. This is, of course, not true; the drag force occurs due to the phenomenon of boundary layer separation mentioned previously but if the fluid motions are very slow a close approximation to the classical motion can be obtained.

2.18 The development of transverse forces

By superimposing a vortex (that is a circulation) on the doublet plus uniform wind considered before (see Fig. 2.33) the development of a transverse force can be demonstrated. Although this may seem an artificial concept, experimental demonstration of the theory developed from it shows that it describes the real situations that occur.

The stream function of the combination is

$$\psi = - Ur\sin\theta + \frac{F\sin\theta}{r} - \frac{\Gamma}{2\pi}\log_e r$$

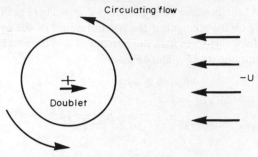

Fig. 2.33

and the velocity potential is

$$\varphi = - Ur \cos \theta - \frac{F \cos \theta}{r} + \frac{\Gamma}{2\pi} \theta$$

The velocity at the cylinder boundary is

$$u_\theta = - \frac{\partial \psi}{\partial r} = 2U \sin \theta + \frac{1}{a} \frac{\Gamma}{2\pi}$$

and, as before $u_r = 0$.

Therefore p at any point on the cylinder is

$$p = p_0 + \tfrac{1}{2}\rho U^2 \left[1 - \frac{1}{U^2} \left(2U \sin \theta + \frac{\Gamma}{2\pi a} \right)^2 \right]$$

$$= p_0 + \tfrac{1}{2}\rho U^2 \left[1 - \left(2 \sin \theta + \frac{\Gamma}{2\pi Ua} \right)^2 \right]$$

As in the case of flow around a cylinder (equation (2.11))

$$F_\mathrm{H} = - a \int_0^{2\pi} p \cos \theta \; d\theta$$

per unit length of cylinder, and (equation (2.12))

$$F_\mathrm{V} = - a \int_0^{2\pi} p \sin \theta \; d\theta$$

By substituting for p into these expressions and integrating it is easy to show that $F_H = 0$.

$$F_V = -a \left\{ \int_0^{2\pi} p_0 \sin\theta \, d\theta + \int_0^{2\pi} \tfrac{1}{2}\rho U^2 \sin\theta \, d\theta \right.$$

$$\left. - \int_0^{2\pi} \tfrac{1}{2}\rho U^2 \left(2\sin\theta + \frac{\Gamma}{2\pi Ua}\right)^2 \sin\theta \, d\theta \right\}$$

The first two of these three integrals can readily be shown to be zero so

$$F_V = +\frac{a}{2}\rho U^2 \left(\int_0^{2\pi} \left(2\sin\theta + \frac{\Gamma}{2\pi Ua}\right)^2 \sin\theta \, d\theta \right)$$

$$= +\frac{a}{2}\rho U^2 \left(\int_0^{2\pi} 4\sin^3\theta \, d\theta + \int_0^{2\pi} \frac{2\Gamma}{\pi Ua}\sin^2\theta \, d\theta + \int_0^{2\pi} \frac{\Gamma^2}{4\pi^2 U^2 a^2}\sin\theta \, d\theta \right)$$

The value of $\int_0^{2\pi} \sin^n\theta \, d\theta$ is always zero if n is odd so

$$F_V = a \left\{ \tfrac{1}{2}\rho U^2 \times \frac{2\Gamma}{\pi Ua} \, [\tfrac{1}{2}\theta - \sin\theta\cos\theta]_0^{2\pi} \right\}$$

$$\therefore \qquad F_V = \rho U \Gamma \text{ units of force/unit length of the cylinder}$$

Thus the force on a cylinder of length $L = \rho U \Gamma L$. This is the Kutta Joukowski equation. This equation explains many phenomena.

2.18.1 The Magnus effect If a ball is sliced when struck it will spin as it travels forwards. Depending upon the direction of the axis of this spin, the ball will experience a force acting perpendicularly to the direction of motion. This force therefore causes the ball to move in a circular path. The spin of the ball imparts a circulation to the adjacent air so creating a value of Γ which will decrease as the ball's spin decreases (due to frictional losses); however U decreases more rapidly than does Γ because of the very much more powerful frictional forces acting caused by boundary layer separation effects. This decrease of the forward velocity causes an even more rapid decrease in the centrally directed acceleration needed to maintain circular motion (this is U^2/r where r is the radius of the ball's path) so the transverse force developed, although decreasing in absolute magnitude, is progressively more capable of curving the ball's path. All players of ball games have discovered this phenomenon themselves. The generation of

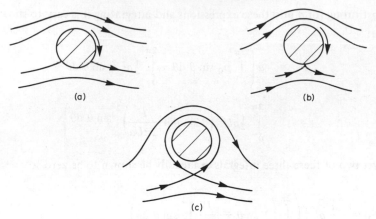

(a)   (b)

(c)

Magnus effect – showing the effect on the flow pattern
of progressively increasing the circulation

Fig. 2.34

such transverse forces by this mechanism is called the Magnus effect (see Fig. 2.34). Hydraulic structures and aerofoils also experience forces transverse to the direction of fluid motion but in such cases it is not so easy to see how the necessary circulation is generated.

The flow around an aerofoil can be obtained by conformally transforming a cylinder (and its associated flow field) into an aerofoil. (Various transforming equations are available and these produce various types of aerofoil.) If the flow around a cylinder without circulation is conformally transformed the result is as shown in Fig. 2.35.

Two stagnation points occur but the downstream one is not located at the trailing edge. This means that a streamline having a very small negative value δ for its stream function would have a bend of extremely small radius at the trailing edge. Such a small radius of a streamline would require a very high pressure gradient (a particle of mass m requires a centrally directed force of mv^2/r to travel in a curve of radius r) and such pressure gradients can only be produced by having an extremely low, possibly negative, absolute pressure at the trailing edge.

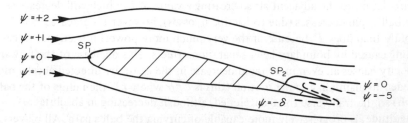

Fig. 2.35

Such negative pressures cannot occur in a gas so the situation cannot exist. In order that the stagnation point occurs at the trailing edge, a flow towards the trailing edge must develop. This in effect is a circulation. Figure 2.36 shows the

Stagnation point

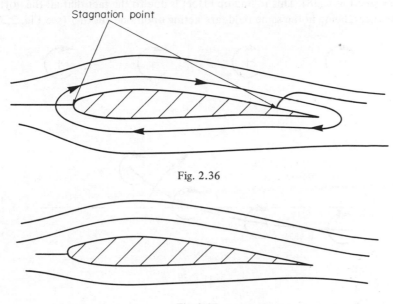

Fig. 2.36

Fig. 2.37

conformally transformed flow around a cylinder plus a circulation. Figure 2.37 shows the resultant flow. The circulation is of just such a magnitude that it moves the stagnation point to the trailing edge.

Some may find this explanation of how a circulation starts unconvincing. It should be realised that it represents a real phenomenon and is not a mathematical 'trick'. The mechanism of generation of circulation may be thought of somewhat differently.

At the moment of starting the flow (take-off in the case of an aeroplane) there are no velocities anywhere so no frictional effects are present. No circulation is present initially. Flow around the trailing edge from the under to the upper edge starts to occur but because pressures at the trailing edge cannot fall low enough to maintain a stable pattern the particles of fluid trying to travel in the necessary curved path separate from the aerofoil and, because they now have

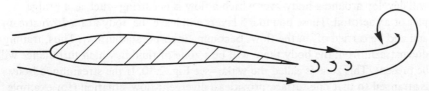

Fig. 2.38

angular momentum, move off downstream in a vortex. More particles attempt to move from the lower to the upper surface and so another vortex is initiated. A sequence of vortices is thus generated (a vortex sheet) and this rolls up into a large vortex (see Fig. 2.38). This rolling-up effect is due to the fact that all the vortices in the sheet, being in the same field, are acting upon one another (see Fig. 2.39).

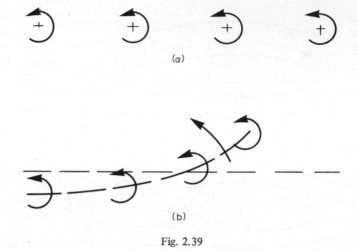

(a)

(b)

Fig. 2.39

The vortex so caused is left behind as the aerofoil moves forward so that anticlockwise angular momentum has been generated in the flow. Now a linear motion cannot generate angular momentum so an equal and opposite angular momentum must be generated around the aerofoil in order that the sum of the momentum in the starting vortex plus that around the aerofoil remains zero. The angular momentum around the aerofoil causes the circulation. When an aeroplane takes off a starting vortex is left behind on the airport runway. Every time the aeroplane increases speed or increases the angle of incidence of its wings another amount of circulation is generated by shedding another starting vortex.

2.19 The wake

Earlier in the chapter mention was made of the boundary layer: the layer in which frictional effects are present and vorticity is not zero. Boundary layers will develop around a body over which a flow is occurring—such as a bridge pier or an aerofoil. These boundary layers will leave the body at its downstream end and be carried off in the flow, becoming incorporated into it. Thus, trailing downstream from the body there will be a zone of fluid in which turbulence will be present. This zone is called the wake—see Fig. 2.40. If the streamlined body is arranged so that one surface provides a divergent flow situation (for example the upper surface of an aerofoil) and a positive pressure gradient occurs, the

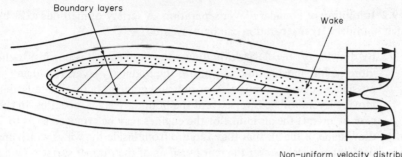

Boundary layers

Wake

Non-uniform velocity distribution
is confined to the wake

Fig. 2.40

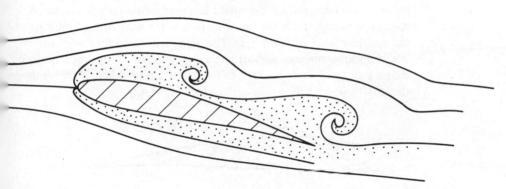

Fig. 2.41

boundary layer may separate from the surface as described earlier in this chapter
—see Fig. 2.41.

Clockwise vortices are shed regularly from the upper surface and the width of
the wake is increased; it also becomes more turbulent. The energy loss in the wake
is the cause of one of the elements of the drag force experienced by the aerofoil,
so when boundary layer separation occurs a sudden very considerable increase in
the drag occurs.

2.19.1 Stalling The clockwise vortices that are shed during boundary layer
separation carry angular momentum away with them so reducing the circulation.
Consequently, the force acting transversely to the body (the lift force) is suddenly
and very drastically reduced. If the body is an aircraft wing this sudden loss of
lift puts it into a dive. The pilot must decrease the angle of incidence of the
aircraft *relative to the plane's direction of motion* as this will cause the boundary
layer to re-attach. By diving the plane will pick up the speed that it lost when the
drag force increased during the period that the boundary layer was detached and
once the angle of incidence has been reduced the lift will be regained and the
pilot can pull the plane out of the dive.

2.19.2 Inhibition of boundary layer separation A variety of methods exist by which boundary layer separation can be prevented.

(1) By drawing off the sluggish fluid near the boundary it is possible to draw more rapidly moving fluid down into the boundary layer. It is then possible to develop a much larger pressure gradient before these faster moving layers will be reversed and the boundary layer separates. In the case of aircraft the air intake of the engines may be arranged so as to draw some of the air that they require from inside the wings which are perforated suitably over the rear portions of their upper surfaces (suction wings).

(2) If the slowly moving air close to the surface can be speeded up by the injection of high speed air, boundary layer separation will be prevented. Air can be brought from the lower surface of the wing through slots to the upper surface (slotted wings). The high pressure on the lower surface communicating through the slot to the low pressure on the upper surface causes high speed flows through to the upper surface where it speeds up the slow flows near the upper surface (see Fig. 2.42).

Slot

Fig. 2.42

2.20 Pressure distribution over an aerofoil

The 'suction' (sub-atmospheric) pressures acting on the upper surface of an aerofo contribute far more lift than do the super-atmospheric pressures on the lower surface (see Fig. 2.43).

Rayleigh's equation can be applied in this situation (see chapter 3). The lift force L is given by

$$L = \rho U l^2 f_1(Re)$$

and the drag force D by

$$D = \rho U^2 l^2 f_2(Re)$$

where f_1 and f_2 indicate different functions.

l^2 represents an area. The one usually used in this case is the plan area of the aerofoil A.

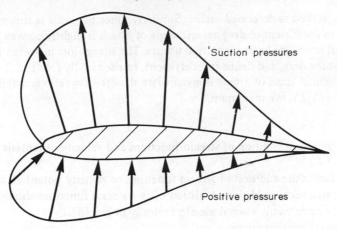

'Suction' pressures

Positive pressures

Fig. 2.43

$f_1(Re)$ is denoted by k_L and $f_2(Re)$ by k_D so

$$L = k_L \rho U^2 A$$

and

$$D = k_D \rho U^2 A$$

Alternatively the lift and drag forces can be written in terms of the dynamic or pilot pressure at the stagnation point, $\frac{1}{2}\rho u^2$. Then

$$L = C_L \tfrac{1}{2} \rho U^2 A$$
$$D = C_D \tfrac{1}{2} \rho U^2 A$$

C_L and C_D are called the coefficients of lift and drag respectively.

The coefficient of lift is approximately proportional to the angle of incidence until boundary layer separation occurs and then it suddenly drops because the

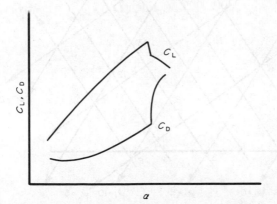

Fig. 2.44

circulation is shed as described earlier. Similarly, since the body is streamlined, the curve of coefficient of drag against angle of attack is slightly convex downwards until boundary layer separation occurs. The energy loss in vortex formation causes the drag, and hence its coefficient, to rise rapidly (see Fig. 2.44).

The optimum angle of attack is given when the lift–drag ratio is maximum or the ratio C_L/C_D is a maximum.

2.21 The graphical addition of stream functions and velocity potentials

As stated earlier the addition of stream functions or velocity potentials is equivalent to a vector addition. Addition of two stream functions should produce the same result when done algebraically or graphically.

Take two stream functions,

$$\psi_1 = 8x + 5y$$

and

$$\psi_2 = -3x + 2y$$

$$\psi_{\text{combined}} = 5x + 7y \text{ by addition algebraically}$$

What has happened is that velocity components in the x and y directions have been added.

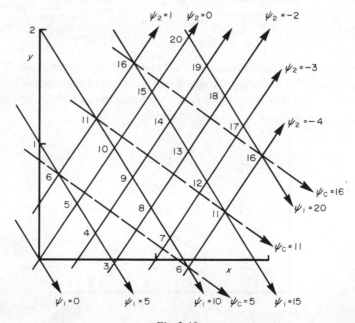

Fig. 2.45

For stream function ψ_1,

$$u_1 = \partial\psi/\partial y = 5 \text{ and } v_1 = -\partial\psi/\partial x = -8$$

Similarly, for stream function $\psi_2, u_2 = 2, v_2 = +3$.

At the intersections of the two stream function systems there are two values of the stream functions which can be added together as illustrated in Fig. 2.45 If these added values are entered on the diagram, lines of constant ψ_c can be drawn. It can easily be seen that the result for $\psi_c = 6, 11, 16$, etc., as shown, represent the equation $\psi_c = 5x + 7y$ when values of 6, 11, 16, etc. are substituted for ψ_c. Similarly, addition of velocity potentials can be performed graphically.

2.21.1 Addition of a source and a sink

$$\psi_{source} = \frac{Q}{2\pi}\theta; \; \psi_{sink} = -\frac{Q}{2\pi}\theta$$

for $Q = 12$

$$\psi_{source} = \frac{6}{\pi}\theta; \; \psi_{sink} = -\frac{6}{\pi}\theta$$

θ	$\frac{\pi}{6}$	$\frac{\pi}{3}$	$\frac{\pi}{2}$	$\frac{2\pi}{3}$	$\frac{5}{6}\pi$	π
ψ_{source}	1	2	3	4	5	6
ψ_{sink}	-1	-2	-3	-4	-5	-6

The streamlines of the combined stream functions are arcs of circles (see Fig. 2.46).

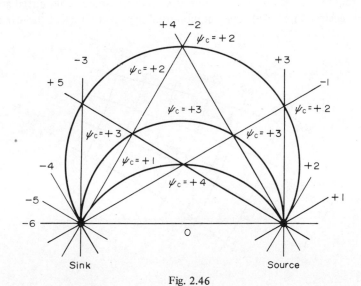

Fig. 2.46

A further example of the technique of the graphical addition of stream functions can be obtained by the addition of a uniform wind and a source—a semi-streamlined body (see Fig. 2.47). Very complex combinations of stream functions which may be too complex for analysis can be dealt with by graphical methods following the techniques illustrated above.

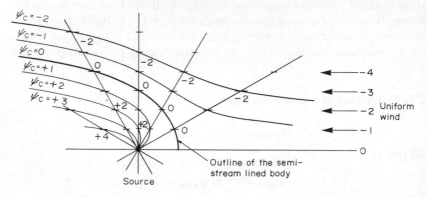

Fig. 2.47

2.22 The Flow Net

To generate a flow net draw streamlines within a particular flow for constant increments $\delta\psi$ of the stream function and then draw lines of constant velocity potential (equipotential lines) for increments $\delta\varphi$ of the velocity potential and make $\delta\psi = \delta\varphi$ (see Fig. 2.48). The equipotential lines will intersect the stream-

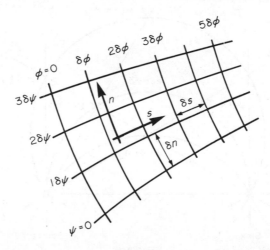

Fig. 2.48

lines orthogonally, that is at right angles, as shown earlier (p. 42). Using natural coordinates s and n

$$\partial\psi/\partial n = v \qquad \text{and} \qquad \partial\varphi/\partial s = v$$

so

$$\delta\psi = v\,\delta n \qquad \text{and} \qquad \delta\varphi = v\,\delta s$$

as

$$\delta\psi = \delta\varphi$$

$$v\,\delta n = v\,\delta s \qquad \text{so} \qquad \delta n = \delta s$$

This shows that if the equipotential lines and streamlines are drawn for equal increments of $\delta\psi$ and $\delta\varphi$ and these increments are themselves equal, the net of streamlines and equipotential lines produced must have a square mesh. If the values of $\delta\psi$ and $\delta\varphi$ chosen are extremely small, the squares produced will be true squares but if they are of significant size the squares will be 'curvilinear' squares.

The values of δs and δn are inversely proportional to the local velocity, so if the squares in a flow net are small the velocity is high and conversely if they are large the velocity is low.

If a flow is potential (that is the boundary layers are very thin and the main core of the flow is potential) it is possible to sketch in a system of streamlines which fits the shape of the boundaries. If a system of equipotential lines can be drawn—orthogonal to these streamlines—and a system of curvilinear squares produced, the resulting flow net is correct (see Fig. 2.49). Velocities throughout the flow can then be accurately estimated. It can be seen that velocities are largest near the inner wall and least at the outer wall of the bend. The process of

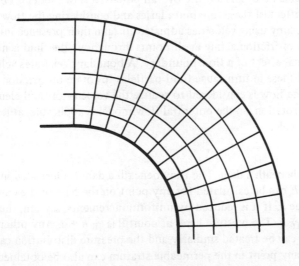

Fig. 2.49

producing a flow net requires a lot of patience, a soft pencil and much use of an eraser. The original estimated streamlines will need modification which will require modification of the equipotential lines. The technique is a trial and error process.

2.23 Percolating flows

The Darcy equation for flow through permeable material is $v = ki$ where i is the hydraulic gradient and v is the velocity of flow ignoring the presence of the particles. Now i can be written as $\Delta E/\Delta s$ where ΔE is the energy loss over the distance Δs measured along the streamline

that is $$v = k\,\Delta E/\Delta s$$

The value ΔE has been assumed to be equal to Δh where h is the head. This is so if $v^2/2g$ is negligibly small and this is certainly true for percolating flows.

Now this means that $kE = \varphi$ as can be seen from

$$v = \partial\varphi/\partial s = k\,\Delta E/\Delta s$$

The flow is therefore a potential flow (and a flow net can be drawn) if the velocity head is negligibly small.

It might seem that the application of potential theory to what is essentially a highly frictional flow must be wrong. The flow of a fluid through a porous medium does not however produce vorticity on a macroscopic scale although it does so on a microscopic scale. Frictional effects occur throughout the fluid but these do not necessarily generate overall velocity gradients. As fluid flows round particles frictional effects do occur, but the overall potential flow model is obtained by ignoring the effect of these tiny boundaries and considering the flow through the space they occupy using velocities adjusted to take their presence into account.

The effect of friction acting at all points throughout the fluid is not the same as the frictional effect of a flow boundary. A boundary generates velocity gradients and these in turn cause fluid particles to rotate but friction distributed throughout the flow is considered to reduce the velocity of fluid elements without causing their rotation, so the potential flow model is applicable, at least on a macroscopic scale.

2.23.1 Flow beneath a dam
The flow beneath a dam is illustrated in Fig. 2.50.

The value E can be calculated for any point on the base of the dam. Upstream it has the value h. If it is reduced by uniform increments, say ten, the value of E at point P is $\frac{5}{10}\,h$. The pressure head at point P is $\frac{5}{10}\,h + d$. Any other point on the dam base can be treated similarly and the pressure distribution can be found. Velocities at any point in the permeable stratum can also be obtained. It should be noted that just downstream of the dam the squares are small so the velocities

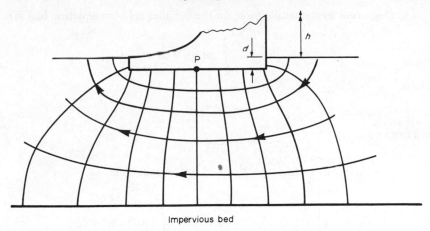

Impervious bed

Fig. 2.50

through the material are high. If they are high enough they may cause particles of sand to lift and be carried away. Once this process starts, the erosion that occurs will shorten the fluid's frictional path so velocities will increase still further causing erosion to occur more and more rapidly. This is a catastrophic sequence of events and eventually the downstream erosion will become so great that the foundations of the dam will be endangered. The material immediately below the dam will suddenly be removed and the dam will collapse. It usually falls upstream rather than downstream—one example is the collapse of certain regulators on the river Nile. This phenomenon is called sand piping.

Worked examples

(1) A long cylinder is located in a flow with its longitudinal axis perpendicular to it. Assuming that the flow over the cylinder is two dimensional and that it can be represented by the zero streamline in a combination of a uniform wind and a doublet, show that the cylinder experiences a force that places it in tension and calculate the magnitude of this force when the cylinder is 0·05 m in diameter and the uniform wind has a velocity of 3 m/s. The fluid is water and the ambient pressure is 100 kN/m² abs.

The equation applicable is (2.10), that is

$$\frac{p_P - p_0}{w} = \frac{U^2}{2g}(1 - 4\sin^2\theta)$$

Now $(p_P - p_0)/w$ is negative when $4\sin^2\theta > 1$, that is when $\sin\theta > \frac{1}{2}$. So $(p_P - p_0)/w$ is negative if θ is greater than 30° and less than 150°. This negative pressure distribution will apply tension to the cylinder.

The magnitude of the tensile force on the cylinder is (from equation (2.12))

$$F_V = - \int_0^\pi pa \sin \theta \ d\theta$$

where

$$a = 0 \cdot 025 \ \text{m}$$

and from equation (2.14)

$$F_V = -a \int_0^\pi (p_0 + \tfrac{1}{2}\rho U^2 (1 - 4 \sin^2 \theta)) \sin \theta \ d\theta$$

$$\therefore \qquad F_V = -a \left[\int_0^\pi (p_0 + \tfrac{1}{2}\rho U^2) \sin \theta \ d\theta - \int_0^\pi 2\rho U^2 \sin^3 \theta \ d\theta \right]$$

The first integral is $2p_0 - \rho U^2$. The second integral

$$I_2 = 2\rho U^2 \int_0^\pi \sin^2 \theta \ d(- \cos \theta)$$

$$= - 2\rho U^2 \int_0^\pi (1 - \cos^2 \theta) \ d(\cos \theta)$$

$$= - 2\rho U^2 \ [\cos \theta - \tfrac{1}{3} \cos^3 \theta]_0^\pi$$

$$= - 2\rho U^2 \ [(-1 - 1) - (-\tfrac{1}{3} - \tfrac{1}{3})] = \tfrac{4}{3}\rho U^2$$

$$F_V = - 2ap_0 + \tfrac{7}{3}a\rho U^2$$

$$\therefore \qquad = - 2 \times 0 \cdot 025 \times 1000 + \tfrac{7}{3} \times 0 \cdot 025 \times 1000 \times 3^2$$

The tensile force acting on the cylinder is thus 475 N/m.

(2) Given that the stream function for flow over a stationary cylinder immersed in a stream flowing at a velocity U is

$$\psi = - Uy + \frac{Ua^2 y}{x^2 + y^2}$$

and using the principle of superposition of velocities to obtain the stream function of a cylinder moving at velocity U through a stationary fluid, develop an equation for the kinetic energy stored in a fluid caused by movement of a cylinder of radius a at velocity through a fluid which is stationary at infinite distances from the cylinder.

An empty steel cylinder of 1 m external diameter and of wall thickness 0·04 m moves in a direction normal to its length. Neglecting all viscosity effects calculate the ratio of the forces required to accelerate the cylinder in sea water ($\rho_{\text{sea water}} = 1070$ kg/m^3) to that required to produce the same acceleration in air ($\rho_{\text{air}} = 1·284$ kg/m^3). ($\rho_{\text{steel}} = 7217$ kg/m^3)

The stream function (ψ_c) of a cylinder moving through a fluid is obtained by adding the stream function of the stationary cylinder over which a uniform wind is flowing to that for a uniform wind moving in the opposite direction i.e.

$$\psi_c = \left(-Uy + \frac{Ua^2 y}{x^2 + y^2} \right) + Uy$$

so

$$\psi_c = \frac{Ua^2 y}{x^2 + y^2} = \frac{Ua^2 \sin \theta}{R}$$

(the second expression is in polar coordinates).

The velocity component of this flow at coordinates R, θ are

$$u_R = \frac{Ua^2 \cos \theta}{R^2} ; \qquad u_\theta = -\frac{Ua^2 \sin \theta}{R^2}$$

so the resultant velocity

$$V = \sqrt{(u_R^2 + u_\theta^2)} = \frac{Ua^2}{R^2} (\cos^2 \theta + \sin^2 \theta)^{1/2}$$

$$V = Ua^2 / R^2$$

The kinetic energy of an element at R, θ is

$$\tfrac{1}{2} mV^2 = \tfrac{1}{2} \rho R \, \delta \theta \, \delta R \, \frac{U^2 a^4}{R^4} = \frac{U^2}{2} \frac{\rho a^4}{R^3} \delta \theta \, \delta R$$

Total KE of the entire fluid field

$$\int_0^{2\pi} \int_a^\infty \frac{U^2}{2} \frac{\rho a^4}{R^3} \delta \theta \, \delta R = - U^2 \pi \rho a^4 \left[\frac{1}{2R^2} \right]_a^\infty$$

$$= \tfrac{1}{2} \pi a^2 \rho U^2$$

∴ KE of flow field = KE of fluid displaced by the cylinder.

As KE = $\int$ Force ds = $\int$ mass × acceleration ds

the kinetic energy is proportional to mass for a given acceleration.

In the case of a cylinder moving through a fluid, the mass of the cylinder must be increased by the mass of the displaced fluid.

$$\text{Force ratio} = \frac{\frac{\pi}{4}(d_e^2 - d_i^2) \times \rho_{steel} + \frac{\pi}{4} d_e^2 \times \rho_{sea\ water}}{\frac{\pi}{4}(d_e^2 - d_i^2) \times \rho_{steel} + \frac{\pi}{4} d_e^2 \times \rho_{air}}$$

$$= \frac{(1^2 - 0\cdot92^2)7217 + 1^2 \times 1070}{(1^2 - 0\cdot92^2)7217 + 1^2 \times 1284}$$

$$\therefore \qquad \text{Force ratio} = 1\cdot97$$

Questions

(1) The stream function of a two-dimensional frictionless incompressible flow against a flat plate normal to the initial velocity is $\psi = -2axy$. Obtain the velocity potential for this flow and show that both the stream function and the velocity potential satisfy the Laplace equation. Given that $a = 0\cdot5$ plot the stream lines defined by $\psi = 1$ and 2 and the equipotentials defined by $\varphi = 1$ and 2. Obtain the velocity at the intersection of the stream line $\psi = 1$ and the equipotential line $\phi = 1$ and given that the pressure at a point $x = 4$ ft [4 m], $y = 0$ is 2 lbf/in^2 [60 N/cm^2] calculate the pressure at this intersection point.

Answer: $v = 1\cdot68$ ft/s [$1\cdot68$ m/s], $p = 2.09$ lbf/in^2 [$60\cdot7$ N/cm^2].

(2) The flow around a cylinder situated in a uniform flow can be considered to be the same as that produced by a combination of a doublet and a uniform flow if the viscosity is negligible.

(a) Derive an expression that describes the pressure distribution around such a cylinder and draw a graph in cartesian coordinates of pressure against θ.
(b) Indicate on this graph how the behaviour of a real fluid differs from its theoretical behaviour.
(c) Plot on graph paper the stream lines for $\psi = 1$ and $\psi = 2$.

(3) A cylinder is mounted in an air flow with its longitudinal axis transverse to the flow. A small hole is drilled in the surface of this cylinder and this hole is connected to an air–water differential manometer and acts as a pressure tapping. Using this tapping and rotating the cylinder slowly the pressure distribution around the cylinder can be obtained. As the cylinder is rotated it is found that the reading on the manometer varies and over the up-stream face of the cylinder the difference between the maximum and minimum manometer readings is $1\cdot5$ in of water. Working from first principles, specify the locations of the pressure tapping where these maximum and minimum readings are obtained and calculate the undisturbed stream velocity. The specific weight of air is $0\cdot08$ lbf/ft^3.

Answer: At $0°$ and $90°$, $39\cdot6$ s.

(4) A two-dimensional uniform stream of incompressible invixid fluid flowing parallel to the x axis with a velocity of 12 ft/s [4 m/s] combines with a source at the origin. Interpreting this as a flow over a body defined by $\psi = 0$ of 3 ft [1 m] width at $x = -\infty$ sketch carefully the flow over the body. Calculate

(a) the distance from the source to the stagnation point,
(b) the distance from the source to the body surface at an angle of 90° to the x axis,
(c) the magnitude of the velocity and its direction relative to the x axis of the fluid along the surface at the 90° angle.

Answer: (a) 0·477 ft   (b) 0·75 ft   (c) 14·2 ft/s at $32\frac{1}{2}°$
       (a) [0·159 m]   (b) [0·25 m]   (c) [4·73 m/s at $32\frac{1}{2}°$].

(5) Define a source and sink. If a source and sink of equal strength are separated by a distance of $2a$ show that the stream lines will be circular arcs. Taking the distance a as 1 cm draw the streamline on one side of the axis only having a radius of 5 cm and calling the value of the stream function for this line unity insert the streamlines corresponding to the values 2 and 4.

(6) Explain the meaning and use of the stream function in dealing with problems of two-dimensional fluid motion. A line source of strength $50\,\pi$ ft s units is situated in a uniform stream flowing at 40 ft/s. At a distance of 2 ft downstream from the source there is an equal sink. Locate two points of zero velocity in the resultant field of flow and show how to trace the streamlines passing through these two points.

(7) Find the expression for the stream functions for a point source of strength m and for a point vortex of circulation Γ and hence obtain the stream function for the combined flow when the source is at the centre of the vortex. Find the expressions for the radial and tangential velocities and hence show that the streamlines make a constant angle with the radius vector, also show that the locus of all points having constant velocity V is the circle

$$R = \frac{\sqrt{(m^2 + \Gamma^2)}}{2\pi V}$$

(8) Calculate the circulation around a cylinder of radius 1 m if the velocity in metres per second is

$$V = 7 \sin \theta + 7$$

Also determine the lift per metre length if the free stream velocity is 3 m/s and the fluid density is 1·25 kg/m³.

Answer: $\Gamma = 14\pi\,m^2/s$, $L = 165$ N/m.

(9) A 90° bend of rectangular cross section with inner radius 2 m and external radius 6 m carries a flow of gas. Ignoring frictional effects and assuming that no boundary layer separations occur sketch a flow net for the flow and indicate a method of obtaining the pressure profile across the line of symmetry of the bend.

3 Dimensional Analysis

The solution of any problem in the physical sciences is usually expressed in the form of a relationship between a number of variables. Normally, such a relationship is written as an equation which describes how the dependent variable varies with the independent variables. The methods of establishing such relationships fall into two classes.

In the first method, a mathematical model of the phenomenon is developed. This consists of a statement of the physical laws that are operating expressed in the form of equations. Almost always some assumptions have to be made that simplify the situation to a point at which the resulting mathematics can be performed. Such a simplified mathematical model may represent the situation quite accurately but sometimes the simplifying assumptions are so drastic that the model bears little relation to the physical problem.

The second method is used when either the mathematical approach does not give good results, or the problem is so complex that no model can be developed whatever assumptions are made. Sometimes, such problems involve so many variables, or require mathematics of such difficulty, that no analytic solution is possible. In this case a different approach is required. One approach is to construct a physical model which is capable of reproducing the phenomenon—usually in miniature. Measurements of parameters of interest can be taken from this model and scaling laws are then used to derive values applicable to the real situation (the prototype). This approach is particularly useful in fluid mechanics. Much money and effort has been invested in developing models to represent the forces acting upon hydraulic structures, or to predict the magnitudes and directions of fluid motions.

Dimensional analysis forms the basis for physical modelling. This technique allows measurements taken from a model to be scaled up to apply to the prototype. Dimensional analysis also allows the variables in an equation to be arranged into a smaller number of dimensionless groups of variables. In problems involving fluids this can be of great importance.

Dimensional analysis is concerned with the fundamental relationships between variables. Examples of different variables are velocity, length, acceleration and

82

mass density. These quantities are fundamentally different from one another although they possess certain qualities in common. It is reasonable to assume that all variables can be derived from quite a small number of certain basic entities. These are called *dimensions* and should not be confused with the quantities usually described by this word, for example length, breadth and height.

Two groups of dimensions can be used, one comprising mass [M], length (L), and time (T). There is no reason to think that mass is a more basic quantity than force and so the other set of dimensions that can be used is force (F), length (L) and time (T).

The dimensions of all other variables can be expressed in terms of either of these sets of dimensions. For instance velocity must have the dimensions of L/T as it is the ratio of distance travelled to the time taken. Acceleration (velocity/time) will have the dimensions of L/T^2, force (mass x acceleration) the dimensions of ML/T^2 and stress (force/area) the dimensions of M/LT^2. Table 3.1 gives the dimensions of the quantities that will be used in this book.

Table 3.1.

Quantity	Symbol	Dimensions
Mass	M	M or FT^2/L
Length	L	L
Time	T	T
Velocity	v	L/T
Acceleration	a	L/T^2
Force	F	ML/T^2 or F
Stress or pressure	f or p	M/LT^2 or F/L^2
Any modulus	E or K	M/LT^2 or F/L^2
Mass density	ρ	M/L^3 or FT^2/L^4
Weight density	w	M/L^2T^3 or F/L^3
Dynamic viscosity	μ	M/LT or FT/L^2
Kinematic viscosity	v	L^2/T
Flow	Q	L^3/T
Energy or work	W	ML^2/T^2 or FL
Power	P	ML^2/T^3 or FL/T
Surface tension	σ	M/T^2 or F/L

Any problem can be expressed as a relationship between a group of variables, that is

$$f(V_1, V_2, V_3, V_4, \ldots V_i, \ldots V_n) = 0$$

where V_i denotes some variable and f denotes 'a function of'.

Consider the problem presented by the need to predict the resistance to motion R experienced by a body of size l when travelling at velocity v through a fluid of mass density ρ and viscosity μ. There must be a function such that

$$f(R, \rho, l, v, \mu) = 0$$

It is possible to express such a relationship in the form of an infinite series so that in general

$$R = A\,(\rho^a l^b v^c \mu^d) + B\,(\rho^e l^f v^g \mu^h) + \ldots \qquad (3.1)$$

every term on the right hand side must represent an element of force as the left hand side is a force and it is impossible to equate or to add unlike things. Thus the dimensions of each of the terms on the right hand side are those of R. Consider then the first term of the right hand side and substitute the dimensions of each variable for that variable. The dimensions of the resulting group of dimensions must reduce to those of the left hand side, that is force. Therefore

$$\mathrm{MLT}^{-2} = \mathrm{M}^a \mathrm{L}^{-3a} \mathrm{L}^b \mathrm{L}^c \mathrm{T}^{-c} \mathrm{M}^d \mathrm{L}^{-d} \mathrm{T}^{-d}$$

Equating the indices of the three dimensions

$$\text{M:} \quad 1 = a + d \qquad (3.2)$$

$$\text{T:} \; {-2} = -c - d \qquad (3.3)$$

$$\text{L:} \quad 1 = -3a + b + c - d \qquad (3.4)$$

From equation (3.2) $a = 1 - d$. From equation (3.3) $c = 2 - d$. Substituting for a and c into equation (3.4) gives

$$1 = -3 + 3d + b + 2 - d - d$$

hence
$$b = 2 - d$$

(Note that the constant A was not considered when writing the dimensions of the variables as it is a pure number, that is a ratio, with no dimensions.)

The process performed above could now be carried out for the second term also and the result would be

$$e = 1 - h$$
$$f = 2 - h$$
$$g = 2 - h$$

If these values of the indices are now substituted back into equation (3.1) the series will be:

$$R = A(\rho^{1-d} l^{2-d} v^{2-d} \mu^d) + B(\rho^{1-h} l^{2-h} v^{2-h} \mu^h) + \ldots$$

A common group $\rho l^2 v^2$ occurs in every term and this can be taken out so that

$$R = \rho v^2 l^2 \left(A \left(\frac{\mu}{\rho v l} \right)^d + B \left(\frac{\mu}{\rho v l} \right)^h + \ldots \right)$$

The infinite series in the brackets is a function of the group $\mu/\rho v l$. The reciprocal, $\rho v l/\mu$, of this group is more usually used, so the series then becomes

$$R = \rho v^2 l \; \varphi \; (\rho v l/\mu)$$

If the dimensions of the variables in the group $\rho vl/\mu$ are substituted for the variables themselves it will be found that they will all cancel, that is the group is dimensionless. It is known as Reynolds number. The group $R/\rho v^2 l^2$ is also dimensionless; this can be seen when the equation is re-arranged:

$$\frac{R}{\rho v^2 l^2} = \varphi\left(\frac{\rho vl}{\mu}\right)$$

As the group on the right hand side is dimensionless the one on the left hand side must also be dimensionless.

In place of a relationship between a set of variables that describe a physical phenomenon a relationship between two dimensionless groups has been obtained. The number of variables has been reduced from five to two if each dimensionless groups is regarded as an individual variable. Without such a dimensional analysis it would be necessary to establish how R varies with each of ρ, v, l, and μ. Each variable would have to be altered in turn while the others were kept constant. Such an investigation would be prohibitively laborious and also very expensive, as it would involve changing the working fluid to investigate the effect of ρ and μ. On the other hand, a quite simple experiment can be performed to investigate the variation of $R/\rho v^2 l^2$ with $\rho vl/\mu$ and a single curve relating these two variables would then convey all the information that would have been obtained in the previous experiment. Moreover, the variable $\rho vl/\mu$ can be changed simply by altering the value of v; there is no need to alter any of the other variables. Since the dimensional analysis has reduced the five variables involved to two the number of graphs required has been reduced from four to one.

In general, the process of dimensional analysis will reduce the number of variables in a problem by the number of dimensions involved in each variable, for example, six variables would be reduced by three if mass length and time are all involved in each variable. The reduction will be correspondingly less if one of the dimensions is not involved in a particular variable.

The rather tedious process of dimensional analysis illustrated above can be greatly simplified.

3.1 The Buckingham π theorems

The ideas discussed above can be formalised as two theorems. In the western world these are usually ascribed to Buckingham.

3.1.1 First theorem A relationship between a set of physical variables can be stated as a relationship between a set of independent dimensionless groups made up from suitably chosen variables.

3.1.2 Second theorem The number of dimensionless groups required to specify completely the relationship is the number of variables, n, minus the number of dimensions, m, involved in the variables. (These dimensionless groups are often called π groups). Although this may be intuitively obvious it is very difficult to prove with mathematical rigour.

These two theorems will be used in the problem discussed above. Initially an intuitive judgment is made as to the variables involved. As before, the resistance to motion R is judged to depend upon the density and viscosity of the fluid, ρ and μ, through which the object is moving, its size l, and its velocity v. If this judgment had been wrong and a variable inserted which was not in fact having any influence upon the resistance, the dimensional analysis would produce an additional group which would be inapplicable in the circumstances. Conversely, if a variable had been omitted a dimensionless group of significance would not be generated. The Buckingham π theorems permit the equation

$$f(\pi_1, \pi_2) = 0 \qquad (3.5)$$

to be written. (Only two groups are involved as $n = 5$ and $m = 3$ so the number of π groups $= n - m = 5 - 3 = 2$.) All that remains to be done is to develop expressions for these π groups. It is possible to write a large number of π groups that will all be dimensionless and which will therefore be suitable for use. However, some can be immediately discarded. For example if two groups are produced which are not independent of one another, one or the other can be discarded. Then some groups will be less suitable than others. It will be shown below that many groups represent ratios of real quantities, for example Reynolds number is a ratio of the inertia forces acting on a body to the viscous forces acting on it. The group $R/\rho v^2 l^2$ is the ratio of the resistance force experienced to the inertia force. Other groups however, have no such physical interpretation and it is therefore logical to eliminate these.

Consider the flow between any two adjacent streamlines. If the area of flow is A, the momentum flow (or inertia) is $\rho A v^2$ and the resistance experienced by this flow is proportional to $\mu A \, dv/dy$ (by Newton's law of viscosity). Now the distance apart of two adjacent streamlines depends upon the scale of the object creating the motion, in other words the value l. The area A depends upon l^2 and dy depends upon l.

The ratio of inertial force/viscous force acting at any point in the field of flow is $\rho A v^2 / \mu A \, dv/dy$. Now dv depends upon v—the larger v the larger dv so

$$\frac{\text{inertial force}}{\text{viscous force}} = \frac{\rho A v^2}{k \mu A v/l} \propto \frac{\rho v l}{\mu}$$

that is, Reynolds number is proportional to the ratio inertial force/viscous force.

3.2 Construction of π groups

The technique of constructing a π group will now be described. Initially the decision must be made as to what type of group is required. Say the group is to be a ratio of inertia force to some other type of force. Inertia force is proportional to the rate of flow of momentum, $\rho A v^2$ or $\rho l^2 v^2$. Thus, if the inertia force is to be involved in the π group the variables ρ, v, and l must also occur in the group. A group can be formed by combining any four variables which between them contain the dimensions of mass, length and time. If three of the variables chosen are of the form mass density, velocity and length, the resulting group will be a ratio of inertia force to some other force. For instance, if the variables selected are ρ, v, l and μ the resulting group will be the ratio of inertia force to viscous force, that is Reynolds number. Thus

$$\pi_1 = \rho^a v^b l^c R$$

The second π group can be formed similarly, the repeating variables ρ, v, l, being combined with the other variable μ

$$\pi_2 = \rho^a v^b l^c \mu$$

The method of processing these π groups can now be demonstrated. Each group must be dimensionless, thus for π_1

$$M^a L^{-3a} L^b T^{-b} L^c MLT^{-2} = |0|$$

Equating indices of M, L and T respectively to zero gives

$$M: a + 1 = 0$$
$$T: -b - 2 = 0$$
$$L: -3a + b + c + 1 = 0$$

Thus $a = -1$, $b = -2$ and $c = -2$. Hence

$$\pi_1 = \rho^{-1} v^{-2} l^{-2} R$$

Similarly

$$\pi_2 = \rho^a v^b l^c \mu$$

Thus

$$M^a L^{-3a} L^b T^{-b} L^c ML^{-1} T^{-1} = |0|$$

$$M: a + 1 = 0$$
$$T: -b - 1 = 0$$
$$L: -3a + b + c - 1 = 0$$

Thus $a = -1$, $b = -1$ and $c = -1$,

$$\pi_2 = \rho^{-1} v^{-1} l^{-1} \mu = \mu/(\rho v l)$$

In most problems, the value of $\mu/\rho vl$ is very small. In a pipe flow it could be 10^{-6} for instance. This is inconvenient and the reciprocal of this number is used instead. This is of course quite valid as it is also dimensionless. By substituting these expressions (3.5)

$$f\left(\frac{R}{\rho v^2 l^2}, \frac{\rho vl}{\mu}\right) = 0$$

or

$$\frac{R}{\rho v^2 l^2} = f\left(\frac{\rho vl}{\mu}\right) \qquad (3.6)$$

or

$$R = \rho v^2 l^2 f(\rho vl/\mu) \qquad (3.7)$$

This last form is called the Rayleigh equation.

(Note: The reader is recommended to follow the sequence indicated when writing the equations for the indices of M, L, and T, as he will then find that they solve explicitly as demonstrated.)

3.3 The physical significance of some commonly used groups

As stated before, the dimensionless groups commonly used all have physical significance as ratios of various physical quantities. The groups encountered in fluid mechanics are listed in Table 3.2.

The precise meaning of these groups is worth closer examination. In many circumstances, the behaviour of a body is controlled by two physical effects. For instance, the motion of a billiard ball depends upon its inertia and also upon

Table 3.2.

Group	Name	Symbol	Meaning
$\rho vl/\mu$	Reynolds number	Re, N_R	inertial stress/viscous stress
$v^2/(gl)$	Froude number	Fn, N_F	inertial stress/gravitational stress
$\rho v^2 l/\sigma$	Weber number	We, N_W	inertial stress/surface tension stress
v/c or $\sqrt{(\rho v^2/K)}$			
	Mach number	Mn, N_M	local velocity of fluid/local velocity of sound
$p/(\rho v^2)$	Euler number	En, N_E	pressure/inertial stress
$\rho V^2/E$	Cauchy number	Cn, N_C	(local velocity/wave velocity)^2

There are other numbers such as Petrov's number which arises in lubrication theory and Peclet's number which arises in heat transfer theory. Additionally there are numbers such as $\rho g^{0.5} h^{1.5}/\mu$ and $Q/g^{0.5} h^{2.5}$ which arise in free surface flow through apertures such as orifices, notches, and weirs. These are variant forms of the Reynolds and Froude numbers. Dimensionless numbers arising in the theory of rotodynamic machines such as Q/ND^3, $gH/N^2 D^2$, the efficiency and the Reynolds number $\rho ND^2/\mu$ establish the modelling criteria for such machines.

the frictional forces it experiences when travelling over the table's surface. If it is travelling rapidly, the frictional effects can almost be ignored and its behaviour in the immediate future can be predicted from its inertia (momentum) and the laws of impact. However, if it is travelling slowly, the frictional effects may well dominate the motion. A type of Reynolds number could be written for the ball and this, as before, would be the ratio of inertial force to frictional force. If this number is large, the frictional force is negligible and if it is small the frictional force would very largely determine the motion. Similarly in the case of fluid motion, if Reynolds number is large, frictional effects are relatively small and the motion is governed largely by the momentum of the fluid; conversely if it is small, the motion will be dominated by friction.

The Froude number appears in problems in which gravitational forces act upon and control, to some extent, the fluid motion. This happens when circumstances are such that free surfaces exist. Examples are wave motions on a liquid surface, river and channel flows and flow under gravitational action through orifices, notches and weirs. In many cases, frictional effects are present as well as gravitational effects; a dimensional analysis will then yield both the Reynolds and the Froude numbers. For example, in the case of a ship or sub-marine travelling on or near the surface of an ocean, the resistance to motion will be due partly to the energy it loses in creating waves upon the ocean surface and partly to the energy loss caused by friction between the ship's hull and the adjacent slower moving water. Its resistance will not only depend upon ρ, v, l and μ as described earlier but also upon g—acceleration due to gravity. A unit volume of fluid raised through a height h by the action of a wave will need $\rho g h$ units of energy; this indicates that g must be included in the dimensional analysis.

Therefore,

$$f(R, \rho, v, l, \mu, g) = 0$$

There are six variables and three dimensions so three π groups can be expected. Hence,

$$f(\pi_1, \pi_2, \pi_3) = 0$$

Now π_1 and π_2 will be the same as in the previous analysis and

$$\pi_3 = \rho^a v^b l^c g$$

Therefore,

$$M^a L^{-3a} L^b T^{-b} L^c L T^{-2} = |0|$$

$$M: a = 0$$

$$T: -b - 2 = 0. \quad \text{Therefore, } b = -2$$

$$L: -3a + b + c + 1 = 0. \quad \text{Therefore, } c = +1$$

Therefore,

$$\pi_3 = gl/v^2 \text{—this is inverted to give } v^2/gl$$

Therefore,

$$\frac{R}{\rho v^2 l^2} = f\left(\frac{\rho v l}{\mu}, \frac{v^2}{gl}\right)$$

$$v^2/gl = Fn = \text{Froude number}$$

or,

$$R = \rho v^2 l^2 f(Re, Fn)$$

The Froude number is the ratio: inertial force/gravitational force. This is Rayleigh's equation for the resistance of a body when both gravitational and frictional effects are present.

The Weber number arises in problems that are concerned with surface tension. Ripples (as opposed to waves) are a phenomenon caused by surface tension, so the Weber number can be expected to occur in the equations that describe ripple motions. At very low pressure heads flows over, Vee and rectangular notches are not as predicted by the discharge equations. This is because surface tension is significant under such circumstances and the theory ignores surface tension. (At pressure heads of approximately $\frac{1}{8}$ inch the liquid may stand on the edge of a notch and there is no flow at all.)

Normally the coefficient of discharge of a notch (this is a type of Froude number) is plotted against a form of Reynolds number. At low heads it would be more sensible to plot this coefficient against the Weber number as surface tension effects are more important than friction effects. This is not done in practice as there is little interest in the behaviour of flow-measuring devices at such low heads.

3.4 Models

As the basis of most modelling techniques is dimensional analysis, it is appropriate to describe this basis at this point.

Usually a model of the hydraulic structure under examination is built. The scale of this model may be greater than full size, but normally it is less than full size. (The hydraulic structure under investigation is called the prototype.) The model is usually built to an undistorted geometric scale. This is not always true, as in some river models different scales are used for horizontal and vertical distances. Measurements can be taken from such a model and a suitable scaling law applied which will predict their values for the prototype. These scaling laws are usually derived from dimensional considerations.

To illustrate how such scaling laws can be obtained from dimensional analysis, we use the result obtained earlier (equation (3.6)) for the resistance experienced by a body moving through a fluid

$$R/(\rho v^2 l^2) = \varphi(\rho v l/\mu)$$

This equation applies whatever the size of the body. This means that l defines the size of the body; it can be thought of as defining a leading dimension, all other dimensions of the body bearing some fixed relationship to l. Thus all geometrically similar bodies will be described by the same function φ irrespective of the fluid used. This can be demonstrated by inserting into the analysis another length—say d. There will now be an additional π group which must be either d/l or l/d.

Equation (3.6) will then become

$$R/(\rho v^2 l^2) = \varphi\ (\rho v l/\mu, d/l) \tag{3.8}$$

If the value of d/l is the same for two bodies of different sizes—that is they are geometrically similar—then equation (3.8) above reduces to equation (3.6). This result must be true for both the model and the prototype. (If the nature of the function φ were known there would be no need to build a model at all, but as the problem cannot be analytically analysed (except in certain specific ranges of the value of Reynolds number) this function is *not* known.)

Write the relationship for the model and the prototype using subscripts m for the model and p for the prototype

$$\frac{R_m}{\rho_m v_m^2 l_m^2} = \varphi\left(\frac{\rho_m v_m l_m}{\mu_m}\right)$$

and

$$\frac{R_p}{\rho_p v_p^2 l_p^2} = \varphi\left(\frac{\rho_p v_p l_p}{\mu_p}\right)$$

Dividing the first by the second equation

$$\frac{R_m/(\rho_m v_m^2 l_m^2)}{R_p/(\rho_p v_p^2 l_p^2)} = \frac{\varphi(\rho_m v_m l_m/\mu_m)}{\varphi(\rho_p v_p l_p/\mu_p)}$$

We can now get no further unless an assumption is made. If it is assumed that $\rho_m v_m l_m/\mu_m$ is equal to $\rho_p v_p l_p/\mu_p$ then $\varphi(\rho_m v_m l_m/\mu_m) = \varphi(\rho_p v_p l_p/\mu_p)$ which gives

$$\frac{R_m/(\rho_m v_m^2 l_m^2)}{R_p/(\rho_p v_p^2 l_p^2)} = 1$$

so

$$\frac{R_m}{R_p} = \frac{\rho_m v_m^2 l_m^2}{\rho_p v_p^2 l_p^2} \tag{3.9}$$

or

$$\lambda_R = \lambda_\rho \lambda_v^2 \lambda_l^2$$

where λ here denotes 'scale of'. This is the scaling law of resistance.

The equality of the Reynolds numbers of the model and prototype was an essential requirement in the development of this result. The implications of this assumption must next be examined.

If

$$Re_m = Re_p \qquad (3.10)$$

then

$$\frac{\rho_m v_m l_m}{\mu_m} = \frac{\rho_p v_p l_p}{\mu_p}$$

so

$$\frac{v_m}{v_p} = \frac{\rho_p}{\rho_m} \frac{\mu_m}{\mu_p} \frac{l_p}{l_m} \qquad (3.11)$$

or

$$\lambda_v = \frac{\lambda_\mu}{\lambda_\rho \lambda_l} \qquad (3.12)$$

The symbol v (nu) is often used to represent μ/ρ and this is called the kinematic viscosity of the fluid as opposed to μ—the dynamic viscosity of the fluid.

Equation (3.11) can thus be written

$$\frac{v_m}{v_p} = \frac{\nu_m}{\nu_p} \frac{l_p}{l_m}$$

or

$$\lambda_v = \lambda_\nu / \lambda_l$$

Substituting this result back into equation (3.9) gives

$$\frac{R_m}{R_p} = \frac{\rho_m}{\rho_p} \left(\frac{\nu_m}{\nu_p} \frac{l_p}{l_m}\right)^2 \left(\frac{l_m}{l_p}\right)^2$$

Therefore,

$$\frac{R_m}{R_p} = \frac{\rho_m}{\rho_p} \left(\frac{\nu_m}{\nu_p}\right)^2$$

or

$$\lambda_R = \lambda_\rho (\lambda_\nu)^2$$

This means that, given the value of R_m (obtained by direct measurement from the model), the force experienced by the prototype can be predicted, but only when it is moving at a velocity that produces the same Reynolds number as that of the model when the R_m measurement was taken. That is to say that R_p can be predicted from

$$R_p = \frac{\rho_p}{\rho_m} \left(\frac{v_p}{v_m}\right)^2 R_m$$

and the result will apply when the prototype is travelling at a velocity

$$v_p = \frac{v_p}{v_m} \frac{l_m}{l_p} v_m$$

Dynamic similarity between model and prototype is said to exist when the controlling dimensionless group on the right hand side of the dimensionless relationship is the same for both. In equation (3.10) Re_m was equated to Re_p and this condition is now described as that of dynamic similarity. Reynolds number was earlier shown to be the ratio inertia force/viscous force, so that for dynamic similarity

$$\frac{(\text{inertial force})_m}{(\text{inertial force})_p} = \frac{(\text{viscous force})_m}{(\text{viscous force})_p}$$

for all points throughout the fluid in both the model and prototype. This means that fluid paths (streamlines in steady flow) in model and prototype are geometrically similar.

If the problem is that of a ship traversing a water surface, wave motions would be the dominating influence and, as explained before, g would have to be introduced into the analysis. As before,

$$R/(\rho v^2 l^2) = \varphi(Re, Fn)$$

For strict modelling, the Reynolds and Froude numbers of the model and prototype must be made equal to one another if the two functions are to be equal. If this could be done, then true dynamic similarity would be obtained—the streamlines of the flow around the ship would be geometrically similar and at the same time the wave motions around the ship would be geometrically similar. Unfortunately, conditions under which both the Reynolds numbers and Froude numbers for model and prototype are equal cannot exist. For example, if $Re_m = Re_p$ and $Fn_m = Fn_p$

$$\frac{v_m}{v_p} = \frac{v_m}{v_p} \frac{l_p}{l_m}; \quad \frac{v_m}{v_p} = \sqrt{\left(\frac{l_m}{l_p}\right)}$$

equating these values of v_m/v_p gives

$$\frac{v_m}{v_p} = (l_m/l_p)^{3/2}$$

The possibility of using a different liquid for the model from that used in the prototype, is not really feasible. Modelling techniques involve very large volumes of liquids and the use of a liquid other than water would be prohibitively expensive. Further, many liquids that might be used are so offensive and dangerous that they would not be suitable, even if expense were no object. The impossibility of simultaneous Reynolds and Froude-number modelling is well recognised and causes both the ship designer and the river hydraulicist much trouble and work. Generally, when the gravitational effects present are important, the model is operated on the Froude- number scaling laws and the effects of failing to model for Reynolds number are ignored. This means there may be errors in the predictions for the prototype; this is called *scale effect*. Special formulae have been developed to correct such errors which are unfortunately inevitable.

At this point it is as well to realise that Reynolds-number modelling also involves difficulties. The results previously derived for the moving body problem are

$$\frac{R_m}{R_p} = \frac{\rho_m}{\rho_p} \left(\frac{v_m}{v_p}\right)^2$$

and

$$\frac{v_m}{v_p} = \frac{\nu_m}{\nu_p} \frac{l_p}{l_m}$$

Now if the modelling fluid is the same as that occurring in the prototype situation

$$v_m/v_p = l_p/l_m$$

This suggests that if the model is to be smaller than the prototype, its operating speed must be greater, in inverse ratio to the scale used. Also if the fluid is the same ($\rho_m = \rho_p$ and $\nu_m = \nu_p$) then $R_m = R_p$. To provide apparatus that is capable of meeting these requirements can be formidably difficult. For instance, a common application of Reynolds-number modelling is in aircraft design. If the prototype speed is to be 400 mph and the model is to be one tenth the size of the prototype, then its speed should be 4000 mph. This places the model speed into the hypersonic range where compressibility effects—hitherto ignored—will be of far greater significance than the frictional effects described by Reynolds-number modelling. Furthermore the model will have to withstand the same forces as would be experienced by the prototype. Supports for the model will not be easy to design under such circumstances. If the model is tested in air which is pressurised to, say, 30 atmospheres, the density of the air will be increased by a factor of 30 but its dynamic viscosity will be only marginally affected. Thus ν_m will be reduced by a factor of approximately 30 and ρ_m increased correspondingly. Then,

$$v_m = 1/30 \times 10 \times 400 = 133\tfrac{1}{3} \text{ mph}$$

and

$$R_m = 30 \times (1/30)^2 \times R_p = R_p/30$$

A pressurised wind tunnel needs strong and expensive structures and its operating costs are very high, largely due to its great power demands. Only large commercial undertakings and national governments can build such tunnels.

Pure Froude-number modelling is relatively uncomplicated and as free surface flows are much more the province of the civil engineer than the mechanical engineer, the former is more fortunate than his colleague who frequently has to use Reynolds-number modelling. The civil engineer's problems really start when a problem involves Froude-number modelling but frictional effects are not negligible.

3.5 Examples of dimensional analysis

In this section, various problems that are discussed elsewhere in this book are analysed by dimensional analysis. The section may be omitted on a first reading but may be found helpful later. Reference will be made to this section when such problems arise.

3.5.1 Orifices, notches and weirs

Flow through orifices, notches and weirs are all examples of free-surface flows. The flow Q will depend upon the following parameters.

(1) The diameter of the orifice, or the breadth of the weir whichever is appropriate.
(2) The head, h, over the device.
(3) The fluid parameters: mass density ρ, and the dynamic viscosity μ.
(4) The intensity, g, of the earth's gravitational field.

That is,

$$Q = f(l, \rho, g, h, \mu)$$

or,

$$f(Q, l, \rho, g, h, \mu) = 0$$

There are six variables and three dimensions are involved so the number of π groups is three. That is,

$$f(\pi_1, \pi_2, \pi_3) = 0$$

Now it was stated earlier that ρ, v, and l should be chosen to be repeating variables, but in this problem there is no variable v. However, v is implicit in Q and l—as the ratio of Q to l^2 is a velocity. Alternatively a velocity can be obtained by combining g and h. In the case of an orifice for example, $v = c_v \sqrt{(2gh)}$. Thus there are two equally valid approaches and both will be given here.

$$\pi_1 = \rho^a Q^b l^c g \qquad \text{or} \qquad \pi_1 = \rho^a g^b h^c Q$$

$$\therefore \ M^a L^{-3a} L^{3b} T^{-b} L^c LT^{-2} = |0| \qquad \text{or} \qquad M^a L^{-3a} L^b T^{-2b} L^c L^3 T^{-1} = |0|$$

$$\text{M: } a = 0 \qquad\qquad\qquad\qquad\qquad a = 0$$

$$\text{T: } -b - 2 = 0 \ \ \therefore \ b = -2 \qquad\qquad -2b - 1 = 0 \ \ \therefore \ b = -\tfrac{1}{2}.$$

$$\text{L: } -3a + 3b + c + 1 = 0 \qquad\qquad -3a + b + c + 3 = 0$$

$$0 - 6 + c + 1 = 0 \qquad\qquad\qquad 0 - \tfrac{1}{2} + c + 3 = 0$$

$$c = 5 \qquad\qquad\qquad\qquad\qquad c = -2\tfrac{1}{2}$$

$$\therefore \qquad \pi_1 = \frac{g l^5}{Q^2} \qquad\qquad\qquad \pi_1 = \frac{Q}{g^{1/2} h^{5/2}}$$

Dimensional analysis cannot differentiate between h and l, as both have the dimensions of length. Bearing this in mind it can be seen that the second π_1 group is the reciprocal of the square root of the first. In effect they are the same.

$$\pi_2 = \rho^a Q^b l^c h \qquad\qquad \pi_2 = \rho^a g^b h^c l$$

$$M^a L^{-3a} L^{3b} T^{-b} L^c L = |0| \qquad M^a L^{-3a} L^b T^{-2b} L^c L = |0|$$

$$\text{M: } a = 0 \qquad\qquad\qquad \text{M: } a = 0$$

$$\text{T: } b = 0 \qquad\qquad\qquad\quad \text{T: } b = 0$$

$$\text{L: } -3a + 3b + c + 1 = 0 \qquad \text{L: } -3a + b + c + 1 = 0$$

$$\therefore \ \ c = -1 \qquad\qquad\qquad \therefore \ \ c = -1$$

$$\pi_2 = h/l \qquad\qquad\qquad\quad \pi_2 = l/h$$

so once again, the second π_2 group is effectively the same as the first, as it is the reciprocal of it.

$$\pi_3 = \rho^a Q^b l^c \mu \qquad\qquad \pi_3 = \rho^a g^b h^c \mu$$

$$M^a L^{-3a} L^{3b} T^{-b} L^c ML^{-1} T^{-1} = |0| \qquad M^a L^{-3a} L^b T^{-2} L^c ML^{-1} T^{-1} = |0|$$

$$\text{M: } a + 1 = 0, \ \therefore a = -1 \qquad \text{M: } a + 1 = 0, \ \therefore \ a = -1$$

$$\text{T: } -b - 1 = 0, \ \therefore b = -1 \qquad \text{T: } -2b - 1 = 0, \ \therefore b = -\tfrac{1}{2}$$

$$\text{L: } -3a + 3b + c - 1 = 0 \qquad \text{L: } -3a + b + c - 1 = 0$$

$$\therefore \ \ c = 1 \qquad\qquad\qquad \therefore \ \ c = -\tfrac{3}{2}$$

$$\pi_3 = \frac{l \mu}{\rho Q} \qquad\qquad\qquad \pi_3 = \frac{\mu}{\rho g^{1/2} h^{3/2}}$$

These two are apparently quite different but both are forms of Reynolds number. (Note: π_1 is a form of Froude number). Taking the reciprocals of both gives:

$$\pi_3 = \frac{\rho Q}{l\mu} \quad \text{and} \quad \pi_3 = \frac{\rho g^{1/2} h^{3/2}}{\mu}$$

The result of the dimensional analysis could be written as

$$\frac{Q^2}{gl^5} = f\left(\frac{\rho Q}{l\mu}, \frac{h}{l}\right)$$

or as

$$\frac{Q}{g^{1/2} h^{5/2}} = f\left(\frac{\rho g^{1/2} h^{3/2}}{\mu}, \frac{h}{l}\right) \qquad (3.13)$$

and both are valid. The second form is a little more useful than the first because Q is usually unknown and h is the known measured value. If Q appears in the group for Reynolds number, it is not possible to substitute and obtain its value, but with the second form this difficulty does not arise.

It is not necessary to perform such a double solution. The second form is obtainable from the first and vice versa. Suitable combination of the groups gives

$$\left[\left(\frac{Q}{gl^5}\right)\left(\frac{l\mu}{\rho Q}\right)^2 \left(\frac{l}{h}\right)^3\right]$$

Upon simplification this becomes

$$\mu/(\rho g^{1/2} h^{3/2})$$

and taking the reciprocal

$$\rho g^{1/2} h^{3/2}/\mu$$

This introduces a new idea. A combination of dimensionless groups can be arranged to give a new dimensionless group and this is just as valid as the original groups. *Any* combination of groups can be manipulated as shown, provided that in the end the number of groups specified by the first Buckingham theorem—no more and no less—is still available and that these final groups are independent of one another. For instance it would not be permissible to use the square of one group as a new group while retaining the first group.

Consider equation (3.13). This may be written

$$Q = g^{1/2} h^{5/2} f\left(\frac{\rho g^{1/2} h^{3/2}}{\mu}, \frac{h}{l}\right) \qquad (3.14)$$

This result is applicable to an orifice, a notch or a weir. h and l cannot be distinguished and consequently anywhere h appears l could be substituted. The velocity through a small orifice is not variable across it, so the flow through it must be proportional to l^2 where l represents the orifice diameter. Therefore the left hand side of equation (3.14) needs modification. Similarly the Reynolds number needs modification so that the leading dimension l can be introduced.

From the normal analysis for an orifice

$$Q = C_d \frac{\pi}{4} d^2 \sqrt{(2gh)}$$

$Q/g^{1/2}h^{5/2}$ multiplied by $(h/l)^2$ gives

$$\frac{Q}{g^{1/2}h^{1/2}l^2}$$

$\rho g^{1/2}h^{3/2}/\mu$ by l/h gives $\rho l\sqrt{(gh)}/\mu$.

Except for the omission of $\sqrt{2}C_v$ this group is $\rho v l/\mu$ (for an orifice $v = C_v\sqrt{(2gh)}$ which is, of course, the Reynolds number. Therefore,

$$Q = l^2\sqrt{(gh)}f(\rho v\sqrt{(gh)}l/\mu)$$

If l denotes the orifice diameter this result shows that the normal analysis is correct, but in addition it shows that C_d is a function of Reynolds number. By comparing the two results

$$\frac{\pi}{4}\sqrt{2}C_d = f\left(\frac{\rho\sqrt{(gh)}d}{\mu}, \frac{d}{h}\right)$$

so

$$C_d = \psi\left(\frac{\rho\sqrt{(gh)}d}{\mu}, \frac{d}{h}\right)$$

Thus the dimensional analysis has revealed more about the nature of the coefficient of discharge than was obtainable from the conventional analysis.

Conventional analysis for a Vee notch gives

$$Q = C_d\frac{8}{15}\sqrt{(2g)} \tan\frac{\theta}{2}h^{5/2}$$

If l is interpreted as the surface breadth of the notch, l/h will be 2 tan $\theta/2$, so when the dimensional-analysis result is applied to a notch the l/h group must be interpreted as the tan $\theta/2$ term in the conventional solution.

Thus

$$C_d = \psi\left(\frac{\rho g^{1/2}h^{3/2}}{\mu}, \frac{l}{h}\right)$$

Conventional analysis for a rectangular notch gives

$$Q = C_d\frac{2}{3}\sqrt{(2g)}bh^{3/2}$$

Interpreting l as the breadth of the notch, b, and multiplying the $Q/g^{1/2}h^{5/2}$ group by h/b gives $Q/g^{1/2}bh^{3/2}$ and so the result becomes

$$Q = g^{1/2}bh^{3/2}f\left(\frac{\rho g^{1/2}h^{3/2}}{\mu}, \frac{b}{h}\right)$$

so

$$C_d = \psi\left(\frac{\rho g^{1/2}h^{3/2}}{\mu}, \frac{b}{h}\right)$$

These results show that in the case of orifices and rectangular notches the coefficient of discharge varies with both the Reynolds number and the value of l/h and as neither can be regarded as constant as h varies, it is possible to predict that C_d must vary with h. In the case of the Vee notch, C_d only varies with the Reynolds number, so providing that the head is large enough for the surface tension effects to be negligible, the C_d of a Vee notch should be less affected by head variation than are the coefficients of discharge of rectangular notches. This is found in practice to be true.

3.5.2 The dimensional analysis of pipe flow

Flow in pipes is an example of a surface being exposed to frictional shear stresses caused by a fluid moving over it; the surface is the inside face of the pipe. This frictional stress must depend upon the velocity of the fluid; the diameter of the pipe must have an influence too, as if all the fluid is being constrained to travel in a small diameter pipe it must be exposed to the frictional effects from the boundary to a greater extent than it would be if it were travelling at the same velocity in a larger diameter pipe. The nature of the fluid must also affect the frictional shear stress at the pipe wall, that is the mass density and the coefficient of dynamic viscosity must affect the shear stress magnitude.

The roughness of the surface must also influence the flow and this parameter will be denoted by k which will signify the mean height of the roughnesses. It is possible for the fluid compressibility to be involved in the problem, for example, when gases are flowing through a pipe compressibility effects are extremely important, and when flow is unsteady and pressure transients occur, the size of such transients is governed to a very great extent by the compressibility of the fluid. Thus

$$\tau = f(\rho, d, v, l, \mu, k, K) \text{ or } f(\tau, \rho, d, v, l, \mu, k, K) = 0$$

where τ denotes the frictional shear stress at the pipe wall, ρ denotes the mass density of the fluid, v denotes the mean velocity of the flow, l denotes the length of the pipeline, μ denotes the coefficient of dynamic viscosity of the fluid, k denotes the mean height of the roughnesses on the pipe wall and K denotes the bulk modulus of the fluid.

The dimensions of τ are those of stress—M/LT^2, the dimensions of k are L, and the dimensions of K are stress i.e. M/LT^2. There are $8 - 3 = 5$ groups, so

$$f(\pi_1, \pi_2, \pi_3, \pi_4, \pi_5) = 0$$

$$\pi_1 = \rho^a v^b d^c \tau$$

therefore,

$$M^a L^{-3a} L^b T^{-b} L^c M L^{-1} T^{-2} = |0|$$

$$M: a + 1 = 0, \quad \therefore a = -1$$

$$T: -b - 2 = 0, \therefore b = -2$$

$$L: -3a + b + c - 1 = 0, \quad \therefore c = 0$$

so

$$\pi_1 = \frac{\tau}{\rho v^2}$$

$$\pi_2 = \rho^a v^b d^c l$$

therefore,

$$M^a L^{-3a} L^b T^{-b} L^c L = |0|$$

$$M: a = 0$$

$$T: -b = 0$$

$$L: -3a + b + c + 1 = 0; \quad \therefore c = -1$$

so

$$\pi_2 = l/d$$

$$\pi_3 = \rho^a v^b d^c \mu$$

therefore,

$$M^a L^{-3a} L^b T^{-b} L^c M L^{-1} T^{-1} = |0|$$

$$M: a + 1 = 0, \quad \therefore a = -1$$

$$T: -b - 1 = 0, \quad \therefore b = -1$$

$$L: -3a + b + c - 1 = 0, \quad \therefore c = -1$$

Therefore,

$$\pi_3 = \frac{\mu}{\rho v d} \quad \text{or} \quad \frac{\rho v d}{\mu}$$

that is, Reynolds number.

$$\pi_4 = \rho^a v^b d^c k$$

k has the dimensions of length so $\pi_4 = k/d \approx$ the roughness number

$$\pi_5 = \rho^a v^b d^c K$$

Therefore,

$$M^a L^{-3a} L^b T^{-b} L^c ML^{-1} T^{-2} = |0|$$

$$M: a + 1 = 0, \quad \therefore a = -1$$

$$T: -b - 2 = 0, \quad \therefore b = -2$$

$$L: -3a + b + c - 1 = 0, \quad \therefore c = 0$$

Therefore,

$$\pi_5 = \frac{K}{\rho v^2}$$

Now $K/\rho = c^2$ where c is the velocity of transmission of sound in the fluid in question. Therefore,

$$\pi_5 = (c/v)^2$$

Now take the square root of the reciprocal so that π_5 becomes v/c. This is the well-known Mach number.

So the required result is obtained:

$$f\left(\frac{\tau}{\rho v^2}, \frac{l}{d}, \frac{\rho v d}{\mu}, \frac{k}{d}, \frac{v}{c}\right) = 0$$

or

$$\tau = \rho v^2 f\left(\frac{l}{d}, \frac{\rho v d}{\mu}, \frac{k}{d}, \frac{v}{c}\right)$$

The pressure difference between the ends of the pipe, p, must be related to the pipe wall shear stress by the equation

$$\Delta p A = \tau l P$$

where A is the cross sectional area of the pipe, and P is the wetted perimeter of the pipe.

Therefore

$$\Delta p = \frac{\tau l}{A/P}$$

But A/P is called the hydraulic mean radius of the pipe (see Chapter 6) and is usually denoted by m (in American practice by R) therefore

$$\Delta p = \frac{\tau l}{m}$$

and the fluid head equivalent to this pressure difference is $h = p/w$ as $w = \rho g$. Therefore,

$$h_f = \frac{p}{w} = \frac{\tau l}{\rho g m}$$

Denote

$$f\left(\frac{1}{d}, \frac{\rho v d}{\mu}, \frac{k}{d}, \frac{v}{c}\right) \text{ by } f/2$$

then

$$h_f = \frac{flv^2}{2gm}$$

(In American practice a value of f four times this size is used; thus the American expression is $h_f = flv^2/8gm$).

This approach shows that f must depend upon the Reynolds number, the roughness number, the l/d ratio and the Mach number.

In almost all civil engineering practice the Mach number is so small as to suggest that compressibility effects are quite negligible. (This is not true in unsteady flow phenomena). If the pipe length l is large in comparison with d, its effect upon the problem also ceases to be important. (This is equivalent to saying that the pipe entry length is very small in comparison with the pipe length—see Chapter 7 for a detailed explanation). Thus, these two numbers can reasonably be omitted in most pipe flow problems and then

$$f = \varphi(Re, k/d)$$

This justifies the way in which Nikuradse, Stanton and Pannel have presented their results in graphical form.

3.6 Units

By considering the dimensions of a variable it is possible to

(1) give names to the units in which variables are expressed. A simple example of this facility is velocity. The dimensions are L/T so the units of velocity are ft/s or cm/s or miles/hour. Again, the dimensions of dynamic viscosity are M/LT, so the units are slugs/ft or lbs/ft s or grams/cm s. The last unit is the one applicable to the cgs system and is usually called the poise in honour of Poiseuille, who did much of the early work on laminar flow.

(2) convert from one system of units to another. Consider the coefficient of dynamic viscosity again. If it is required to calculate the number of poises in one slug/ft s then

$$X \times \frac{1}{1 \times 1} = \frac{32 \cdot 2 \times 454}{2 \cdot 54 \times 12 \times 1} = 479 \cdot 6$$

What has been done here is as follows:

$X \times 1$ gram/(1 cm $\times$ 1 s) has been equated to the number of grams in one slug (that is 32·2 $\times$ 454)/(the number of centimetres in one foot (that is 2·54 $\times$ 12) $\times$ 1 second). Thus there are 479 poise in 1 slug/ft s.

Similarly

$$1 \text{ kg/m s} = 10 \text{ poise}$$

and

$$1 \text{ slug/ft s} = 47 \cdot 96 \text{ kg/m s}$$

or

$$1 \text{ kg/m s} = 0 \cdot 0209 \text{ slugs/ft s}$$

3.7 Dimensional homogeneity of equations

The reader should now be quite clear that it is only possible to equate like things. Thus, if a mathematical equation has been developed, a check can be made upon its dimensional accuracy by ascertaining that all the terms in the equation have the same dimensions as one another and as the left hand side of the equation. This process will often reveal algebraic slips that have occurred in the process of developing an equation and the reader is strongly recommended to adopt this procedure as a standard routine in all his work; the process applies in all areas of study and is not, of course, confined to fluid mechanics. An equation which is dimensionally homogeneous is called a rational equation but some empirically derived equations are not dimensionally homogeneous, for example the Manning expression for the Chezy C, and these are called irrational.

Worked examples

(1) The flow through a sluice gate set into a dam is to be investigated by building a model of the dam and sluice at 1:20 scale. Calculate the head at which the model should work to give conditions corresponding to a prototype head of 20 metres. If the discharge from the model under this corresponding head is 0·5 m³/s, estimate the discharge from the prototype dam.

The variables involved are Q—the flow rate, d—a leading dimension of the sluice, h—the head over the sluice, ρ—the mass density of the water, g—the gravity field intensity and μ—the viscosity of water, that is,

$$f(Q, \rho, g, h, d, \mu) = 0$$

There are $6 - 3 = 3\pi$ groups.

These are

$$\rho^a g^b h^c Q$$
$$\rho^a g^b h^c d$$
$$\rho^a g^b h^c$$

$$\pi_1 \to M^a L^{-3a} L^b T^{-2b} L^c L^3 T^{-1} = |0|$$

M: $a = 0$

T: $-2b - 1 = 0, \quad \therefore b = -1/2$

L: $-3a + b + c + 3 = 0, \quad \therefore c = -3 - b + 3a = -2\frac{1}{2}$

$$\therefore \pi_1 = \frac{Q}{g^{1/2} h^{5/2}}$$

$$\pi_2 \to M^a L^{-3a} L^b T^{-2b} L^c L = |0|$$

by inspection, π_2 can be seen to be d/h.

$$\pi_3 \to M^a L^{-3a} L^b T^{-2b} L^c M L^{-1} T^{-1} = |0|$$

M: $a + 1 = 0, \quad \therefore a = -1$

T: $-2b - 1 = 0, \quad \therefore b = -0.5$

L: $-3a + b + c - 1 = 0$

$$\therefore c = 1 + 3a - b = 1 - 3 + \tfrac{1}{2} = -\tfrac{3}{2}$$

$$\therefore \pi_3 = \frac{\mu}{\rho g^{1/2} h^{3/2}}$$

The solution is thus:

$$\frac{Q}{g^{1/2} h^{5/2}} = \varphi \left(\frac{\rho g^{1/2} h^{3/2}}{\mu}, \frac{d}{h} \right)$$

(Note that dimensionless groups can be inverted as they remain dimensionless).

As flow through a sluice resembles flow through a large orifice, this expression requires slight amendment.

Multiply $Q/g^{1/2} h^{5/2}$ by $(h/d)^2$

that is

$$\frac{Q}{g^{1/2} d^2 h^{1/2}}$$

Then

$$Q = d^2 \sqrt{(gh)} \varphi \left(\frac{\rho g^{1/2} h^{3/2}}{\mu}, \frac{d}{h} \right)$$

This type of problem involves both the Reynolds and the Froude numbers.

$$\frac{Q^2}{d^2\sqrt{(gh)}} \propto \frac{Q}{A\sqrt{(gh)}}$$

as A is a function of d^2, but $Q/A = v$, so

$$\frac{Q}{d^2\sqrt{(gh)}} \propto \frac{v}{\sqrt{(gh)}} \propto \frac{v^2}{gh}$$

that is, the Froude number.

As has already been demonstrated, it is not possible to obtain Reynolds and Froude-number modelling simultaneously. If the Reynolds number for the model is large, the value for the prototype will be even larger and the effect of the Reynolds-number variation will be minimal. Therefore, it is only necessary to model for the Froude number and the d/h value.

$$d_m/h_m = d_p/h_p$$

$$\therefore \qquad h_p/h_m = d_p/d_m = 20$$

$$\therefore \qquad h_m = 20/20 = 1 \text{ m}$$

$$Q_p/Q_m = (h_p/h_m)^{5/2} = 20^{5/2}$$

$$\therefore \qquad Q_p = \tfrac{1}{2} 20^{5/2} = 894 \text{ m}^3/\text{s}$$

(2) A model of an aeroplane built to $\frac{1}{10}$th scale is to be tested in a wind tunnel which operates at a pressure of 20 atmospheres. The aeroplane is expected to fly at a speed of 500 km/h. At what speed should the wind tunnel operate to give dynamic similarity between model and prototype. The drag measured on the model is 33·75 newtons. What power will be needed to propel the aircraft at 500 km/h.

Rayleigh's formula (equation 3.7) applies

$$R = \rho v^2 l^2 \, \varphi(Re)$$

For dynamic similarity

$$Re_m = Re_p$$

$$\frac{\rho_m v_m l_m}{\mu_m} = \frac{\rho_p v_p l_p}{\mu_p}$$

$$\therefore \qquad v_m = \frac{\rho_p}{\rho_m} \frac{\mu_m}{\mu_p} \frac{l_p}{l_m} v_p$$

Now the coefficient of dynamic viscosity is not significantly altered by pressure changes unless the pressure change is large, so:

$$\mu_m = \mu_p$$

Due to the compressibility of air,

$$\rho_m = 20\,\rho_p$$

$$v_m = \tfrac{1}{20} \times \tfrac{1}{1} \times 10 \times v_p = 250 \text{ km/hr}$$

As

$$Re_m = Re_p$$

$$R_p/R_m = 20 \times 4 \times 10^2$$

$$R_p = 8000\,R_m$$

Power needed to propel the aircraft

$$P = v_p R_p = \frac{500 \times 1000}{3600} \times \frac{8000\,R_m}{1000} \text{ kW}$$

$$= \frac{40\,000}{36} \times 33.75$$

$$P = 37.5 \text{ MW}$$

Questions

(1) A sphere when placed in water moving at a velocity of 5 ft/s [1·6 m/s] experiences a drag of 1 lbf [4 N]. Another sphere of twice the diameter is placed in a wind tunnel. Find the velocity of the air which will give dynamically similar conditions. Also find the drag in this case. $v_{air}/v_{water} = 13$. $\rho_{air} = 0.08$ lb/ft³ [1·28 kg/m³].

Answer: 32·5 ft/s [10·4 m/s], 0·217 lbf [0·865 N].

(2) Explain briefly the use of the Reynolds number in the interpretation of tests on the flow of liquid in pipes. Water flows through a ¾ in [2 cm] dia. pipe at 5 ft/s [1·6 m/s]. Calculate the Reynolds number and find also the speed required to give the same Reynolds number when the pipe is transporting air. Obtain the ratio of the pressure drops in the same length of pipe in both cases.

$v_{water} = 1.41 \times 10^{-5}$ ft²/s [1·31 × 10⁻⁶ m²/s]

and $\rho_{water} = 62.4$ lb/ft³ [1000 kg/m³]

$v_{air} = 16.2 \times 10^{-5}$ ft²/s [15·1 × 10⁻⁶ m²/s]

$\rho_{air} = 0.075$ lb/ft³ [1·19 kg/m³]

Answer: 22 200 [24 400], 57·5 ft/s [18·4 m/s], 0·159 [0·157].

(3) Explain what is meant by a non-dimensional coefficient and show that Reynolds number vd/ν is non-dimensional. If the discharge Q through an orifice is a function of the diameter d, the pressure p causing flow, the density ρ, and the viscosity of the fluid μ, show that

$$Q = Cp^{1/2}d^2/\rho^{1/2}$$

where C is some function of the non-dimensional number $d\rho^{1/2}p^{1/2}/\mu$.

(4) Show that one form of the law governing the frictional loss in a pipe can be expressed in the form

$$\tau/\rho v^2 = \phi(Re)$$

and also show that $\tau/\rho v^2$ is dimensionless, τ being the frictional force per unit area of wetted surface. The frictional loss in a pipe carrying air is to be estimated from a suitable test on a similar pipe carrying water. Using the particulars given in the table, find the value of v and the ratio of the pressure drops per unit length of pipe, water to air.

Fluid	Pipe dia. inches [cm]	Velocity ft/s [m/s]	Density lb/ft^3 [kg/m^3]	Viscosity lb/ft s [kg/m s]
Water	2 [5]	v	62·4 [1000]	5·38 x 10^{-4} [8 x 10^{-4}]
Air	4 [10]	40 [12]	0·078 [1·24]	1·24 x 10^{-5} [1·84 x 10^{-5}]

Answer: Velocity 4·35 ft/s [1·3 m/s], pressure ratios = 18·9 [18·9].

(5) A cylinder 6 inches [0·16 m] in diameter is to be mounted in a stream of water in order to estimate the force on a tall chimney of 36 inches [1 m] diameter which is subjected to a wind of 100 ft/s [33 m/s]. Calculate (a) the speed of the stream necessary to give dynamic similarity between model and chimney (b) the ratio of the forces per ft [m] run of height.

Chimney $\rho = 0·07$ lb/ft^3, $\mu = 10·6 \times 10^{-6}$ lb/ft s

$$[1·12 \text{ kg/m}^3], \quad [16 \times 10^{-6} \text{ kg/m s}]$$

Model $\rho = 62·3$ lb/ft^3, $\mu = 5·4 \times 10^{-4}$ lb/ft,

$$[1000 \text{ kg/m}^3], \quad [8 \times 10^{-4} \text{ kg/m s}]$$

Answer: (a) 34·3 ft/s [11·2 m/s]; (b) 0·0572 [0·0567].

(6) In the rotation of similar discs in a fluid in which the motion of the fluid is turbulent, show by the method of dimensions that a rational formula for the frictional torque T of a disc of diameter d rotating at a speed N in a fluid of viscosity μ and density ρ is:

$$T = \rho N^2 d^5 \ f(\rho N d^2/\mu)$$

Hence show that in similar discs rotating in the same fluid the frictional torques at the corresponding speeds vary as the diameter of the discs. What is the ratio of the corresponding speeds?

Answer: $N_1/N_2 = (d_2/d_1)^2$.

(7) If the resistance to the motion of a sphere through a fluid is a function of the density and viscosity of the fluid, and the radius r and velocity v of the sphere, show that the resistance R is given by

$$R = \frac{\mu^2}{\rho}\, \varphi\left(\frac{\rho v r}{\mu}\right)$$

Hence show that if at very low velocities the resistance R is proportional to the velocity v than $R = k\mu r v$ where k is a dimensionless constant. A fine granular material of specific gravity 2·5 is in uniform suspension in still water of depth 3·3 m. Regarding the particles as spheres of diameter 0·002 cm find how long it will take for the water to clear. Take $k = 6\pi$ and $\mu = 0{\cdot}0013$ kg/m s.

Answer: 218 mins 39·3 s.

(8) Prove that the flow of a liquid over a 90° Vee notch can be expressed as $Q = g^{1/2}h^{5/2}\, \varphi(gh^3/v^2)$ where h is the head and v is the kinematic viscosity. Experiments on the flow of water over such a notch show that very nearly $Q = 1{\cdot}3\, h^{2 \cdot 48}$ using metre, second units. Show that for a fluid the viscosity of which is n times that of water the corresponding formula is $Q = 1{\cdot}3\, n^{0 \cdot 0133}h^{2 \cdot 48}$.

(9) It is found that the rate of discharge of fluids of viscosity μ and density ρ flowing over a Vee notch is given by $Q = g^{1/2}h^{5/2}\psi(\rho g^{1/2}h^{3/2}/\mu)$ where h is the head and ψ is some unknown function. It is required to find the flow of fluid of specific gravity 0·8 and viscosity eight times that of water over a Vee notch with a depth of 0·3 m. It is found that when water is run over a similar notch the flow is given by $Q = 0{\cdot}828\, h^{5/2}$ m³/s, where h is the depth in metres. Determine the depth at which water should be run over this test notch and hence calculate the probable rate of flow over the working notch.

Answer: 6·46 cm   0·0408 m³/s.

(10) A fluid dynamometer is to be used to absorb the power developed by an engine when under test. To check the design of the dynamometer a one fifth scale model was built and tests were carried out on it. Test results were as follows.

Model speed = 2500 rpm; torque developed = 5·23 lbf ft [7 Nm]

The kinematic viscosity of the oil to be used in the prototype dynamometer is to be ten times that used in the model and the specific gravity will be the same. At what speed should the prototype dynamometer be run to obtain dynamic similarity to these model conditions and what torque and power will the prototype dynamometer then be developing.

Answer: 2615 lb ft [3500 Nm], 497·9 hp [367 kW], speed for dynamic similarity = 1000 rpm.

(11) A model submarine is tested in an air stream at an air pressure of 25 atmospheres. The velocity of the air flow is 40 ft/s [12 m/s]. The model is built to a one tenth scale. At the test speed the model experiences a drag of 30 lbf [120 N]. At what speed will the prototype be travelling when dynamic similarity exists between it and the model and what horsepower will it dissipate at this speed? The kinematic viscosity of air at atmospheric pressure is thirteen times that of water and its density at atmospheric pressure is 0·079 lb/ft^3 [1·26 kg/m^3].

Answer: 7·7 ft/s [2·3 m/s], 49·2 hp [32·2 kW].

4 The Basic Equations of Engineering Fluid Mechanics

In Chapter 2 the mathematical approach to fluid mechanics was outlined. It is undoubtedly a very powerful technique but it is not the only approach available. Engineers use essentially the same methods in a rather different form. For example the continuity equation, the force equation and the energy equation, all of which were given in cartesian coordinates and in differential form in Chapter 2 are used in their natural coordinate integrated forms in engineering practice.

4.1 Continuity equation

Consider the flow between two streamlines (Fig. 4.1). In time δt element ABCD will move to position $A'B'C'D'$. Now no flow can cross streamlines as these are by definition lines drawn in the fluid parallel to the direction of motion of the fluid particles.

The amount entering across AB in time δt must be $\rho_1 a_1 V_1 \delta t$ and the amount leaving across CD must be $\rho_2 a_2 V_2 \delta t$. Now $\delta x_1 = V_1 \delta t$ and $\delta x_2 = V_2 \delta t$ so the amount stored must be stored within the volume $A'B'C'D'$. If the flow is steady, no storage can occur, so

$$\rho_1 a_1 V_1 = \rho_2 a_2 V_2$$

or

$$\rho a V = \text{a constant}$$

If the fluid is incompressible ρ is constant and then

$$a V = \text{constant}$$

110

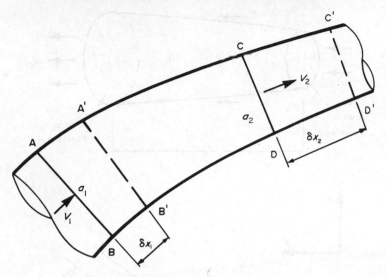

Fig. 4.1

By differentiating these equations, the differential forms given in Chapter 2 can readily be obtained:

$$\delta\rho/\rho + \delta a/a + \delta V/V = 0$$

4.2 The force equation

The force acting on an element causes a change of momentum according to Newton's second law. Let the total force be F, the component in the x direction F_x and in the y direction F_y. From Fig. 4.2

The force

$$F_x = \frac{d}{dt}(MV\cos\alpha) = M\frac{d}{dt}(V\cos\alpha)$$

$$= M\left(V\frac{\partial V}{\partial x} + \frac{\partial V}{\partial t}\right)\cos\alpha$$

but for the case of steady flow

$$\frac{\partial V}{\partial t} = 0$$

so

$$F_x = MV\frac{\partial V}{\partial x}\cos\alpha$$

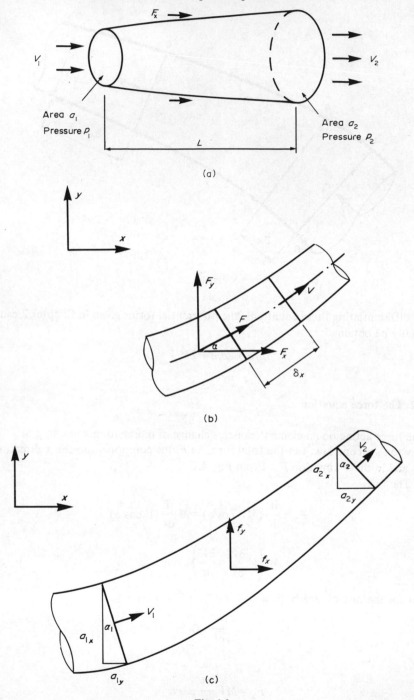

Fig. 4.2

But

$$M = \rho a \, \delta x$$

so

$$F_x = \rho (aV) \, \delta V \cos \alpha$$
$$F_x = \rho Q \, \delta u$$

Similarly the application of Newton's second law in the y direction gives

$$F_y = \rho Q \, \delta v$$

In natural coordinates

$$F = \rho Q \, \delta V$$

This approach can be used as it stands remembering that F must include pressure forces acting upon the element plus any external forces.

Alternatively a further developed equation can be used, as will now be shown.

$$Q \, \delta u = a_1 V_1 \, \delta u$$
$$= a_1 V_1 (u_2 - u_1)$$

but as

$$a_1 V_1 = a_2 V_2$$
$$Q \, \delta u = a_2 V_2 u_2 - a_1 V_1 u_1$$

Consider the projections of areas a_1 and a_2 onto planes perpendicular to the x direction. Call these a_{1x} and a_{2x} (see Fig. 4.2c). Then

$$a_{1x} = a_1 \cos \alpha_1 \text{ and}$$
$$a_{2x} = a_2 \cos \alpha_2$$

But

$$u_1 = V_1 \cos \alpha_1 \text{ and}$$
$$u_2 = V_2 \cos \alpha_2$$

∴

$$a_1 V_1 u_1 = \frac{a_{1x}}{\cos \alpha_1} V_1 V_1 \cos \alpha_1 = a_{1x} V_1^2$$

and

$$a_2 V_2 u_2 = \frac{a_{2x}}{\cos \alpha_2} V_2 V_2 \cos \alpha_2 = a_{2x} V_2^2$$

If we denote the external forces as f_x and f_y and resolve in the x direction

$$p_1 a_{1x} - p_2 a_{2x} + f_x = a_{2x} V_2^2 - a_{1x} V_1^2$$

$$\left(\frac{p_1}{w} + \frac{V_1^2}{g} \right) a_{1x} - \left(\frac{p_2}{w} + \frac{V_2^2}{g} \right) a_{2x} + \frac{f_x}{w} = 0 \qquad (4.1)$$

By an exactly similar process, the force equation in the y direction is

$$\left(\frac{p_1}{w} + \frac{V_1^2}{g}\right) a_{1y} - \left(\frac{p_2}{w} + \frac{V_2^2}{g}\right) a_{2y} + \frac{f_y}{w} = 0 \tag{4.2}$$

Note: both f_x and f_y are external forces acting on the element (in the x and y directions respectively). The pressure forces acting on the ends of the element were included in equations (4.1) and (4.2) but the pressures on the sides of the element were not included: these contribute to the forces f_x and f_y.

4.3 The energy equation

Consider an element of fluid between two streamlines (Figs. 4.3a, b). Assume that the area, velocity and pressure increase with s. (This means there are no complications with signs in the following equation). Pressure forces act upon *all*

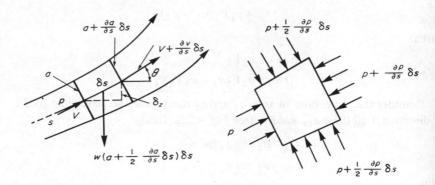

Fig. 4.3

sides of the element. If the element is small the mean pressure on the sides of the element is closely approximated by $p + \frac{1}{2}(\partial p/\partial s)\,\delta s$. The force acting in the direction s is

$$pa + \left(p + \frac{1}{2}\frac{\partial p}{\partial s}\delta s\right)\frac{\partial a}{\partial s}\delta s - \left(p + \frac{\partial p}{\partial s}\delta s\right)\left(a + \frac{\partial a}{\partial s}\delta s\right) - w\left(a + \frac{1}{2}\frac{\partial a}{\partial s}\delta s\right)\delta s \sin\theta$$

The second term is the component of pressure force acting in the s direction caused by the pressures on the sides of the elements. Remember that pressures act equally in all directions and that the projection of the side area of the element onto a plane perpendicular to the s direction is $\partial a/\partial s\,\delta s$. The fourth term is the component of the weight of the element acting in the s direction.

Expanding, this expression gives

$$\text{Force} = pa + p\,\frac{\partial a}{\partial s}\,\delta s + \frac{1}{2}\frac{\partial p}{\partial s}\frac{\partial a}{\partial s}(\delta s)^2$$

$$-pa - a\,\frac{\partial p}{\partial s}\,\delta s - p\,\frac{\partial a}{\partial s}\,\delta s - \frac{\partial p}{\partial s}\frac{\partial a}{\partial s}(\delta s)^2$$

$$-wa\,\delta s\,\sin\theta - \frac{w}{2}\frac{\partial a}{\partial s}(\delta s)^2\sin\theta$$

Ignoring terms containing $(\delta s)^2$ and simplifying gives

$$\text{Force} = -a\,\frac{\partial p}{\partial s}\,\delta s - wa\,\delta s\,\sin\theta$$

Now

$$\delta s\,\sin\theta = \delta z$$

$$\therefore \qquad \text{Force} = -a\,\frac{\partial p}{\partial s}\,\delta s - wa\,\delta z$$

This force causes the acceleration of the element according to Newton's second law: force = mass x dV/dt. Now mass = $\rho a\,\delta s$ and $dV/dt = V\,\partial V/\partial s + \partial V/\partial t$

$$\therefore \qquad -a\,\frac{\partial p}{\partial s}\,\delta s - wa\,\delta z = \rho a\,\delta s\left(V\frac{\partial V}{\partial s} + \frac{\partial V}{\partial t}\right)$$

$$\therefore \qquad \frac{\partial p}{\partial s} + \rho\left(V\frac{\partial V}{\partial s} + \frac{\partial V}{\partial t}\right) + w\frac{\partial z}{\partial s} = 0$$

If the flow is steady $\partial V/\partial t = 0$ so

$$\frac{\partial p}{\partial s} + \rho V\frac{\partial V}{\partial s} + w\frac{\partial z}{\partial s} = 0$$

Integrating with respect to s gives

$$p + \rho V^2/2 + wz = \text{constant}$$

Dividing through by w

$$p/w + V^2/2g + z = \text{constant}$$

This is called Bernoulli's equation. Each of the terms in this equation represents a component of energy per unit weight of the fluid.

When developing the Bernoulli equation no external forces were taken into account. Such external forces could be generated by friction between the element and adjacent fluid and this would cause energy loss so that Bernoulli's equation would not apply. Bernoulli's equation can *only* be applied if no energy losses are occurring.

4.4 Flow through small orifices

Energy losses as fluid moves down a streamtube as shown in Fig. 4.4 are very small, so the application of Bernoulli should be appropriate. The centre line of the tube is defined as the line joining ① and ②, so

$$\frac{p_1}{w} + \frac{V_1^2}{2g} + z_1 = \frac{p_2}{w} + \frac{V_2^2}{2g} + z_2 \tag{4.3}$$

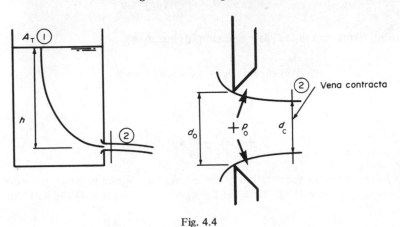

Fig. 4.4

To solve this equation for one variable, the values of all the others must be known. p_1 is atmospheric pressure, V_1 is very small if the orifice is small and the tank large so $V_1^2/2g$ is negligible, $z_1 - z_2$ is the height of the water surface above point 2. Unless p_2 can be specified nothing further can be done. If the point 2 is assumed to be *in* the plane of the orifice p_2 is *not* atmospheric pressure because the streamlines are heavily curved. These curvatures mean that the pressure at the centre of the orifice should be greater than atmospheric pressure so that outwardly directed pressure gradients can exist which will provide the centrally directed forces for the required centrally directed accelerations. At the first point where the streamlines are parallel in the effluxing jet the pressure is atmospheric so this is the position chosen for point 2. This point where the sides of the jet first became parallel is called the *vena contracta* and the pressure here must be atmospheric pressure. Then if p_a is the atmospheric pressure equation (4.3) becomes

$$p_a/w + 0 + h + z_2 = p_a/w + V_2^2/2g + z_2$$

so

$$V_2^2/2g = h$$

so

$$V_2 = \sqrt{(2gh)}$$

$$= \sqrt{\left(2\frac{p}{\rho}\right)}$$

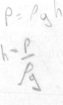

$p = \rho g h$

$h = \dfrac{p}{\rho g}$

This is not perfectly true because there is a small energy loss which was not taken into account. A coefficient can be introduced so that this expression gives an accurate result

$$V = C_v \sqrt{(2gh)}$$

The coefficient C_v is called the *coefficient of velocity* of the orifice. For a sharp-edged small orifice (that is $d_0 \ll h$)C_v is approximately 0·98 which shows how well the energy approach works. The flow through the orifice is given by the product of the velocity and the area of the jet at the vena contracta. That is

$$Q = VA_{vc}$$

$$Q = C_v A_{vc} \sqrt{(2gh)}$$

Unfortunately, the area of the vena contracta is not easily calculated by a theoretical method and an element of empiricism must be introduced. The ratio of the area of the vena contracta to the area of the orifice is called the *coefficient of contraction* of the orifice C_c.

$$C_c = A_{vc}/A_0$$

so

$$Q = C_c A_0 C_v \sqrt{(2gh)} = C_d A_0 \sqrt{(2gh)}$$

the coefficient C_d is called the *coefficient of discharge* and is equal to the product $C_v C_c$. The relationship between the three coefficients and h is shown in Fig. 4.5.

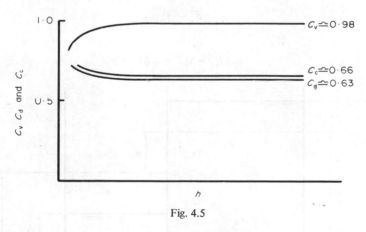

Fig. 4.5

4.4.1 Time of emptying a tank through an orifice

Let the cross sectional area of the tank be A_T. The continuity equation is

$$-A_T \, dh = Q \, dt$$

(tank volume reduction = outflow rate x time in which the volume reduction occurs). h is the surface elevation above the orifice at time t and the negative sign is introduced because dh is a depth reduction.

$$\therefore \qquad -A_T\, \mathrm{d}h = C_d A_0 \sqrt{(2gh)}\, \mathrm{d}t$$

$$\therefore \qquad \mathrm{d}t = \frac{-A_T\, \mathrm{d}h}{C_d A_0 (2gh)^{1/2}}$$

Integrating

$$t = \frac{2A_T}{C_d A_0 (2g)^{1/2}}\, [H^{1/2} - 0]$$

$$\therefore \qquad t = \frac{2A_T H^{1/2}}{C_d A_0 (2g)^{1/2}}$$

(H is the initial depth at time $t = 0$).

4.4.2 Time of equalisation of levels in two tanks Denote the initial levels in two tanks, referred to the plane of the orifice, by H_1 and H_2 and the level difference by H (see Fig. 4.6). Denote levels at a time t by h_1 and h_2 and level difference by h.

Efflux through the orifice = $Q = C_d A_0 (2gh)^{1/2}$

$$-A_1\, \delta h_1 = Q\, \mathrm{d}t$$

also

$$-A_1\, \delta h_1 = A_2\, \delta h_2$$

and

$$-\delta h_1 + \delta h_2 = -\delta h$$

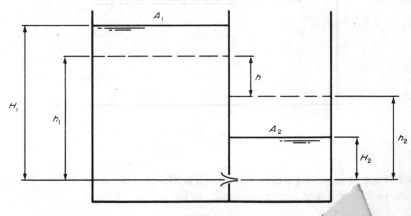

Fig. 4.6

(remember that a decrease in level has a negative value)

$$\therefore \qquad \delta h_2 = \delta h + \delta h_1 = -\frac{A_1}{A_2}\delta h_1$$

so

$$\delta h_1 = \frac{\delta h}{1 + A_1/A_2}$$

$$\therefore \qquad \frac{-A_1 A_2}{A_1 + A_2}\delta h = Q\,\delta t = C_d A_0 (2gh)^{1/2}\,\delta t$$

$$\therefore \qquad dt = -\frac{A_1 A_2}{C_d A_0 (2g)^{1/2}(A_1 + A_2)}h^{-1/2}\,\delta h$$

$$t = \frac{2A_1 A_2}{(A_1 + A_2)C_d A_0 (2g)^{1/2}}[h^{1/2}]_0^H$$

$$t = \frac{2H^{1/2}}{(1/A_1 + 1/A_2)C_d A_0 (2g)^{1/2}}$$

The value A_2 = infinity corresponds to the emptying of a single tank.

4.5 The venturimeter

The venturimeter is a flow-measuring device used for pipelines. It consists of a tapered convergent pipe section followed by a slowly divergent pipe section as illustrated in Fig. 4.7. Pressure tappings at the meter entry and the throat each

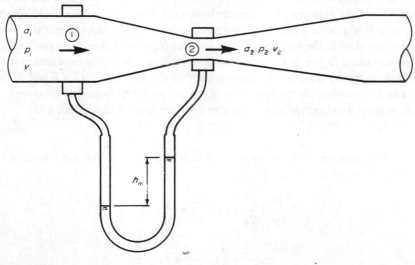

Fig. 4.7

consist of three holes drilled into the tube at the cross section concerned, the holes being located at 120° angular spacings around the section. These three holes connect to a gallery which is, in turn, tapped for the manometer connection. The way in which the holes are drilled is important. It is vital that the internal surfaces adjacent to the holes should be smooth and flat. In large meters they can be scraped down but in small meters it is desirable that the holes should be

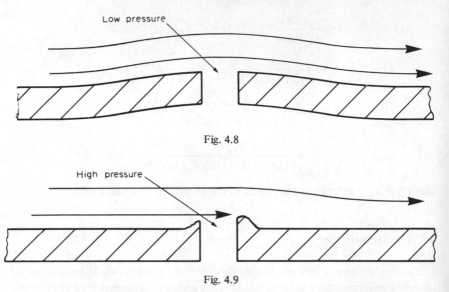

Low pressure

Fig. 4.8

High pressure

Fig. 4.9

drilled either through the pipe wall into a plug or through the pipe wall and then into and through the diametrically opposite pipe wall. The entry hole should be plugged and the diametrically opposite hole used. These techniques avoid roughen' of the hole edge which may cause reduced pressures at the tapping due to curvilinear flow in the boundary layer. A metal fragment adhering to the hole edge may cause higher or lower pressures at the hole depending on whether it is located on the downstream or upstream side of the hole (see Figs. 4.8 and 4.9).

The downstream divergent section of the venturimeter is used to reconvert high velocity heads at the throat back to pressure head with minimal loss.

4.5.1 Venturimeter analysis Apply Bernoulli's equation to the entry section 1 and to the throat section 2

$$\frac{p_1}{w} + \frac{v_1^2}{2g} + z_1 = \frac{p_2}{w} + \frac{v_2^2}{2g} + z_2$$

$$\therefore \qquad \frac{v_2^2 - v_1^2}{2g} = \frac{p_1 - p_2}{w} + z_1 - z_2$$

But by continuity $a_1v_1 = a_2v_2$,
so

$$v_2 = \frac{a_1}{a_2}v_1$$

$$\therefore \quad \left[\left(\frac{a_1}{a_2}\right)^2 - 1\right]\frac{v_1^2}{2g} = \frac{\Delta p}{w} + z_1 - z_2$$

$$\therefore \quad v_1 = \left[\frac{2g(\Delta p/w + z_1 - z_2)}{(a_1/a_2)^2 - 1}\right]^{1/2}$$

$$\therefore \quad Q = a_1v_1 = a_1a_2\left[\frac{2g(\Delta p/w + z_1 - z_2)}{a_1^2 - a_2^2}\right]^{1/2}$$

Because small energy losses occur in the convergent section a coefficient of discharge must be introduced into Bernoulli's equation.

$$\therefore \quad Q = C_d a_1\left[\frac{2g(\Delta p/w + z_1 - z_2)}{(a_1/a_2)^2 - 1}\right]^{1/2}$$

C_d will be 0·98 if

(1) The Reynolds number for the pipeline is such that flow is turbulent.

(2) The pressure tappings are constructed as described above.

(3) No bends or valves occur in the pipeline within a length of twenty diameters upstream of the meter. Bends cause secondary flows (see Section 5.5) generating helical flow patterns which must die away before the flow enters the meter.

(4) The local pressure never falls below the pressure at which gas is released from the water, that is about 8 ft [2.3 m] head absolute.

A typical C_d curve is shown in Fig. 4.10.

The largest effect causing C_d to be less than unity is the non-uniformity of the velocity distribution across the pipe. This means that the assumption that

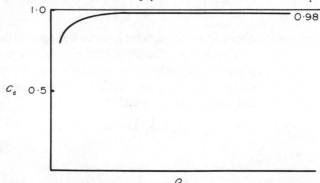

Fig. 4.10

$v^2/2g$ represents the mean kinetic energy per unit weight of flow is not strictly accurate. This is itself a friction effect of course.

The energy loss in the convergent cone can be calculated if the C_d value is known. The Bernoulli equation is written including an allowance for friction loss, that is

$$\frac{p_1}{w} + \frac{v_1^2}{2g} + z_1 = \frac{p_2}{w} + \frac{v_2^2}{2g} + z_2 + h_f$$

where h_f is the head loss due to friction.

Then as before

$$Q = a_1 \left(\frac{2g(\Delta p/w + z_1 - z_2 - h_f)}{(a_1/a_2)^2 - 1} \right)^{1/2}$$

from before

$$Q = C_d a_1 \left(\frac{2g(\Delta p/w + z_1 - z_2)}{(a_1/a_2)^2 - 1} \right)^{1/2}$$

∴
$$C_d^2(\Delta p/w + z_1 - z_2) = (\Delta p/w + z_1 - z_2 - h_f)$$

∴
$$h_f = (\Delta p/w + z_1 - z_2)(1 - C_d^2)$$

but

$$\frac{\Delta p}{w} + z_1 - z_2 = \frac{Q^2}{C_d^2 2g a_1^2} \left[\left(\frac{a_1}{a_2} \right)^2 - 1 \right] = \frac{1}{C_d^2} \left(\frac{v_2^2 - v_1^2}{2g} \right)$$

so

$$h_f = \left(\frac{1}{C_d^2} - 1 \right) \frac{v_2^2 - v_1^2}{2g} \text{ or}$$

$$h_f = \left(\frac{1}{C_d^2} - 1 \right) \frac{Q^2}{2g} \left(\frac{1}{a_2^2} - \frac{1}{a_1^2} \right)$$

As the result stands it is suitable for use with a meter equipped with pressure gauges to measure pressures p_1, and p_2. Usually venturimeters are equipped with some form of differential manometer rather than pressure gauges. In these circumstances the manometer will correct automatically for non-horizontality of the meter—see Fig. 4.11, and p_A will be equal to p_B.

∴
$$p_1 + w_f s + w_f h_m = p_2 + w_f \Delta z + w_f s + w_m h_m$$

but $\Delta z = z_2 - z_1$,
so

$$\frac{p_1 - p_2}{w_f} + z_1 - z_2 = \left(\frac{w_m}{w_f} - 1 \right) h_m = \left(\frac{s_m}{s_f} - 1 \right) h_m$$

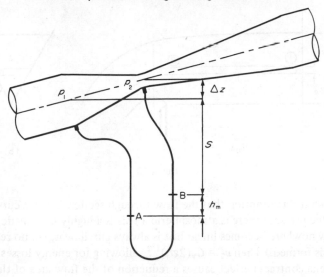

Fig. 4.11

In effect, the lack of horizontality of the meter does not affect the manometer reading h_m,
so

$$Q = C_d a_1 \left\{ \frac{2g(s_m/s_f - 1)h_m}{(a_1/a_2)^2 - 1} \right\}^2$$

4.6 Notches

A notch is a device which permits water to flow through it, developing a free surface as it does so. Notches may have a number of different shapes—triangular, circular, trapezoidal, rectangular or hyperbolic.

The model that has been developed to predict the relationship between the flow Q and the head H over the base of the notch is a poor one and many people think that it should no longer be used. While accepting that the theory is based on unrealistic assumptions, I feel that it is worth including so that comparisons can be made with the results obtained from dimensional considerations (Chapter 3).

4.6.1 The Vee or triangular notch
The Vee notch is illustrated in Figs 4.12a and b.

It can be argued that a vena contracta is formed at section 2 and that Bernoulli's equation can be applied between sections 1 and 2. If the cross sectional area at 1 is large by comparison with that at 2 the velocity there will be small and $v_2^2/2g$ will be negligible.
Hence

$$p_1/w = h + p_a/w = p_2/w + v_2^2/2g$$

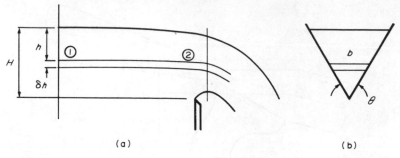

Fig. 4.12

We make the assumption that the flow through section 2 is *not* curvilinear and thus the pressure there is atmospheric. (This is a highly unrealistic assumption as the flow nowhere becomes linear but is always curvilinear, i.e. no real vena contracta is formed.) Then $v_2 = C_v(2gh)^{1/2}$, allowing for energy losses. Assuming that the vena contracta effect causes a reduction of the flow area of the element so that it is given by $C_c b \delta h$

$$\delta Q = C_v C_c b \, \delta h (2gh)^{1/2}$$

But

$$b = 2(H - h) \tan \theta/2$$

∴

$$\delta Q = 2C_c C_v \tan \theta/2 (2g)^{1/2}(H - h)h^{1/2} \, \delta h$$

Integrating between 0 and H

$$Q = 2C_d \tan \theta/2 (2g)^{1/2} [\tfrac{2}{3}Hh^{3/2} - \tfrac{2}{5}h^{5/2}]_0^H$$

∴

$$Q = \tfrac{8}{15}C_d \tan \theta/2 (2g)^{1/2}H^{5/2}$$

It will be seen that the coefficients of contraction and velocity for the element that have been introduced have been assumed to be constant for all elements. It is improbable that this is true. In any case, as has been said before, no true vena contracta can be said to occur in the flow.

It is surprising to find that the equation works reasonably well if an experimentally derived value for C_d is used. The result obtained from dimensional analysis is

$$Q = g^{1/2}H^{5/2} \, \varphi \, (b/H, \rho g^{1/2}H^{3/2}/\mu, \rho g H^2/\sigma)$$

that is,

$$Q = g^{1/2}H^{5/2} \, \varphi(b/H, Re, We)$$

The effect of b/H in the function can be interpreted as $\tan \theta/2$. Thus the coefficient of discharge is given by

$$C_d = \frac{8}{15}\sqrt{2} \, \varphi(Re, \tan \theta/2, We)$$

As b/H is fixed for all values of H it can be predicted that C_d should vary from notch to notch depending upon the value of the vertex angle but that for notches having the same vertex angle it should only depend upon Re if depths are such that surface tension effects can be neglected. At high Reynolds numbers it would be reasonable to expect the coefficient of discharge to become constant and invariant with Reynolds number. This in fact occurs as shown in Fig. 4.13.

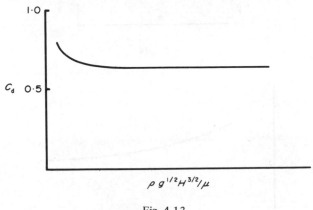

Fig. 4.13

The Vee notch is thus an excellent notch to use without calibration providing it is worked at high levels.

For a 90° sharp-edged Vee notch $Q = 2.48H^{2.48}$ where Q is in ft^3/s and H is in ft is an excellent formula. A satisfactory approximation for this formula when using SI units would be $Q = 1.37H^{2.48}$. The nearness of the index 2.48 to the theoretical value of 2.5 shows how very nearly constant the coefficient of discharge becomes over the working range of pressure head. It is important that a Vee notch working with water should not be operated under a head less than 0.75 inches if this formula is to be used; this is due to the increasing significance of surface tension at low heads.

4.6.2 The rectangular notch The criticisms of the theory advanced for the Vee notch applies with equal force to the rectangular notch. This is illustrated in Fig. 4.14. As for the Vee notch

$$\delta Q = C_d b \, \delta h \sqrt{(2gh)}$$

so

$$Q = C_d b (2g)^{1/2} \int_0^H h^{1/2} \, dh$$

$$Q = \tfrac{2}{3} C_d b (2g)^{1/2} H^{3/2}$$

The curve corresponding to this equation is shown in Fig. 4.15.

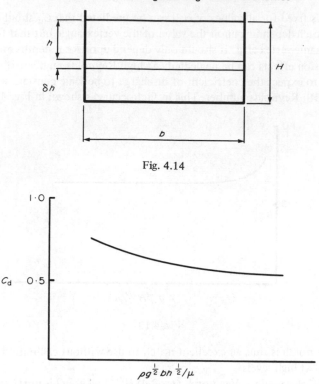

Fig. 4.14

Fig. 4.15

Compared with the Vee notch the rectangular notch needs calibration more because its C_d value varies more over the operating range. C_d values for Vee notches range from 0·62 to 0·59 as the θ value is reduced from 180° to 90°.

A formula advanced by Francis for rectangular notches which attempts to account for approach conditions is

$$Q = 3 \cdot 33\ (b - 0 \cdot 1\ nH)H^{3/2} \quad \text{where } Q \text{ is in ft}^3/\text{s and}$$
$$H \text{ is in ft.}$$

or

$$Q = 1 \cdot 84\ (b - 0 \cdot 1\ nH)H^{3/2} \quad \text{where } Q \text{ is in m}^3/\text{s and } H \text{ is in m.}$$

n is the number of *side* contractions. A side contraction is illustrated in Fig. 4.16. The vena contracta of a notch such as that in Fig. 4.16 has a shape like that in Fig. 4.17 (here there are two side contractions.)

If the notch is at the end of a channel and one side of the notch is located at the side of the channel (Fig. 4.18a) then the contraction at that side will be inhibited (Fig. 4.18b) and $n = 1$.

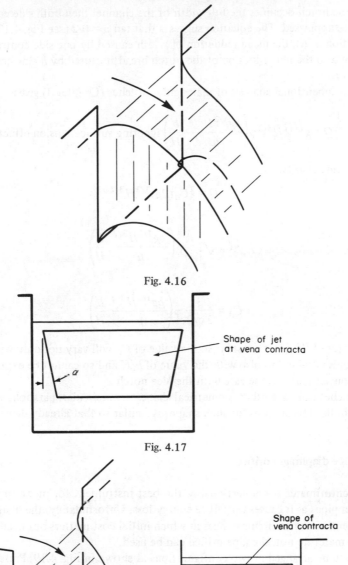

Fig. 4.16

Shape of jet
at vena contracta

α

Fig. 4.17

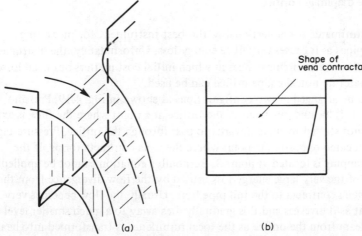

Shape of
vena contracta

(a)   (b)

Fig. 4.18

If the notch occupies the full width of the channel then both side contractions will be suppressed. The equation suggests that $\tan \alpha = 0.2$ (see Fig. 4.17 for the definition of α), the mean reduction of width caused by one side contraction is $\frac{1}{2}H \tan \alpha$ so the net reduction of the notch breadth caused by n side contractions is $0.1\,nH$.

The dimensional analysis of rectangular notches (Chapter 3) gives

$$Q = g^{1/2}bH^{3/2}\,\varphi\left(\frac{\rho g^{1/2}H^{3/2}}{\mu},\frac{b}{H}\right) \text{ ignoring surface tension effects}$$

Comparing results:

$$Q = \tfrac{2}{3}C_d(2g)^{1/2}H^{3/2}$$

so

$$\tfrac{2}{3}C_d\sqrt{2} = \varphi\left(\frac{\rho g^{1/2}H^{3/2}}{\mu},\frac{b}{H}\right)$$

and

$$C_d = \frac{3}{2\sqrt{2}}\,\varphi\left(\frac{\rho g^{1/2}H^{3/2}}{\mu},\frac{b}{H}\right)$$

in this case b/H is *not* constant so the value of C_d will vary not only with the Reynolds number but also with the value of b/H and so cannot be expected to be a constant, as was the case with the Vee notch.

Notches can have other geometrical shapes—semicircular, parabolic and hyperbolic. The analysis for such shapes is similar to that already demonstrated.

4.7　Pipe diaphragm orifices

The venturimeter is unquestionably the best instrument for measuring flow in pipes as it causes very little energy loss. Unfortunately, the instrument is expensive and in circumstances in which initial cost matters but head losses in the meter do not, the pipe orifice can be used.

It can be arranged in two configurations as shown in Fig. 4.19. Pressure tapping 1 is located upstream of the orifice at a point where the flow is axial and has not started to curve inwards to pass through the orifice. Pressure tapping 2 may be located opposite the point where the vena contracta forms. If the second tapping is located at point 3, Bernoulli's equation cannot be applied because of the very large energy loss caused by the flow divergence from the narrow vena contracta to the full pipe bore. Usually the flow becomes very turbulent as it diverges and this gradually dies away to a much smaller level downstream from the orifice as the local turbulence is transformed into heat. The two configurations must be analysed separately.

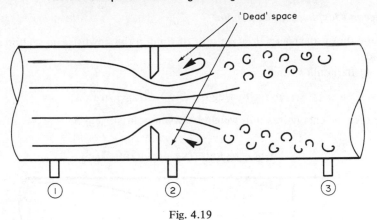

Fig. 4.19

(1) Second tapping located at the vena contracta

Bernoulli's equation can be applied reasonably accurately to points 1 and 2.

$$p_1/w + v_1^2/2g = p_2/w + v_2^2/2g$$

and

$$a_1 v_1 = a_2 v_2$$

Then

$$v_2^2 - v_1^2 = 2g(p_1 - p_2)/w$$

and

$$v_2 = \frac{a_1}{a_2} v_1$$

so

$$v_1 = \sqrt{\left\{ \frac{2g(p_1 - p_2)/w}{(a_1/a_2)^2 - 1} \right\}}$$

and

$$Q = a_1 v_1 = C_v a_1 \sqrt{\left\{ \frac{2g(p_1 - p_2)/w}{(a_1/a_2)^2 - 1} \right\}}$$

This result is the same as the venturimeter result. But

$$a_2 = C_c a_0, \text{ and denoting } a_0/a_1 \text{ by } 1/m$$

$$Q = C_v a_1 \sqrt{\left\{ \frac{2g(p_1 - p_2)/w}{(m/C_c)^2 - 1} \right\}}$$

(2) Second tapping located at point 3

An estimate for the energy loss per unit of fluid in the downstream section must be made.

From Bernoulli's equation

$$p_2/w + v_2^2/2g = p_3/w + v_3^2/2g + h_f$$

h_f denotes the energy loss/unit weight between points 2 and 3.
Then

$$(p_3 - p_2)/w = (v_2^2 - v_3^2)/2g - h_f$$

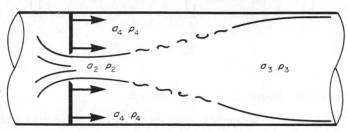

Fig. 4.20

An 'external' force is acting upon the fluid in the divergent section due to the pressure of the fluid in the 'dead' space. (See Fig. 4.20). The force equation is

$$\left(\frac{p_2}{w} + \frac{v_2^2}{g}\right)a_2 - \left(\frac{p_3}{w} + \frac{v_3^2}{g}\right)a_3 = -\frac{p_4(a_3 - a_2)}{w}$$

An assumption about the value of p_4 must now be made. As the flow at point 2 (the vena contracta) is parallel, the pressure just outside the area must be the same as the pressure within the area, i.e. $p_4 = p_2$. The assumption that $p_4 = p_2$ and that p_4 is constant over the area leads to a result which can be justified experimentally.
So

$$p_2 a_2 + a_2 w v_2^2/g - p_3 a_3 - a_3 w v_3^2/g + p_2(a_3 - a_2) = 0.$$

$$\therefore \qquad (p_3 - p_2)/w = (a_2 v_2^2 - a_3 v_3^2)/ga_3$$

$$\therefore \qquad \frac{v_2^2 - v_3^2}{2g} - h_f = \frac{(a_2/a_3)v_2^2 - v_3^2}{g}$$

$$\therefore \qquad \frac{v_2^2 - v_3^2}{2g} - h_f = \frac{2(a_2/a_3)v_2^2 - 2v_3^2}{2g}$$

$$\therefore \qquad h_f = (v_2^2 - 2(a_2/a_3)v_2^2 - v_3^2 + 2v_3^2)/2g$$

but

$$(a_2/a_3)v_2 = v_3$$

so

$$h_f = \frac{v_2^2 - 2v_2v_3 + v_3^2}{2g}$$

∴

$$h_f = (v_2 - v_3)^2/2g$$

Applying Bernoulli from points 1 to 3

$$p_1/w + v_1^2/2g = p_3/w + v_3^2/2g + (v_2 - v_3)^2/2g$$

∴

$$\frac{p_1 - p_3}{w} = \frac{(v_2 - v_3)^2}{2g} \quad \text{as } v_1 = v_3$$

so

$$\frac{p_1 - p_3}{w} = \left(\frac{a_3}{a_2} - 1\right)^2 \frac{v_3^2}{2g}$$

∴

$$Q = a_3v_3 = C_v a_3 \sqrt{\left\{\frac{2g(p_1 - p_2)/w}{(m/C_c - 1)^2}\right\}}$$

This result is very different from the result for the first configuration. (Note that an assumption was made that no energy loss occurred between points 1 and 2.)

4.8 The pitot tube

The pitot tube is a device for measuring local velocities in a flow. It consists of two concentrically arranged tubes bent to form a right-angle bend as illustrated in Fig. 4.21. One leg of the bend is aligned with the direction of flow. The

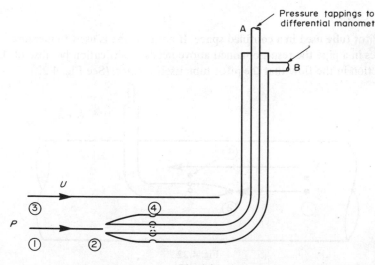

Fig. 4.21

inner tube has an open end and the sealed outer tube has a streamlined shape. A number of small holes (usually 6 or 8) is drilled through the outer tube a small distance from the nose of the device (approximately 2 inches). The pressure at the stagnation point located at the upstream tip of the device is thus transmitted to tapping A while the ambient pressure of the flow is transmitted to tapping B. The difference in pressure between tappings A and B can be measured by any suitable differential manometer.

Applying Bernoulli's equation to points 1 and 2 gives

$$p_1/w + U^2/2g = p_2/w + 0 \quad \Rightarrow \quad \left(\frac{p_2 - p_1}{w}\right) = \frac{u^2}{2g}$$

(the velocity at point 2 is zero as this is a stagnation point). Applying the same equation to points 3 and 4

$$p_3/w + U^2/2g = p_4/w + U^2/2g$$

$$\therefore \qquad p_3 = p_4$$

But p_3 also equals p_1

so

$$p_1 = p_4$$

$$\therefore \qquad (p_2 - p_4)/w = U^2/2g$$

$$w = \rho g$$
$$p = \rho g h$$
$$h = \frac{p}{\rho g} = \frac{p}{w}$$

The pressure head $(p_2 - p_4)/w$ is the pressure head measured by the manometer h, therefore $U = (2gh)^{1/2}$ where h is the head measured by the manometer after it has been converted into head of working fluid; that is if h_m is the manometer head $h = (s_m/s_f - 1)h_m$ (where s_m is the specific gravity of the manometer fluid).

A coefficient of velocity must be introduced because of small energy losses.

$$\therefore \qquad U = C_v[2g(s_m/s_f)h_m]^{1/2} \tag{4.4}$$

C_v for a well designed pitot tube lies between 0·98 and 1·0.

4.8.1 Pitot tube used in a confined space
If a pitot tube is used to measure velocities in a pipe the result obtained above needs modification because of the constriction in the flow that the pitot tube itself creates. (See Fig. 4.22).

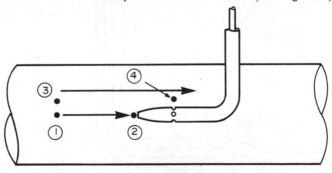

Fig. 4.22

As before

$$p_1/w + U^2/2g = p_2/w$$

and

$$p_3/w + U^2/2g = p_4/w + v_4^2/2g$$

$$v_4 = \frac{UA}{A - a}$$

where A is the pipe area and a is the pitot tube cross section. But

$$p_3 = p_1 \text{ as before}$$

∴

$$\frac{p_1}{w} + \frac{U^2}{2g} = \frac{p_4}{w} + \left(\frac{A}{A - a}\right)^2 \frac{U^2}{2g}$$

∴

$$\frac{p_1 - p_4}{w} = \left[\left(\frac{A}{A - a}\right)^2 - 1\right] \frac{U^2}{2g}$$

and

$$\frac{p_2 - p_1}{w} = \frac{U^2}{2g}$$

Eliminating p_1 from these equations gives

∴

$$\frac{p_2 - p_4}{w} = \left(\frac{A}{A - a}\right)^2 \frac{U^2}{2g}$$

if $a \to 0$ or $A \to \infty$ the previous result is obtained

∴

$$U = C_v \left(\frac{A - a}{A}\right) \bigg/ \sqrt{\left\{2g\left(\frac{s_m}{s_f} - 1\right)h_m\right\}}$$

It is desirable to make a as small as possible compared with A, if the simpler result (4.4) is to be used.

4.9 Applications of the force equation

4.9.1 Force on a tapered bend A tapered bend is illustrated in Fig. 4.23. The force equation is

$$\left(\frac{p_1}{w} + \frac{v_1^2}{g}\right) a_{1x} - \left(\frac{p_2}{w} + \frac{v_2^2}{g}\right) a_{2x} = -\frac{f_x}{w}$$

In the x direction

$$a_{1x} = A_1 \qquad a_{2x} = A_2 \cos \alpha$$

$\therefore$
$$\left(\frac{p_1}{w} + \frac{v_1^2}{g}\right) A_1 - \left(\frac{p_2}{w} + \frac{v_2^2}{g}\right) A_2 \cos \alpha = -\frac{f_x}{w}$$

In the y direction

$$a_1 = 0, \qquad a_{2y} = A_2 \sin \alpha,$$

$\therefore$
$$-\left(\frac{p_2}{w} + \frac{v_2^2}{g}\right) A_2 \sin \alpha = -\frac{f_y}{w}$$

Now f_x and f_y are the forces that the bend exerts upon the fluid in the x and y direction so as to change its direction. The forces that the fluid exerts upon the bend, P_x and P_y, are equal in magnitude but opposite in direction by Newton's third law so

$$P_x = \left(p_1 A_1 + \frac{w}{g} A_1 v_1^2\right) - \left(p_2 A_2 \cos \alpha + \frac{w}{g} A_2 v_2^2 \cos \alpha\right)$$

and

$$P_y = -\left(p_2 A_2 \sin \alpha + \frac{w}{g} A_2 v_2^2 \sin \alpha\right)$$

Applying the energy equation

$$\frac{p_1}{w} + \frac{v_1^2}{2g} = \frac{p_2}{w} + \frac{v_2^2}{2g}$$

if the bend is in the horizontal plane so that $z_1 = z_2$ and if energy losses are negligible. Also

$$A_1 v_1 = A_2 v_2 \quad \text{(continuity equation)}$$

$\therefore$
$$p_2 = w\left(\frac{p_1}{w} + \left[1 - \left(\frac{A_1}{A_2}\right)^2\right] \frac{v_1^2}{2g}\right)$$

By substitution

$$P_x = (p_1 A_1 + \rho A_1 v_1^2) - \left[\left(p_1 + \rho\left(1 - \left(\frac{A_1}{A_2}\right)^2\right)\frac{v_1^2}{2}\right) A_2 \cos \alpha\right.$$
$$\left. + \rho A_2 \cos \alpha\, v_2^2\right]$$

Now substituting for v_2

$$P_x = p_1(A_1 - A_2 \cos \alpha) + \rho v_1^2 \left\{ A_1 - \frac{1}{2}\left(1 - \left(\frac{A_1}{A_2}\right)^2\right)A_2 \cos \alpha - \frac{A_1^2}{A_2}\cos \alpha \right\}$$

Also

$$P_y = -\left[p_1 - \left(1 - \left(\frac{A_1}{A_2}\right)^2\right)\rho\frac{v_1^2}{2}\right]A_2 \sin \alpha - \rho\left(\frac{A_1}{A_2}\right)^2 A_2 v_1^2 \sin \alpha$$

$$= -p_1 A_2 \sin \alpha - \rho v_1^2 \left[\left(1 - \left(\frac{A_1}{A_2}\right)^2\right)\frac{A_2 \sin \alpha}{2} + \left(\frac{A_1}{A_2}\right)^2 A_2 \sin \alpha\right]$$

$$= -\rho v_1^2 A_2 \sin \alpha \left[\left(1 - \left(\frac{A_1}{A_2}\right)^2\right)\Big/2 + \left(\frac{A_1}{A_2}\right)^2\right] - p_1 A_2 \sin \alpha$$

$$= -\frac{\rho v_1^2}{2}A_2 \sin \alpha \left(1 + \left(\frac{A_1}{A_2}\right)^2\right) - p_1 A_2 \sin \alpha$$

P_x and P_y act in the directions shown in Fig. 4.23.

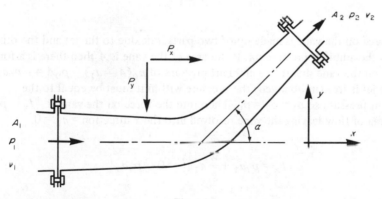

Fig. 4.23

4.9.2 Forces acting on vanes

Jet impact on a flat vane (Fig. 4.24)

$$-\frac{f_x}{w} = \left(\frac{p_1}{w} + \frac{v_1^2}{g}\right)a_1 - \left(\frac{p_2}{w} + \frac{v_2^2}{g}\right)a_2$$

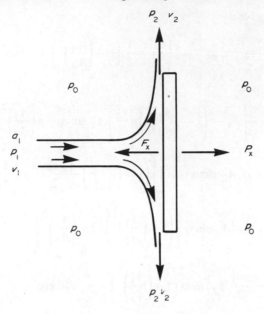

Fig. 4.24

The force on the vane is made up of two parts, one due to the jet and the other due to the ambient pressure p_0. If the area of the vane is A then there is a force acting on the vane due to the ambient pressure of $p_0(A - a_1) - p_0 A = -p_0 a_1$. As the jet is free (unbounded) the pressure within it must be equal to the ambient pressure so $p_1 = p_2 = p_0$. Therefore the force on the vane = $-f_x - p_1 a_1$. The area of flow leaving the vane resolved into the x direction = $a_2 = 0$.

$$\therefore \qquad P_x = p_1 a_1 + \frac{w}{g} a_1 v_1^2 - 0 - p_1 a_1$$

$$P_x = \rho a_1 v_1^2$$

(Note that $\rho a v^2 = \rho a v \times v$ = mass flow $\times$ velocity

$$= \text{rate of momentum transfer.})$$

If the ambient pressure had been taken as zero this result would have been obtained immediately. It is clear that in the case of free jets impinging upon vanes the ambient pressure can produce no net force upon the vanes and this method of making the ambient pressure zero is legitimate.

Jet impact upon a curved vane (Fig. 4.25)

In the x direction

$$\frac{f_x}{w} = \frac{v_1 a_{1x}^2}{g} - \frac{v_2^2 a_{2x}}{g}$$

$$a_{1x} = A_1, \quad a_{2x} = A_2 \cos \gamma$$

$$\therefore \quad \frac{f_x}{w} = \frac{A_1 v_1^2}{g} - \frac{A_2 \cos \gamma_1 v_2^2}{g}$$

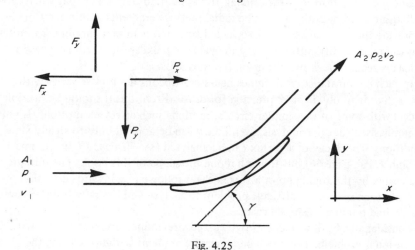

Fig. 4.25

Now $A_1 v_1 = A_2 v_2$ (continuity equation) and because of friction between the fluid and the vane v_2 is less than v_1. Assume

$$v_2 = k v_1 \quad \text{where} \quad k \approx 0.85$$

$$\therefore \quad -\frac{f_x}{w} = \frac{A_1 v_1^2}{g}(1 - k \cos \gamma)$$

$$P_x = -f_x = \rho A_1 v_1^2 (1 - k \cos \gamma)$$

In the y direction

$$a_1 = 0 \qquad a_{2y} = A_2 \sin \gamma$$

$$\therefore \quad \frac{f_y}{w} = -\frac{A_2 \sin \gamma_1 v_2^2}{g} = -\frac{k A_1 v_1^2 \sin \gamma}{g}$$

$$\therefore \quad P_y = -f_y = -\rho A_1 v_1^2 k \sin \gamma$$

The negative sign means that P_y is in the direction of y decreasing.

4.10 The variation of the Bernoulli constant across stream lines

Bernoulli's equation is

$$\frac{p}{w} + \frac{v^2}{2g} + z = E$$

where E can be regarded as the total energy of a unit weight of fluid and has the dimension of distance. Up to this point it has been treated as a constant along a stream line if the flow is not frictional. It was pointed out earlier that when flow is along a curved path there must be radial pressure gradients within the fluid to provide the necessary centrally directed forces to maintain curvilinear motion. This can make it difficult to specify the value of p except in circumstances in which a stream tube is straight or on free surfaces.

In such circumstances it becomes necessary to consider how E varies with radius, that is to obtain an expression for dE/dr. From this it should be possible to deal with some of the simpler circular motions such as vortex motions. It must be emphasised that the analysis given below can be applied only to steady state conditions. The decay of a vortex can be analysed (see *Viscous Flow*, volume 1 by Shih J. Pai, page 66) but this is beyond the scope of this book. The initiation of a vortex by the rolling up of a vortex sheet followed by inwards diffusion of vorticity is an extremely difficult problem to analyse and has only been achieved in a limited number of special cases.

Consider steady flow between two curved streamlines, treating it as a two dimensional problem. There will be a pressure gradient within the flow, the pressure increasing in the radial direction (See Fig. 4.26).

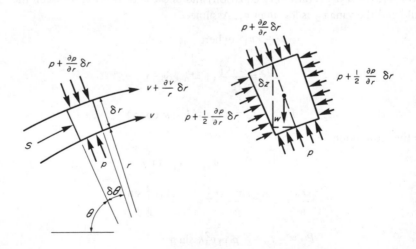

Fig. 4.26

Resolving forces acting upon the element in the radial direction:

(1) Pressure forces

The pressure force acting outwards is

$$pr\,\delta\theta - \left(p + \frac{\partial p}{\partial r}\,\delta r\right)\left(r + \delta r\right)\delta\theta + \left(p + \frac{1}{2}\frac{\partial p}{\partial r}\,\delta r\right)\delta r\,\delta\theta$$

Note that the last term in this expression is the component of the pressure force acting on the sides of the element resolved in the radial direction.

Simplifying and ignoring second order small quantities reduces the expression to

$$-\frac{\partial p}{\partial r}r\,\delta r\,\delta\theta$$

(2) Weight forces

The weight of the element is $wr\,\delta\theta\,\delta r$. Its component in the outward radial direction is

$$-wr\,\delta\theta\,\delta r\sin\theta$$

but $\delta r\sin\theta$ is δz, the change in elevation from the lower to the higher streamline, so the weight force acting in the outward radial direction is

$$-wr\,\delta z\,\delta\theta$$

The total of these two forces creates the required centrally directed acceleration v^2/r

so

$$-\left(-\frac{\partial p}{\partial r}\,\delta r\,\delta\theta\,r - wr\,\delta z\,\delta\theta\right) = \frac{w}{g}r\,\delta\theta\,\delta r\,\frac{v^2}{r}$$

Simplifying

$$\frac{\partial p}{\partial r} + w\frac{dz}{dr} + \frac{w}{g}\frac{v^2}{r} = 0$$

Differentiating Bernoulli's equation gives

$$\frac{dE}{dr} = \frac{d(p/w)}{dr} + \frac{v}{g}\frac{dv}{dr} + \frac{dz}{dr}$$

$$= \frac{1}{w}\left(\frac{dp}{dr} + \frac{w}{g}v\frac{dv}{dr} + w\frac{dz}{dr}\right)$$

Now under conditions of steady two-dimensional flow $\partial p/\partial r = \mathrm{d}p/\mathrm{d}r$ so substituting for $\mathrm{d}p/\mathrm{d}r$

$$\frac{1}{w}\left(-w\frac{\mathrm{d}z}{\mathrm{d}r} - \frac{w}{g}\frac{v^2}{r} + \frac{w}{g}v\frac{\mathrm{d}v}{\mathrm{d}r} + w\frac{\mathrm{d}z}{\mathrm{d}r}\right) = \frac{\mathrm{d}E}{\mathrm{d}r}$$

Rearranging and cancelling gives

$$\frac{\mathrm{d}E}{\mathrm{d}r} = \frac{v}{g}\left(\frac{\mathrm{d}v}{\mathrm{d}r} + \frac{v}{r}\right) \tag{4.5}$$

This equation can be used to solve the steady states of two vortices which are of common occurrence.

4.11 The free vortex

In a flow which is not significantly affected by friction any perturbation of the flow will generate rotational motion and the vortices so created are known as free vortices. The free vortex is not self sustaining and the model of a steady-state free vortex is unrealistic as even a small amount of friction in the flow would cause it to decay. However, it is an extremely useful mathematical concept because when it is combined with an inwardly directed flow (a sink) a self sustaining flow results and here the mathematical solution is a very good description of such natural phenomena as the bath tub vortex, the tornado, the hurricane, the cyclonic flow and the whirlpool. When the flow is outwardly directed the combination of a free vortex and a source describes the behaviour of anticyclones.

To start a free vortex no more energy than is already in the flow is required, but it is necessary to introduce some rotational motion. By adding a 'swirl' to a sink flow a free spiral vortex can be produced for instance. If the production of the rotation introduces no angular momentum to the flow then two vortices of equal and opposite sign will be generated. An example of this is the twin vortices found upstream of undershot sluice gates.

It is a common statement that in the northern hemisphere a free spiral vortex will rotate in an anticlockwise direction and conversely, in the southern hemisphere it will rotate in a clockwise direction. This statement is true if no initial predisposing whirl causes the vortex to rotate in an opposite direction. The effect of the earth's rotation upon the direction of the vortex is small and almost any initial whirl or geometrical asymmetry will be sufficient to mask the earth's effect.

The way in which the earth influences the direction of a free vortex can be understood as follows. Imagine a large water-filled circular tank placed exactly above the north pole of the earth. The water in the tank will rotate once a day in an anticlockwise direction viewed from above, as does the earth. If a central plug is now removed, water from the periphery will more inwards and because no energy or angular momentum is added it will travel more rapidly around the centre in order to keep its angular momentum constant as its distance from the centre of rotation is reduced. As the initial direction of rotation was the same as that of the earth the vortex will be moving in an anticlockwise direction. As no energy has been added to the flow the energy of any particle must remain constant so the increase in its kinetic energy that occurs as its radius of motion decreases must be at the expense of its positional and pressure energy, so its depth must decrease.

4.11.1 The analysis of the free vortex The free vortex, ideally, dissipates no energy. Thus throughout it, E is constant and $dE/dr = 0$.

$\therefore$
$$dv/dr + v/r = 0$$

so
$$dv/v = - \, dr/r$$

$\therefore$
$$\log v + \log r = A$$

and
$$vr = C$$

or
$$v = C/r$$

The constant C is called the *strength of the vortex*. In Chapter 2 it is shown that C is related to the circulation Γ by the equation $\Gamma = 2\pi C$.

As the energy of the flow field is invariant Bernoulli's equation can be applied across streamlines as well as along them.

Considering two streamlines at radii R and r and applying Bernoulli's equation

$$\frac{p_R}{w} + \frac{v_{Rt}^2}{2g} + z_R = \frac{p_r}{w} + \frac{v_{rt}^2}{2g} + z_r$$

$\therefore$
$$\frac{p_R - p_r}{w} + z_R - z_r = \frac{v_{rt}^2 - v_{Rt}^2}{2g} \tag{4.6}$$

(The second subscript t denotes the tangential direction). Substituting

$$v_{rt} = C/r \quad \text{and} \quad v_{Rt} = C/R$$

$$\frac{p_R - p_r}{w} + z_R - z_r = \frac{C^2}{2g} \left(\frac{1}{r^2} - \frac{1}{R^2} \right)$$

But, from Fig. 4.27 it can be seen that

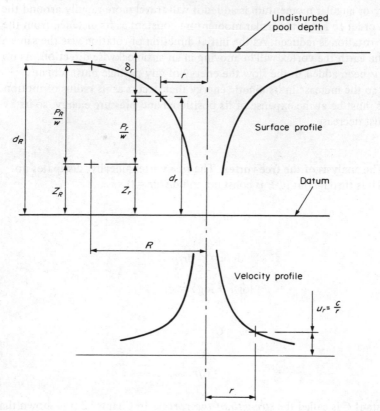

Fig. 4.27

$$\left(\frac{p_R}{w} + z_R\right) - \left(\frac{p_r}{w} + z_r\right)$$

is equal to

$$d_R - d_r$$

so

$$d_R - d_r = \frac{C^2}{2g}\left(\frac{1}{r^2} - \frac{1}{R^2}\right)$$

and this is the equation of a second order hyperboloid.

It R is made equal to infinity $v_R \to 0$ and $d_r \to d_\infty$ which is the undisturbed pool depth, so

$$\delta_r = \frac{C^2}{2gr^2}$$

Of course, in a real vortex the velocity gradients near the centre become so large that viscosity effects prevent the development of the central infinite velocity and infinite value of δ_r that the theory predicts. Points in flows at which such infinite values are predicted are called singular points.

4.12 Radial flow

A flow of this type is either a sink or a source and in the case when no external energy is added to the flow and none is taken from it by such mechanisms as friction, Bernoulli's equation can be applied along radii. Denoting velocities by v_{Rr} and v_{rr} where the subscript r denotes the radial direction

$$\frac{p_R - p_r}{w} + z_R - z_r = \frac{v_{rr}^2 - v_{Rr}^2}{2g}$$

Note the similarity between this δ and equation (4.6). The plot of p against r is called Barlow's curve.

4.13 The free spiral vortex

As the analysis of the free cylindrical vortex and radial flow just demonstrated have both been based on Bernoulli's equation it is equally reasonable to think that Bernoulli's equation can be applied to their combination—the free spiral vortex.

Denote the velocity at a radius R by v_R, its tangential component by v_{Rt} and its radial component by v_{Rr} and similarly for the velocity at radius r. Then

$$\frac{p_R}{w} + \frac{v_R^2}{2g} + z_R = \frac{p_r}{w} + \frac{v_r^2}{2g} + z_r$$

But

$$v_R^2 = v_{Rt}^2 + v_{Rr}^2 \quad \text{and} \quad v_r^2 = v_{rt}^2 + v_{rr}^2$$

So

$$\frac{p_R - p_r}{w} + z_R - z_r = \frac{v_{rt}^2 - v_{Rt}^2}{2g} + \frac{v_{rr}^2 - v_{Rr}^2}{2g}$$

but if only unit thickness of the flow is considered and flow in a direction perpendicular to the plane under consideration does not occur then

$$v_{rr} = \frac{q}{2\pi r} \quad \text{and} \quad v_{Rr} = \frac{q}{2\pi R}$$

where q is the flow through the vortex per unit thickness, and as

$$v_{rt} = C/r \quad \text{and} \quad v_{Rt} = C/R$$

$$\frac{p_R - p_r}{w} + z_R - z_r = d_R - d_r = \frac{C^2 + (q/2\pi)^2}{2g}\left(\frac{1}{r^2} - \frac{1}{R^2}\right)$$

This result applies quite accurately to many naturally occurring free-spiral vortices.

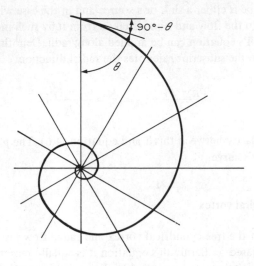

Fig. 4.28

The path of a fluid particle passing through such a vortex is an equiangular spiral as can readily be seen from the equation for the angle θ that the velocity vector v_r makes with the radius.

$$\tan \theta = \frac{v_{rt}}{v_{rr}} = \frac{C/r}{q/2\pi r} = \frac{2\pi C}{q}$$

that is the value of θ is constant for all radii (see Fig. 4.28). It is interesting that the spiral arms of galaxies are very similar in shape to this particle path.

4.14 The forced vortex

If a stirring device is placed in a fluid and rotated the fluid is constrained to rotate with it and must obey the law of solid body rotation:

$$v_{rt} = r\Omega$$

where Ω is the angular velocity of the rotation. As the free vortex velocity distribution is not obeyed the energy distribution throughout the fluid field is not uniform.

Now

$$v/r = \Omega \quad \text{and} \quad dv/dr = \Omega$$

so

$$\frac{dE}{dr} = \frac{v}{g}\left(\frac{v}{r} + \frac{dv}{dr}\right) = \frac{r\,\Omega\,2\,\Omega}{g}$$

so

$$\frac{dE}{dr} = \frac{2r\,\Omega^2}{g}$$

$$\therefore \qquad E = r^2\,\Omega^2/g + A$$

where A is an arbitrary constant.

Now

$$\frac{r^2\,\Omega^2}{g} = \frac{v_{rt}^2}{g} \quad \text{and} \quad E = \frac{p_r}{w} + \frac{v_r^2}{2g} + z_r$$

so

$$\frac{p_r}{w} + z_r = \frac{v_{rt}^2}{g} - \frac{v_{rt}^2}{2g} + A = \frac{v_{rt}^2}{2g} + A$$

but

$$p_r/w + z_r = d_r$$

$$\therefore \qquad d_r = r^2\,\Omega^2/2g + A$$

This is the equation of a paraboloid (see Fig. 4.29). When $r = 0$, $d_0 = A$, so

$$d_r = d_0 + r^2\,\Omega^2/2g$$

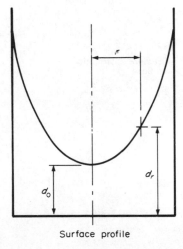

d_0

Surface profile

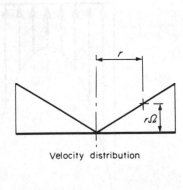

Velocity distribution

Fig. 4.29

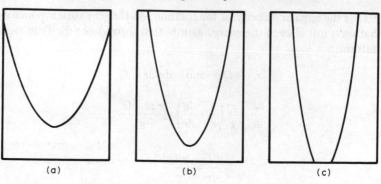

(a)   (b)   (c)

Fig. 4.30

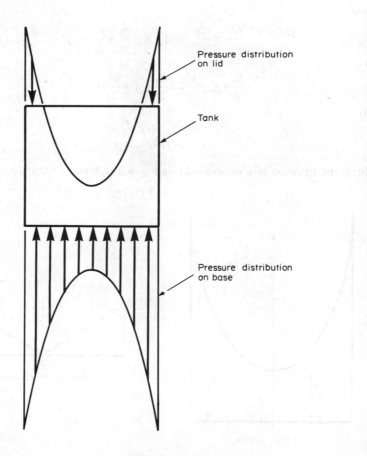

Pressure distribution on lid

Tank

Pressure distribution on base

Fig. 4.31

If a part-filled cylindrical vertical tank is rotated about its axis the contained fluid will eventually move as a forced vortex under the action of viscous shears. If the tank is closed a number of different modes of motion are possible (see Fig. 4.30).

(a) When the surface does not touch the lid.
(b) When the surface touches the lid but does not uncover the bottom or vice versa.
(c) When the surface touches the lid and also uncovers the bottom.

In all cases the surfaces are paraboloids or parts of paraboloids. The necessary condition for solution of all three types of motion is that the liquid volume does not change so the air volume above the liquid does not change either.

Pressure distributions on the lid can be obtained by continuing the curve of the paraboloid up to its point of intersection with the extension of the tank wall (Fig. 4.31). The pressure distribution on the base is a mirror image of the surface profile.

The total forces acting on the lid and base can be simply obtained by obtaining the appropriate volume of revolution of the pressure profile and multiplying by the specific weight of the fluid. It is not necessary to calculate both volumes of revolutions as once the force on the lid has been obtained the force on the base can very simply be obtained by adding the weight of the fluid contained within the tank to the force on the lid.

4.15 The Rankine vortex

When a free cylindrical vortex is started it rapidly decays because at its centre there are high velocities and high velocity gradients. These velocity gradients create large viscous shears which cause energy losses and hence the vortex decays.

The viscous shears so created also tend to make the vortex core rotate as a solid body so the situation becomes that of a forced vortex surrounded by a free vortex. This is called a Rankine vortex. The surface profile and velocity distribution diagrams are combinations of the forced and free vortex diagrams (Fig. 4.32).

The point at which the two velocity distributions join is not a perfect discontinuity as theory predicts but a more gradual variation giving rise to a transition zone in which conditions correspond neither to the forced nor to the free vortex models. Assuming theoretical conditions at the radius at which the two vortices join, r_c
Then

$$v_c = r_c \Omega \quad \text{and} \quad v_c = C/r_c$$

$$r_c \Omega = C/r_c$$

$$r_c = \sqrt{(C/\Omega)}$$

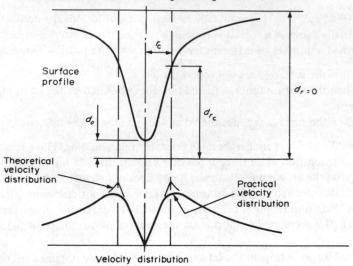

Fig. 4.32

For the forced vortex

$$d_{rc} - d_0 = \frac{v_c^2}{2g} = \frac{r_c^2 \, \Omega^2}{2g}$$

For the free vortex

$$d_{r=\infty} - d_{rc} = \frac{C^2}{2g} \left(\frac{1}{r_c^2} - \frac{1}{\infty^2} \right)$$

substituting r_c from above

$$d_{rc} - d_0 = C \, \Omega / 2g$$

and

$$d_{r=\infty} - d_{rc} = C \, \Omega / 2g$$

so

$$d_{rc} - d_0 = d_{r=\infty} - d_{rc}$$

Therefore the forced vortex core occupies half of the total depth of the vortex.

This vortex is of particular interest because almost all so-called free vortices seen in nature are actually Rankine vortices or free spiral vortices. The vortices seen in a Karman vortex trail are always Rankine vortices. The pure free cylindric: vortex cannot actually occur, the nearest approximation being a Rankine vortex.

An application of Rankine-vortex theory is in centrifugal pumps under no-flow conditions. The flow within the impeller can be thought of as an example of a forced vortex and that within the volute as a free vortex and as the rotationa speed of the impeller is specified the head developed by the pump can be calculated including the head gain in the volute.

Well-developed secondary flows may be associated with vortices (see Chapter :

4.16 Vorticity

In Chapter 2 vorticity ζ was defined and shown to be given by the equation

$$\zeta = v/r + dv/dr$$

From equation (4.5) it can now be seen that

$$\frac{dE}{dr} = \frac{v}{g}\zeta$$

Therefore if E is invariant with r, dE/dr is zero so ζ must also be zero. If this is so, potential flow theory applies. If dE/dr is not zero then ζ has value and this can be used to establish the limits of boundary layers. The dynamic tapping of a pitot tube gives the value of E so if a pitot tube is traversed across a flow that portion in which E changes has vorticity within it. On this a practical technique for delimiting boundary layers can be based.

Worked examples

(1) A tank of 5 m^2 plan area is fitted with a sharp-edged orifice of 5 cm diameter in its base. The coefficient of discharge of the orifice is 0·62. Calculate the time taken for the level in the tank to fall from 2 metres depth to 0·5 metres depth. If water is now admitted to the tank at a rate of 0·01 m^3/s calculate the rate at which the surface will be rising when the depth in the tank is 1 metre and the depth in the tank when the level becomes steady.

$$\text{Time} = \frac{2A}{C_d a_0 \sqrt{(2g)}} (h_2^{1/2} - h_1^{1/2})$$

$$= \frac{2 \times 5}{0{\cdot}62 \times \dfrac{\pi}{4}({\cdot}05)^2 \times \sqrt{19{\cdot}62}} (2^{1/2} - 0{\cdot}5^{1/2})$$

$$= 1{\cdot}31 \times 10^3 \text{ s}$$

Rate of rise of surface

$$A\frac{dh}{dt} = Q - C_d a_0 \sqrt{(2gh)} \qquad \text{where } Q = \text{inflow}$$

$$A\frac{dh}{dt} = Q - 0{\cdot}62 \times \frac{\pi}{4} \times ({\cdot}05)^2 \sqrt{(19{\cdot}62 \times 1{\cdot}0)}$$

$$\frac{dh}{dt} = 0{\cdot}922 \times 10^{-3} \text{ m/s}$$

Level in steady state

$$Q = C_d a_0 \sqrt{(2gh_s)}$$

$$h_s = \frac{Q^2}{C_d^2 a_0^2 2g}$$

$$= 3 \cdot 44 \text{ m}$$

(2) A venturimeter of throat diameter 4 cm is fitted into a pipeline of 10 cm diameter. The coefficient of discharge is 0·96. Calculate the flow through the meter when the reading on a mercury-water manometer connected across the upstream and throat tapping is 25 cm. If the energy loss in the downstream divergent cone of the meter is $10 \, v_p^2/2g$ per unit weight of fluid, calculate the head loss across the meter. (v_p is the velocity in the pipeline.)

$$Q = C_d a_p a_t \sqrt{\left\{ 2g \frac{\left(\dfrac{p_1 - p_2}{w} + z_1 - z_2 \right)}{a_p^2 - a_t^2} \right\}}$$

$$Q = 0 \cdot 96 \times \frac{\pi}{4} \times (0 \cdot 10)^2 \sqrt{\left\{ \frac{19 \cdot 62 \times 0 \cdot 25 \times 12 \cdot 6}{(10/4)^4 - 1} \right\}}$$

$$Q = 9 \cdot 61 \times 10^{-3} \text{ m}^3/\text{s}$$

Note that the manometer reading is related to the value $(p_1 - p_2) + z_1 - z_2$ by the relation

$$\frac{p_1 - p_2}{w} + z_1 - z_2 = \left(\frac{w_{hg}}{w_w} - 1 \right) h_m$$

$$= (13 \cdot 6 - 1) h_m$$

Energy loss/unit wt across the convergent cone

$$= \left(\frac{1}{C_d^2} - 1 \right) \left(\frac{v_t^2 - v_p^2}{2g} \right)$$

Therefore total energy loss across the meter

$$= \left(\frac{1}{C_d^2} - 1 \right) \left(\frac{v_t^2 - v_p^2}{2g} \right) + 10 \frac{v_p^2}{2g}$$

$$= 1 \cdot 01 \text{ m}$$

(3) Two pipe lengths are connected by a tapered 70° bend which lies in a horizontal plane. The pressure head (gauge) upstream of the bend is 25 m where the flow is 0·6 m³/s, and the pipe diameter is 0·5 m. Downstream of the bend the pipe diameter is 0·25 m. Calculate the force acting upon the fluid specifying its magnitude and direction.

Either of the two approaches described in the text may be used.

First, h_2 must be calculated. Assuming no energy loss occurs between points 1 and 2 (Fig. 4.23)

$$h_1 + \frac{v_1^2}{2g} = h_2 + \frac{v_2^2}{2g} \quad \text{and} \quad \frac{\pi}{4} \times 0.5^2 \times v_1 = \frac{\pi}{4} \times 0.25^2 \times v_2$$

$$\therefore \qquad h_2 = h_1 + \frac{v_1^2 - v_2^2}{2g}$$

but

$$v_2 = 4v_1$$

so

$$h_2 = h_1 - 15(v_1^2/2g)$$

but

$$v_1 = \frac{0.6}{\pi/4 \times 0.5^2} = 3.06 \text{ m/s}$$

$$h_2 = 17.863$$

In the x direction

$$\left(\frac{p_1}{w} + \frac{v_1^2}{g}\right)A_1 - \left(\frac{p_2}{w} + \frac{v_2^2}{g}\right)A_2 + \frac{F_x}{w} = 0$$

$$A_1 = \frac{\pi}{4} \times 0.5^2 = 0.196; \quad A_2 = 0.0491 \times \cos 70° = 0.0168$$

$$v_2 = 4 \times v_1 = 12.2$$

$$\therefore F_x = -9810\left[\left(25 + \frac{3.056^2}{9.81}\right) \times 0.19635 - \left(17.863 + \frac{12.224^2}{9.81}\right) \times 0.01679\right]$$

$$= -44.54 \text{ kN}$$

In the y direction

$$A_1 = 0, \quad A_2 = 0 \cdot 0491 \times \sin 70° = 0 \cdot 0461$$

$$\therefore \quad F_y = -9810\left[\left(\frac{p_1}{w} + \frac{v_1^2}{g}\right) \times 0 - \left(17 \cdot 863 + \frac{12 \cdot 224^2}{9 \cdot 81}\right) \times 0 \cdot 0461\right]$$

$$= +14 \cdot 967 \, kN$$

P_x, the force acting on the bend in the x direction is the reaction to F_x so

$$P_x = 44 \cdot 54 \text{ kN}$$

Similarly

$$P_y = -14 \cdot 967 \text{ kN}$$

The resultant force $R = \sqrt{(P_x^2 + P_y^2)} = 46 \cdot 99$ kN. The angle this resultant makes with the extended upstream pipe centre line α is given by

$$\alpha = \tan^{-1}\frac{14 \cdot 967}{44 \cdot 54} = 18 \cdot 57°$$

(4) A cylindrical tank of 0·667 m diameter is situated with its longitudinal axis vertical. It contains water in which floats a solid cylinder of relative density 0·8 and which is 0·333 m in diameter and 0·1667 m in length. When stationary and with the solid cylinder floating with its axis vertical the water surface is 0·333 m above the tank base. If the tank and its contents revolve about the tank's vertical axis at a rotational speed of 40 revs/min calculate the height of the top of the floating solid above the bottom of the tank. Assume that the axes of the solid cylinder and the tank remain coincident and that no water overflows. (See Fig. 4.33)

$$s = v^2/2g$$

$$s = \left(\frac{2\pi N}{60} \times \frac{0 \cdot 333}{2}\right)^2 \Big/ 19 \cdot 62$$

$$= 24 \cdot 8 \times 10^{-3}$$

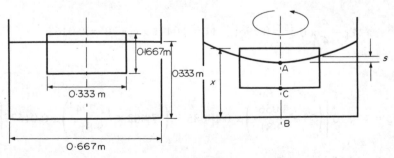

Fig. 4.33

Volume of contained water remains constant and the volume of fluid displaced by the floating solid cylinder must also remain constant.

$$\text{Water volume} = \frac{\pi}{4} \times 0\cdot667^2 \times 0\cdot333 - 0\cdot8 \times 0\cdot1667 \times \frac{\pi}{4} \times 0\cdot333^2$$

$$= 0\cdot105$$

$$\text{Now } 0\cdot105 = AB \times \frac{\pi}{4} \times 0\cdot667^2 + \frac{\pi}{4} \times 0\cdot667^2 \times \frac{1}{2} \frac{(2\pi N/60 \times 0\cdot667/2)^2}{2g}$$

$$-\frac{\pi}{4} \times 0\cdot333^2 \times 0\cdot8 \times 0\cdot1667$$

$$AB = 0\cdot284$$

$$x = 0\cdot284 + 5/2 + 0\cdot2 \times 0\cdot1667$$

$$= 0\cdot3297$$

The height of the top of the solid cylinder above the base is $0\cdot3297$ metres.

Questions

(1) Water discharges from a tank through a convergent–divergent nozzle. The diameters of the throat and exit of the nozzle are $1\frac{1}{2}$ inches [0·04 m] and 3 inches [0·08 m] respectively. The entry to the nozzle is rounded off and losses in the nozzle may be neglected. Calculate the depth of water in the tank over the centre line of the nozzle necessary to cause air release from the water assuming that air release occurs at an absolute head of 8 ft [2·7 m] of water. What will be the discharge under these conditions? If the divergent section of the nozzle is then removed what will be the flow under the same head? The height of the water barometer is 34 ft [10 m].

Answer: $h = 1\cdot733$ ft [0·487 m], $Q_{initial} = 0\cdot5186$ ft^3/s [0·01554 m^3/s], $Q_{final} = 0\cdot1297$ ft^3/s [0·00388 m^3/s]

(2) Show that for a venturimeter possessing a constant coefficient of discharge the loss of head due to friction in the convergent cone can be written as $h_f = kQ^2$ where k is a constant and Q is the rate of flow. Obtain the value of k for a venturimeter for which the inlet and throat diameters are 6 inches [0·15 m] and 3 inches [0·075 m] respectively and the coefficient of discharge is 0·96.

Answer: $0\cdot514$ [208·27]

(3) A closed tank has an orifice of 0·03 m diameter in one of its vertical sides. The tank is filled with oil of specific gravity 0·9 to a depth of 1 m over the centre of the orifice. The air space above the oil is pressurised to 16000 N/m². The coefficient of discharge of the orifice is 0·61. Determine the discharge from the orifice.

Answer: 0·0032 m³/s

(4) A horizontal boiler shell 6 ft [2 m] in diameter and 30 ft [10 m] long is half full of water. Find the time of emptying the shell through a 3 inch [0·08 m] diameter orifice located at the base of the shell for which the coefficient of discharge is 0·8.

Answer: 1206 s [1368·7 s]

(5) A reservoir with vertical walls has a plan area of 60 000 square yards. Discharge from the reservoir takes place over a rectangular weir for which $Q = 30 h^{1·5}$ ft³/s. The sill of the weir is 8 ft above the bottom of the reservoir. Starting with a depth of water of 10 ft in the reservoir and no inflow, what will be the depth of water after free discharge has taken place for one hour.

Answer: 9·54 ft

(6) Derive an expression for the discharge over a sharp-crested rectangular weir taking into account the effects of end contractions. A weir is to be constructed across a stream in which the flow is 0·2 m³/s. (The coefficient of discharge for the notch is 0·623). Find the minimum length of weir necessary if the upstream level is not to rise more than 0·384 m when the stream is discharging five times its normal flow.

Answer: 1·210 m

(7) A venturimeter of throat diameter 3 inches [0·08 m] is fitted in a 6 inch [0·16 m] diameter vertical pipe in which liquid of unit specific gravity flows downwards. Pressure gauges are fitted to the inlet and throat sections, the latter being 3 ft [1 m] below the former. The coefficient of the meter is 0·97. Find the discharge when (a) the pressure gauges read the same (b) when the inlet gauge reads 2·2 lbf/in² [10 kN/m²] higher than the throat gauge.

Answer: 0·688 ft³/5 [0·0224 m³/s], 1·12 ft³/s [0·0318 m³/s]

(8) A jet of water of 10 square inches [0·0062 m²] cross section moving at 40 ft/s [12·0 m/s] is deflected by a curved plate of 135° angle. The plate is moving at 15 ft/s [5 m/s] in the same direction as the jet. Find the amount of work done on the plate per second neglecting friction.

Answer: 2154 ft lb/s [2593 Nm/s]

(9) A tapered bend of 60° deflection angle joins two pipes. The upstream pipe diameter is 4 ft [1·3 m] and the downstream pipe diameter is 2 ft [0·6 m].

When the flow through the pipe line is 107 ft³/s [3 m³/s] the pressure head upstream of the bend is 100 ft [33 m]. Calculate the magnitude and direction of the force acting on the bend and describe a method of resisting this force.

Answer: 71 x 10³ lbf at 16° to the axis of the upstream pipe [394 kN at 13·8°]

(10) The wake of a submarine may be regarded as having a conical shape. The velocity profile across the circular cross section of this cone at a certain distance from the submarine may be treated as a paraboloid, the maximum velocity being on the centre line of the wake its radius being 30 ft [10 m] and the maximum velocity being 15 ft/s [5 m/s]. If the submarine is travelling at 10 ft/s [3 m/s] relative to still water, what power is being used in overcoming drag? Specific gravity of sea water = 1·03.

Answer: 7700 hp [8090 kW]

(11) A hollow cylindrical vessel 2 ft [0·67 m] dia. open at the top, spins about its axis, which is vertical, thus producing a forced vortex motion of the liquid contained in it. Calculate the height of the vessel so that the liquid just reaches the top and just begins to uncover the base when rotating at 120 rpm. If the speed is now increased to 150 rpm what area of the base will be uncovered and if the speed is now reduced until the base is just covered again, how high will the liquid stand at the sides?

Answer: Height = 2·46 ft [0·903 m], area uncovered = 1·14 ft² [0·129 m²],
1·57 ft [0·57 m]

(12) A cylindrical vessel with vertical axis is closed top and bottom with flat ends. Its diameter is 18″ [0·5 m] and the height is 24″ [0·67 m]. It is filled with water to a depth of 20″ [0·6 m], the remainder with air at atmospheric pressure. It is then revolved about its axis at a speed of 200 rpm. Calculate the force on the bottom and top covers.

Answer: Force on top = 71·8 lbs wt [629 N], force on bottom = 255·5 lbf
[1784 N]

(13) A cylindrical vessel 18″ [0·5 m] internal dia. and 30″ [0·9 m] deep, closed top and bottom, can rotate about its axis which is vertical. The vessel contains oil of specific gravity 0·9 to a depth of 20″ [0·6 m] when stationary. It is set in steady rotation and oil rotates with it at the same speed N rpm. Plot on squared paper curves showing (a) the relation between N and y, the depth of the lowest point of the oil surface below the top cover for values of y varying from 10″ [0·3 m] to 30″ [0·9 m].

Answer: $y = 10 + 0·000573 N$ inches up to 132 rpm then $y = 0·152 N$ inches
[$y = 0·3 + 0·0000174 N^2$ m up to 131 rpm then $y = 0·00458 N$ m]

(14) A tube ADBC consists of a straight vertical portion ADB and a curved portion BC, BC being the quadrant of a circle 9″ [0·25 m] radius with centre at D. A is open and the end C is closed. The tube is completely filled with water up to a height of 10″ [0·3 m] above B. Find the speed of rotation about the vertical axis ADB for the pressure at C to be the same as the pressure at B. For this speed find the maximum pressure in BC and its position. Prove any formula used.

Answer: 88 rpm [84 rpm]; 60°, 1·021 ft head [60°, 0·3625 m head]

(15) A horizontal open channel of constant rectangular cross section conveys 600 cusec [20 m³/s] of fluid. At one point in the channel there is a right-angled bend formed by circular arcs, the inner radius being 15 ft [5 m] and the outer 25 ft [8 m]. In the straight portion of the channel the uniform depth of fluid is 6 ft [2 m]. Assuming a perfect fluid with constant specific energy of flow and a free vortex flow around the bend, estimate the depth of flow at the inner and outer radii of the bend.

Answer: Depth at inner radius = 4·65 ft, depth at outer radius = 6·508 ft
[1·552 m and 2·17 m]
Note that the solution of the cubic equation for c is 205 [22·3]

(16) A cylindrical vessel open on top is 2 ft [0·67 m] in diameter and 4 ft [1·33 m] deep and is filled with water. It is now rotated steadily about its axis, which is vertical, at such a speed that the centre of the water surface just touches the bottom of the tank. (a) Develop the formula for the shape of the water surface. (b) Find the speed of rotation. (c) Water is now run into the vessel through a central hole in the bottom at a rate of 10 lbs per second [5 kg/s]. Find the torque required to maintain the steady speed of rotation, stating clearly what assumptions you have made in your solution.

Answer: (a) Paraboloid, (b) 16 radians/s [15·3 rads/s], (c) 5 lbf ft [8·53 Nm]

5 Boundary Layer Theory

In Chapter 2 it was shown that if the motion of a fluid is not affected by friction a velocity potential exists and the Euler equations are obeyed. In some cases, such potential flows can be analysed and pressures and velocities throughout them can be specified.

Unfortunately, there are many other cases which cannot be treated like this because of frictional effects. Sometimes the zones of flow affected by friction are narrow and close to the flow boundaries, but the entire flow may be dominated by friction. When the zone affected by friction is narrow the flow outside the zone may be analysed by potential theory provided that the boundary conditions can be specified.

Prandtl suggested that a flow field should be divided into two parts.

(1) The potential core in which potential theory applies and in which frictional effects are not present.
(2) The boundary layers which are zones of flow situated close to the boundaries in which frictional effects dominate the motion of the fluid.

This division is somewhat arbitrary as boundary layers are not well defined, the two parts of the flow gradually merging into one another. Even so, this concept of the boundary layer has proved extremely useful and provides a powerful tool in the analysis of frictional flows.

5.1 Formation of boundary layers

If a mass of fluid is set in linear motion in a gravity field far from any boundaries there will be no velocity gradients within it. When such a fluid approaches a boundary the fluid close to it is affected by frictional forces which will slow it down. Fluid in contact with the boundary is brought completely to rest—there is no 'slip' at the boundary. (This statement may not apply to gases at very low pressures.) Further away from the boundary, fluid will be slowed but not as much as fluid close to the boundary (see Fig 5.1).

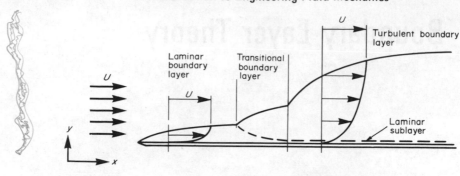

Fig. 5.1

The boundary layer is said to end where the local velocity equals 0·99 U (U being the undisturbed stream velocity). This is arbitrary because frictional effects spread throughout the flow. (The velocity distribution becomes asymptotic to U at large distances from the boundary.) The thickness of the boundary layer, usually denoted by δ, is defined as the distance from the boundary to the point at which the local velocity equals 0·99 U.

As the fluid passes over a greater length of the boundary more fluid will be retarded by friction with the fluid layers near to the boundary, so the boundary layer thickness δ will increase. What is happening here is that the fluid near the top of the boundary layer is dragging the fluid nearer to the boundary along and maintaining it in forward motion. This mechanism may be one of two types.

(1) Normal viscous shearing action produces tractive shear forces which may exert the necessary drag effects upon the slower moving layers deep in the boundary layer. If the boundary layer thickness is small, so that the velocity gradients are large, the tractive shear stress, which is given by $\tau = \mu(\partial u/\partial y)$, can be large enough to account for this effect. As the boundary layer becomes thicker the velocity gradients will decrease and eventually τ will become too small to maintain the slower layers in motion and, if viscous shear stress was the only mechanism available, they would have to come to rest.

(2) When the viscous shear mechanism becomes inadequate to maintain the sluggish layers in forward motion laminar flow will not be possible, deep layers will slow down under the influence of friction from the boundary and between it and faster moving layers higher up in the boundary layer fluid masses will start to rotate. (See Fig. 5.2.) This rotation effect can be considered as the formation of vortex sheets in the boundary layer which roll up to form many small vortices. (See Chapter 2, p. 68). The motion of the fluid particles in this state will rapidly become random and true turbulence will result. Now particles high in the boundary layer which have relatively high velocities and large momenta may move down into the lower layers because of their randomly directed velocities. Conversely particles from the slow moving layers will find their way up into the fast moving layer. The effect is to give a circulation of mass from fast to slow and back to fast layers with a net transport of momentum into the boundary

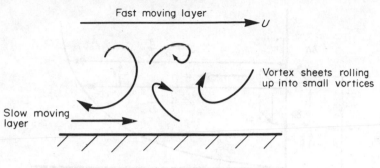

Fig. 5.2

layer. This transfer of momentum will maintain the slower layers in motion and is equivalent to a tractive shear stress.

Thus when a fluid moves over a boundary a boundary layer will develop in which flows will be laminar at the leading edge of the boundary and for some distance downstream of it. As the layer thickens the viscous shear mechanism will become inadequate and flow in the boundary layer will become turbulent. This explains Fig. 5.1. Generally, laminar boundary layers are thin and turbulent boundary layers are relatively thick.

5.1.1 The laminar sub-layer At points very close to the boundary velocity gradients are large and the viscous shear mechanism is powerful enough to transmit the shear stress to the boundary, so the very thin layer adjacent to the boundary is in laminar motion even when the flow in the rest of the boundary layer is turbulent. The sub-layer is only a few thousandths of an inch thick but its presence is vitally important in deciding whether a surface is hydraulically rough or smooth. (See pp. 190 to 196.)

5.2 The Prandtl mixing-length hypothesis

An approach to the development of a theory for turbulent boundary layers is due to Prandtl who applied ideas taken from the kinetic theory of gases to a theory first developed by Reynolds in 1895.

The velocity of any fluid particle can be split into three components U, V, and W in the x, y and z directions. If flow is turbulent these velocity components are not constant with time. However, the components can be regarded as being made up of two parts, one a steady velocity u and the other time varying, u'. So

$$U = u + u'$$
$$V = v + v'$$
$$W = w + w'$$

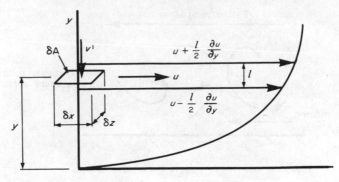

Fig. 5.3

If the flow is unidimensional in the x direction then $U = u + u'$, $V = v'$ and $W = w'$. Consider the flow over a boundary defined by $y = 0$ located in the x, z plane and assume that flow is only occurring in the x direction (Fig. 5.3). The mass crossing the area δA per second $= \rho\,\delta A v'$. The momentum being transferred per second from the fast layer to the slow layer is therefore:

$$\rho\delta A v' \times \text{velocity difference between the two layers}$$

If the two layers are thought of as consisting of rolling balls of fluid moving along with their centres travelling at velocity u (Fig. 5.4) it will be seen that at a fixed point in space on either of the two streamlines the fluid velocity will fluctuate as a ball of fluid passes through it. On streamline 2 the velocity will fluctuate from u up to $u + (l/2)\,\partial u/\partial y$ and back to u so

$$u' = (l/2)\,\partial u/\partial y$$

Similarly on streamline 1 the velocity will fluctuate from u down to $u - (l/2)\,\partial u/\partial y$ and back to u as the rotating fluid mass passes through. The *total* velocity fluctuation between streamlines 1 and 2 is

$$u' = l(\partial u/\partial y)$$

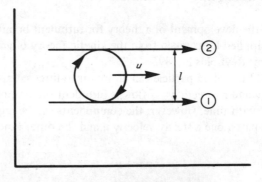

Fig. 5.4

so the momentum transfer per second from one layer to the other is

$$\rho \, \delta A v' l (\partial u / \partial y) = \rho \, \delta A v' u'$$

The rate of transfer of momentum per unit area is the shear stress (Newton's law) so

$$\tau = \rho u' v'$$

This is Reynolds' shear stress equation. Prandtl hypothesised that

$$u' = v' = l (\partial u / \partial y)$$

l which was shown in the diagram as the distance apart of the two layers can now be seen to define the dimensions of the rolling fluid masses. l is usually called the mixing length

$$\therefore \qquad \tau = \rho l^2 (\partial u / \partial y)^2$$

That this mixing length has a physical existence can be seen from the smoke patterns given off by a factory chimney in a steady breeze (Fig. 5.5). The air flow around the chimney is a boundary layer flow, the boundary being the earth's surface. The chimney is a smoke injector and as far as this phenomenon is concerned has no influence on the flow pattern.

Smoke from the chimney fills the centres of the fluid masses rolling over its outlet. As these masses move away downstream of the chimney the smoke gradually diffuses and mixes with all the fluid mass and the eddy appears to increase in size. However the eddy can be observed for some distance downstream before smoke diffuses away from it.

Clearly the size of the eddy must depend upon its location. One situated very near the boundary could not be very large whereas one situated a long way from it could. Prandtl found that the size of the eddies was approximately 0·4 times the distance of its centre from the boundary.

$$\frac{1}{0 \cdot 4 y} \sqrt{\left(\frac{\tau}{\rho} \right)} = \frac{\partial u}{\partial y}$$

$$\therefore \qquad \frac{\partial u}{\partial y} = 2 \cdot 5 \sqrt{\left(\frac{\tau}{\rho} \right)} \frac{1}{y}$$

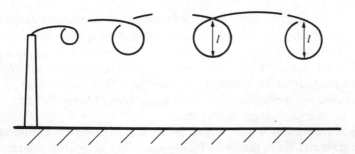

Fig. 5.5

If τ is assumed to have the same value as the bed shear τ_0 and to be invariant with y then

$$u = 2 \cdot 5 \sqrt{(\tau_0/\rho)} \log_e y + C$$

$$\therefore \qquad u = 5 \cdot 75 \sqrt{(\tau_0/\rho)} \log_{10} (y/c)$$

The assumption that τ is constant at τ_0 is difficult to accept. In the case of a free surface flow in which the boundary layer occupies the entire depth of the flow there cannot be a viscous shear stress at the surface itself. However, the logarithmic velocity distribution obtained by making this assumption can be shown experimentally to be closely followed by real fluids. The point is that the precise mode of variation of the shear stress is not very significant deep in the boundary layer where shear stresses have their largest values, but high in the boundary layer where they must be changing rapidly they can have only marginal influence upon the velocity distribution.

Von Karman considered a distribution function for u, expanded it in a Taylor series and from this and other considerations he showed that

$$l = \frac{0 \cdot 4 \; \partial u/\partial y}{\partial^2 u/\partial y^2}$$

and hence

$$\tau = \frac{0 \cdot 16 \; \rho (\partial u/\partial y)^4}{(\partial^2 u/\partial y^2)^4}$$

By making the assumption that τ falls linearly from a value at the boundary to zero at the edge of the boundary layer he obtained an integrable result as below.

$$u = U_{max} + 2 \cdot 5 \; V^* [\sqrt{(1 - y/\delta)} + \log_e \{1 - \sqrt{(1 - y'/\delta)}\}]$$

$$V^* = (\tau_0/\rho)$$

This result is slightly better than Prandtl's result.

5.3 Boundary layer separation

If flow over a boundary occurs in circumstances in which pressure decreases in the direction of flow the fluid will accelerate and the boundary layer will become thinner. Fluid near the boundary will be helped by this negative pressure gradient to maintain its forward motion and there will be no tendency for the boundary layer to separate from the boundary. In convergent flows such negative pressure gradients exist and it is a characteristic of all convergent flows that they are stable and that turbulence decreases within them.

If a positive pressure gradient exists in the fluid the situation is different. Such pressure gradients can be generated by divergent flow or by flow in a curved path; an explanation of the mechanism of this will be given later. Fluid outside the boundary layer will have sufficient momentum to carry it through the adverse pressure gradient; indeed, it is the adverse pressure gradient which slows the flow so that continuity can be maintained in divergent flow. Fluid near the boundary has so little momentum that the adverse pressure gradient will rapidly bring it to rest and may even reverse its motion. In this case the fluid which is conforming to what is usually considered to be boundary layer motion is moved away from the boundary by being displaced by fluid that is moving in the

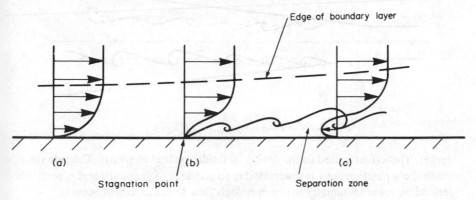

Edge of boundary layer

(a)   (b)   (c)

Stagnation point   Separation zone

Fig. 5.6

reverse direction—this is called boundary layer separation (see Figs. 5.6a, b and c). At the edge of the separated zone a point of inflexion in the velocity distribution occurs (see Fig. 5.7). Flows on either side of the inflexion point are directed in opposite directions. This will generate a vortex sheet which will roll up into discrete vortices. In the separation zone flows are more turbulent due to the vortices present than is the flow in the boundary layer itself, so such separations cause increased energy losses and this is why divergent flow is inherently unstable and why it causes much greater energy loss than normally occurs in parallel or convergent flow.

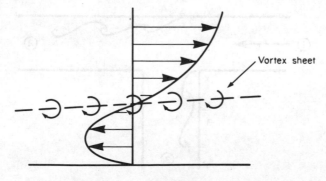

Vortex sheet

Fig. 5.7

5.3.1 Examples of boundary layer separation

Divergent duct or diffuser

This is illustrated in Fig. 5.8. Generally, after a short period the boundary layer separation zone on one side will be suppressed and the other becomes

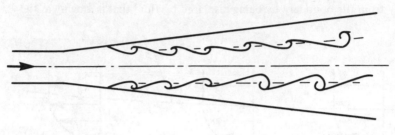

Fig. 5.8

thicker. This effect is used in the design of fluidic control elements. The side on which the separation zone is suppressed is an accident of geometry and is probably decided by some asymmetry in the approach flow to the divergent section.

Tee and Y junctions

These are illustrated in Fig. 5.9. Two boundary layer separations are produced. In separation A, a particle at point 1 may travel through the main pipe to point 2. If the pipe at point 2 has the same area as that at point 1 the velocity at point 2 must be less than that at point 1 because of the flow that has been abstracted through the branch. Therefore the pressure at point 2 must be greater than that at point 1, and a positive pressure gradient must exist along the main pipe. This causes the boundary layer separation at A. The boundary layer on the branch side of the main pipe is drawn off through the branch so no separation can occur

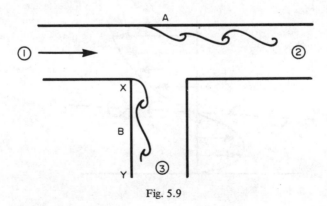

Fig. 5.9

on this side. Next consider separation B. To enter the branch the fluid has to enter a curve of very small radius and needs a centrally directed pressure gradient to provide the necessary centrally directed acceleration. Thus, at X, the pressure must be less than the ambient pressure. When the fluid has passed down the branch it will return to parallel straight motion and the pressure must return to about the same magnitude as the ambient pressure at the junction. Thus along XY there must be a positive pressure gradient and this explains the separation that occurs.

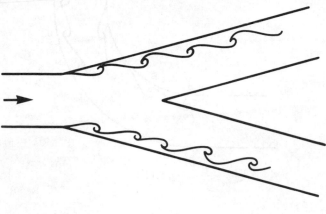

Fig. 5.10

The Tee junction described above is a special case of the Y junction (Fig. 5.10). The essential point here is that junctions such as these produce a net increase in the area of the flow and as such are examples of divergent flow. Downstream of the separation zones the vortices die away and normal pipe flow gradually redevelops; their effect is thus local.

Flow in bends

Two separations occur in flow in bends (Fig. 5.11). Consider separation A. At *a* the pressure is approximately the same as the mean pressure in the pipe at the bend. At b the pressure must be greater than this mean pressure so as to provide the required radially directed acceleration. Thus a positive pressure gradient occurs along ab.

Consider separation B. At c the pressure is less than the mean pressure (if the pressure at b is greater, the pressure at c must be less than the mean bend pressure so that the mean can be maintained). At d the pressure must be the mean value again as the fluid must be in linear motion so a positive pressure gradient must occur along cd—hence the separation B. Flow reattaches after separation A as the pressure gradient becomes negative downstream of it.

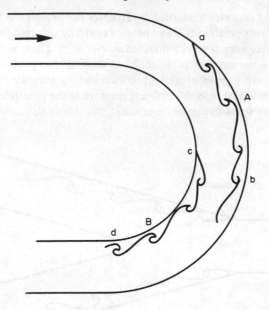

Fig. 5.11

The Karman vortex trail or street

If flow around a cylinder is very slow (Reynolds number $\leqslant 0.5$) the two boundary layers around the cylinder do not detach because the pressure gradients (which depend upon v^2) are very small. (Fig. 5.12). For Reynolds numbers between 2 and 30 the flow boundary layers separate symmetrically producing two attached vortices as shown. From a lower limiting Reynolds number of between 70 and 120 the vortices are shed alternately from each side of the cylinder producing two staggered rows of vortices as illustrated in Fig. 5.13. This is the Karman vortex trail. Each vortex is in the field of every other vortex so if such a system of vortices could exist in a stationary fluid the system would move 'upstream'. If the still-water velocity of the vortex system is v_y the velocity of

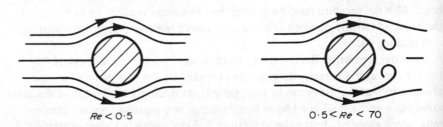

$Re < 0.5$ $0.5 < Re < 70$

Fig. 5.12

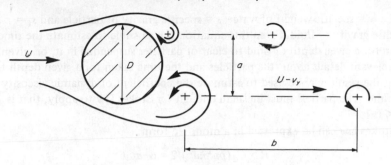

Fig. 5.13

the system in the real flow under consideration will be $U - v_y$. The frequency of the vortex trail must then be

$$f = (U - v_y)/b$$

Von Karman showed that $a/b = 0{\cdot}281$ and G. F. Taylor gave f as

$$f = 0{\cdot}198 \frac{U}{D} \left(1 - \frac{19{\cdot}7}{Re} \right)$$

(note fD/U is the Strouhal number). This result applies up to $Re = 200\,000$.

The Karman vortex trail is the mechanism behind many significant phenomena. The famous failure of the Tacoma narrows suspension bridge was caused by resonance between the Karman vortex trail from the bridge deck and the natural frequency of the bridge. The tramping of power lines is, in essence, caused by the Karman vortex trail phenomenon. Scour *downstream* of a bridge pier is, at least partly, also caused this way. (Remember that most scour effects occur *upstream* of a bridge pier and these are caused by secondary flows—see Section 5.5).

5.4 Drag on spheres

A case of particular importance for civil engineers is that of very small spheres falling through a fluid—the sedimentation situation.

Stokes gave the equation

$$R = 6\pi a \mu v$$

where a is the radius of the sphere, μ is the coefficient of viscosity and v is the velocity of the sphere. R is the drag force or resistance.

If a spherical particle is falling through a liquid the force acting is its weight minus the buoyant effect of the fluid it displaces, so

$$(4/3)\pi a^3 w(s - s_f) = 6\pi a \mu v$$

where w = specific weight of water, s = specific gravity of particle and s_f = specific gravity of fluid. From this equation it is possible to estimate the time taken for a given depth of fluid to clear of particles suspended in it, or, given the relevant details about the particles and the time taken for a given depth to clear, the result may be used to estimate the coefficient of dynamic viscosity of the fluid. The particle must be small enough for Stokes law to apply, that is $Re < 0.1$.

Stokes law can be expressed in a modified form:

$$R = C_d(\rho v^2 \pi a^2)/2 = 6\pi a\mu v$$

$$\therefore \qquad C_d = \frac{12\mu}{\rho av} = \frac{24\mu}{\rho(2a)v} = \frac{24}{Re}$$

This result is accurate, as just stated, up to $Re = 0.1$ but Oseen gives

$$C_d = \frac{24}{Re}(1 + \tfrac{3}{16} Re)$$

which works well up to $Re = 1$. Above 1 and below 100

$$C_d = \frac{24}{Re}(1 + \tfrac{3}{16} Re)^{1/2}$$

5.5 Secondary flow

Generally one dimensional approaches to the analysis of engineering problems give sufficiently accurate solutions. Sometimes two dimensional approaches have to be used. Occasionally results of such analyses are completely wrong because the phenomenon of secondary flow has been ignored.

Secondary flow is a spiral motion created by the interaction of the primary flow and the boundary layer flow. It occurs commonly in curvilinear flow but can also be detected in linear flows. The mechanisms that generate secondary flow in a primary curved flow and in straight flow are different and will be described separately.

5.5.1 Secondary flow in a curvilinear primary flow A flow in a curved path can be considered as part of a vortex of some type (see pp. 138 et seq.). In the radial direction the pressure will alter, reducing towards the centre. Away from any boundary the fluid will be moving with a finite velocity and with a radial pressure gradient such that the force acting on any element of fluid necessary to produce the required centrally directed acceleration for curvilinear flow is provided. Near a boundary, friction will reduce the velocity of an element so that the unchanged pressure gradient will cause an inwardly directed radial velocity component. This causes spiral motions to be superimposed upon the primary flow.

As an example, examine the case of a free surface flow around a channel bend. In a bend such as this a flow closely resembling a portion of a free vortex will be set up (see Fig. 5.14). At Q the longitudinal velocity, v_Q, will be less than that at P, v_p. Hence the value of the acceleration v_Q^2/r required to maintain the

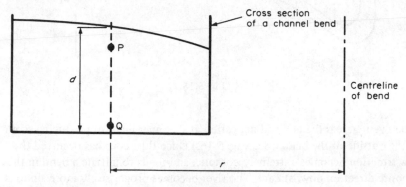

Fig. 5.14

fluid in its curved path parallel to the channel sides will not be as large as that required at P. The transverse pressure gradient acting at P and Q is the same (that is the transverse surface slope) so if the pressure gradient is producing a centrally directed acceleration that causes curvilinear flow at the correct radius r at point P it must be causing a centrally directed acceleration that is too large at Q. This excites a secondary spiral flow as illustrated in Fig. 5.15. As many as three or four such spirals may be generated in any cross section. The boundary layers from the side walls of the channel are not important in this type of secondary flow but they assist in generating it.

One consequence of secondary flow in channel bends is meandering. If the channel is cut in erodible material the secondary helices will cause loose material

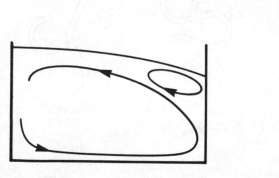

Fig. 5.15

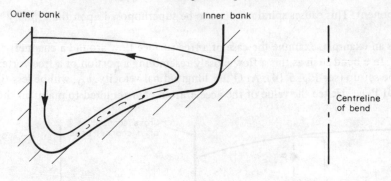

Fig. 5.16

to be swept across the bed and deposited at the inner bank so producing scour
on the outside of the bend. (See Fig. 5.16.) Once this scour has occurred the
flow situation becomes extremely complex and tends to initiate a bend in the
opposite direction downstream. The river becomes progressively more sinuous
and hence longer, its hydraulic mean radius becomes smaller because its wetted
perimeter becomes larger and so the longitudinal surface slope becomes greater.
Eventually the surface rises above the river banks and the meanders are cut off
forming ox-bow lakes. (Fig. 5.17d.) The river may not complete this cycle
however; instead, once the meanders have formed, they themselves may start
travelling downstream as shown in Fig. 5.17b.

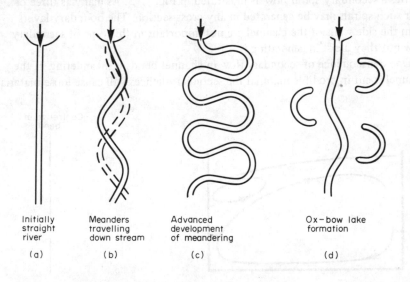

Fig. 5.17

Meandering depends upon the slope of the plain over which the river runs, the erodibility of the bed material and the magnitude of the dominant discharge. For it to occur, both erosion and accretion must be occurring at the same time. If the dominant discharge is too low there will be no bed movement hence no meandering. If it is too large the bed will erode throughout its length and again no meandering will occur. It is extremely difficult to control river meandering and at the moment of writing it is not possible to predict the form or magnitude of the meandering phenomenon except in a most general manner. It can be a most dangerous phenomenon as it can cause a major river to change its course overnight.

Helical flow in pipe bends

Secondary flow occurs in pipe bends for exactly the same reasons as for river bends. Because no free surface is present the secondary flow system consists of two vortices rather than one and these rotate around one another making a rope of vorticity (Fig. 5.18).

These secondary flows cause energy losses over and above normal pipe friction losses and also upset the calibration of instruments such as venturimeters which have been calibrated in straight pipes. Either flow straighteners should be fitted immediately after the bend to eradicate the secondary flow or no instrument should be fitted for a distance of at least 20 diameters downstream of the bend so that the secondary flow may decay.

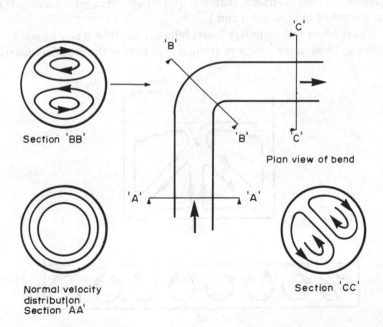

Section 'BB'

Plan view of bend

Normal velocity distribution Section 'AA'

Section 'CC'

Fig. 5.18

Secondary flow in vortices

In a vortex the secondary flow mechanism described for a river bend will operate. As a consequence the secondary spiral will cause suspended particles to be swept in towards the centre at the bottom of the vortex and outwards at the surface. This explains the behaviour of tea leaves when a cup of tea is stirred. It will be found that the tea leaves move in towards the centre at the bottom of the cup.

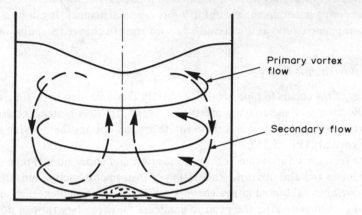

Fig. 5.19

5.5.2 Secondary flow in straight channels In a straight channel secondary flows occur as illustrated in Figs. 5.20a and b.

In a corner where two boundary layers interfere the flow is acted upon by two boundaries and loses more energy in friction than flow in the boundary layers

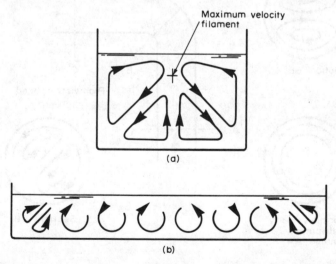

(a)

(b)

Fig. 5.20

away from the corner. Consequently the mass transport mechanism that operates to maintain flow in normal boundary layers (Prandtl mixing-length hypothesis, Section 5.2) must operate even more powerfully to maintain flow in the corners. The resulting increased mass transport into the corners shows itself as a secondary flow as illustrated in Fig. 5.20. In shallow river flows additional spirals may be excited as illustrated in Fig. 5.20b.

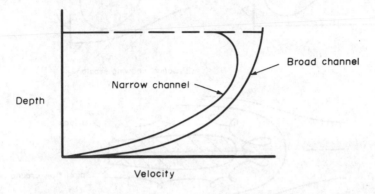

Fig. 5.21

According to most theories of velocity distribution in free surface flows the maximum-velocity filament should be located on the surface and on the centre-line of the channel. In fact it is usually depressed below the surface and secondary flow is the cause of this. It carries the maximum-velocity filament downwards replacing it with slower moving fluid from the sides. Thus in wide channels the maximum-velocity filament can be expected to lie on the surface and in narrow channels it will be displaced downwards by an amount that depends on the strength of the secondary flow. This in turn depends upon the side and bed roughness. (See Fig. 5.21.)

5.5.3 Secondary flows caused by obstructions When an obstruction is placed in a flow a stagnation point will be created somewhere upstream. The potential (pressure head plus elevation above a datum) generated outside the bed boundary layer will be greater than that at the bed by an amount of $v^2/2g$ where v is the free stream velocity so a flow will start towards the bed. In the case of an obstruction in a free surface flow founded on erodible material such as a bridge pier this flow can cause heavy erosion upstream of the bridge pier as shown in Fig. 5.22.

It has been found that by building dwarf walls at approximately $30°$ on either side of the longitudinal axis of the flow the erosion can be greatly reduced (see Figs. 5.23a and b). The explanation for this is that the water between the dwarf

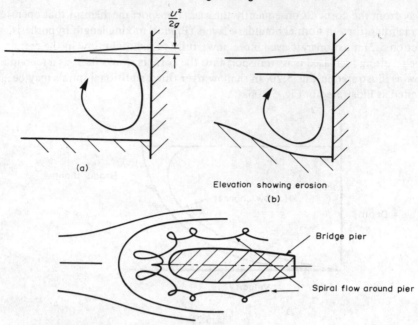

$\dfrac{U^2}{2g}$

(a)

Elevation showing erosion
(b)

Bridge pier

Spiral flow around pier

Plan showing flow at and near the bed

(c)

Fig. 5.22

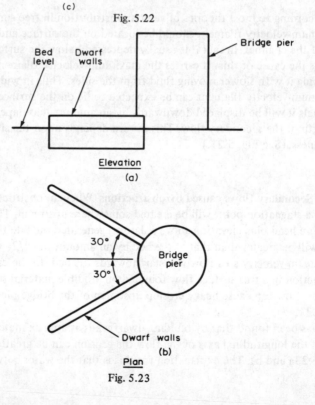

Bed level

Dwarf walls

Bridge pier

Elevation
(a)

Bridge pier

30°

30°

Dwarf walls

Plan
(b)

Fig. 5.23

walls is very nearly stationary so the dangerous secondary current is lifted above the bed by a distance which is of the same order of magnitude as the height of the tops of the dwarf walls above the bed. This protects the bed from being acted upon by the scouring effect.

5.5.4 Undershot sluices

Undershot sluice gates develop twin vortices in the upstream flow. This is due to a secondary flow mechanism. The surface flow along the centreline of the channel is at or near the maximum velocity whereas that on the surface near a side wall has a much lower velocity. When these velocities are both reduced to zero by the water impinging upon the sluice there are changes in flow depth which cause a transverse surface slope. This transverse surface slope causes a secondary flow which is so well developed that visible free spiral vortices may be seen. These may be so strong that an air core may extend down from the upstream water surface to the exit from the sluice. (See Figs. 5.24a, b, and c.)

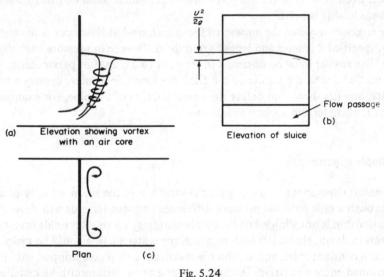

(a)   Elevation showing vortex
          with an air core

Elevation of sluice   (b)

Flow passage

Plan   (c)

Fig. 5.24

6 Pipe Flow

The flow of water through pipes has interested engineers from earliest times.
The problem is far more complex than it would at first seem and it is not
surprising that a definitive solution of the problem has only recently been
attained. Even now there are many research areas which could be investigated
with considerable benefit.

The engineer requires the answer to the question what flow occurs through a
pipe of specified diameter and length given the difference in pressure over the
length. The answer must be obtained in terms of two classes of parameters:
(1) those that define the nature of the fluid, for example its mass density and
viscosity, and (2) those that define the characteristics of the pipe, for example
its length, diameter and surface roughness.

6.1 Simple experiments

The simplest experiment to investigate the variation of the mean velocity of a
flow through a pipe with the pressure difference between its ends will show that
the relationship is not simple but that a discontinuity exists. Reynolds investigated this in detail. He built a tank in which the water surface could be maintained at a constant level, and in which a re-entrant exit pipe equipped with a
bell mouthed entry was fitted. On the centreline of the bell mouth he installed
a dye injector. The arrangement was as illustrated in Fig. 6.1.

If flow is permitted by slightly opening the outlet valve and at the same time
the dye injector control valve is adjusted to admit dye at the same velocity as
the flow, a fine stready dye line will be formed in the pipe. If the flow is
progressively increased and the dye injector correspondingly adjusted, the dye
line will remain stable until a specific condition of flow is reached at which
instabilities in the dye line will start to develop, the line breaking up into a
helical pattern and then reforming. If flow is further increased the dye line will
produce a stable helical pattern and if still further increased the dye line will
break down completely and the dye will be diffused throughout the flow. The
sequence is illustrated in Figs. 6.2, 6.3, 6.4 and 6.5.

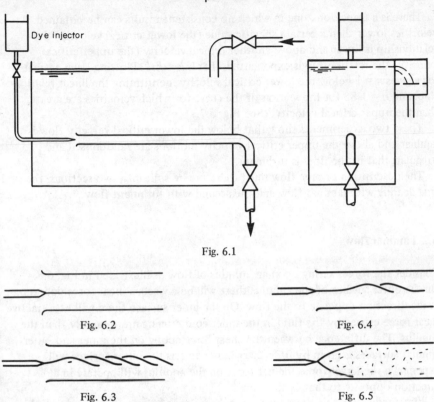

Fig. 6.1

Fig. 6.2

Fig. 6.4

Fig. 6.3

Fig. 6.5

6.1.1 Lower and upper critical velocities It is clear that the nature of the flow changes as the velocity increases from below to above the critical value required to cause the onset of instability of the dye line. A further experiment can be performed to find the relationship between the mean velocity and the pressure difference between the ends of the pipe. The results of such an experiment are graphically represented by a curve such as in Fig. 6.6.

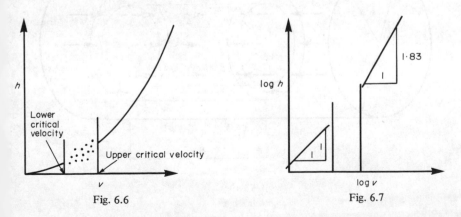

Fig. 6.6

Fig. 6.7

There is a transition zone in which no consistent results can be obtained. At velocities lower than a certain specific value (the lower critical velocity) the relationship is linear and above another critical velocity (the upper critical velocity) the relationship is exponential, that is $h = kv^n$. Plotting $\log h$ against $\log v$ gives $n = 1$ below the lower critical velocity, confirming the linear relationship, and $n = 1 \cdot 83$ for the section of the curve for which velocities are greater than the upper critical velocity. (See Fig. 6.7.)

These two experiments show that below the lower critical velocity flow is laminar and above the upper critical velocity motions are randomised and irregular; that is, the flow is turbulent.

The description of pipe flow must therefore be split into two sections, the first dealing with laminar flow and the second with turbulent flow.

6.2 Laminar flow

Consider the forces acting upon an annulus of flow as illustrated in Fig. 6.8. On the outer surface of the annulus there will be a viscous shear force operating in the direction opposite to the flow. On the inner surface there will be a tractive shear force caused by the fluid in the inner core moving more rapidly than the annulus. The difference between the shear force acting on the inner and outer annular surfaces is given by $\partial/\partial r (2\pi r \tau \delta x) \delta r$ and as the outer surface will experience the larger force the net force on the annulus will operate in a direction opposite to that of the flow.

For steady flow this shear force must be balanced by another force and this is obtained from the pressure difference across the ends of the annulus. In fact the pressure must drop along the pipe if an equality of the shear and pressure

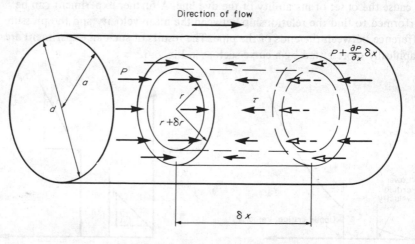

Fig. 6.8

forces is to be obtained but here the pressure gradient will be assumed to be positive and the signs will then look after themselves.

By Newton's first law of motion the force in the direction of flow caused by the pressure changes is equal to the net shear force on the annulus.

$$\therefore \qquad -2\pi r \frac{\partial p}{\partial x} \delta x\, \delta r = \frac{\partial(2\pi r\tau)}{\partial r} \delta r\, \delta x$$

$$\therefore \qquad -\frac{\partial p}{\partial x} r = \frac{\partial(\tau r)}{\partial r}$$

Expanding

$$-\frac{\partial p}{\partial x} r = r\frac{\partial \tau}{\partial r} + \tau$$

$$\therefore \qquad -\frac{\partial p}{\partial x} = \frac{\partial \tau}{\partial r} + \frac{\tau}{r} \qquad\qquad (6.1)$$

This is the differential equation of axi-symmetric flow through circular pipes.

Up to this point no particular assumption has been made about the type of flow in the pipeline other than that it is steady and no assumption has been made about the nature of the fluid, whether it is newtonian or non-newtonian.

If it is assumed that the flow is laminar and that the fluid is newtonian, Newton's law of viscosity can be applied

$$\tau = \mu\frac{\partial u}{\partial y} = -\mu\frac{\partial u}{\partial r}$$

(u is the velocity at a distance y from the wall and $r = a - y$ so $\delta r = -\delta y$). This equation is not limited however. In rheology, equation (6.1) can be applied to any fluid providing the relationship between τ and the shear rate ($\partial u/\partial r$ often denoted by $\dot\gamma$) can be specified. Equation (6.1) would even be applicable to turbulent flow if a relationship linking τ, u and r could be specified.

Assuming laminar flow of a newtonian fluid and substituting for τ

$$-\frac{\partial p}{\partial x} = -\mu\frac{\partial^2 u}{\partial r^2} - \frac{\mu}{r}\frac{\partial u}{\partial r}$$

$$\therefore \qquad \frac{\partial p}{\partial x} = \mu\left(\frac{\partial^2 u}{\partial r^2} + \frac{1}{r}\frac{\partial u}{\partial r}\right)$$

so

$$\frac{r}{\mu}\frac{\partial p}{\partial x} = r\frac{\partial^2 u}{\partial r^2} + \frac{\partial u}{\partial r}$$

Integrating

$$\frac{r^2}{2\mu}\frac{\partial p}{\partial x} = r\frac{\partial u}{\partial r} + A$$

Dividing through by r

$$\frac{r}{2\mu}\frac{\partial p}{\partial x} = \frac{\partial u}{\partial r} + \frac{A}{r}$$

Integrating again

$$\frac{r^2}{4\mu}\frac{\partial p}{\partial x} = u + A\log_e r + B \tag{6.2}$$

When r is zero u is finite (the centreline velocity is finite) but as $\log_e 0$ is $-\infty$, A must be zero. When $r = a$, $u = 0$.

$$\therefore \qquad \frac{a^2}{4\mu}\frac{\partial p}{\partial x} = B$$

$$\therefore \qquad u = \frac{r^2 - a^2}{4\mu}\frac{\partial p}{\partial x}$$

$$= \left(\frac{a^2 - r^2}{4\mu}\right)\left(-\frac{\partial p}{\partial x}\right) \tag{6.3}$$

As r is always less than a and u is positive $\partial p/\partial x$ must be negative, that is the pressure drops along the pipe in the direction of x increasing.

The quantity of flow along the pipe can simply be obtained by integrating

$$Q = \int_0^a 2\pi r u\, dr$$

$$\therefore \qquad Q = -\frac{\partial p}{\partial x}\frac{2\pi}{4\mu}\int_0^a (a^2 r - r^3)\, dr$$

$$= -\frac{\partial p}{\partial x}\frac{\pi a^4}{8\mu}$$

The mean velocity $\bar{v}$ is defined as the flow divided by the area through which it occurs so

$$\bar{v} = \frac{Q}{\pi a^2}$$

and hence

$$\bar{v} = -\frac{\partial p}{\partial x}\frac{a^2}{8\mu}$$

∴

$$\frac{\partial p}{\partial x} = -\frac{8\mu\bar{v}}{a^2}$$

Integrating gives

$$p = -\frac{8\mu\bar{v}x}{a^2} + C$$

When

$$x = 0, p \text{ is } p_1 \text{ (the upstream pressure)}$$

so

$$C = p_1$$

When

$$x = L, p \text{ is } p_2 \text{ (the downstream pressure)}$$

so

$$p_1 - p_2 = \frac{8\mu\bar{v}L}{a^2} = \frac{32\mu L\bar{v}}{d^2}$$

where d is the pipe diameter and equals $2a$. The pressure head equivalent to this pressure difference is

$$h_f = \frac{p_1 - p_2}{w} = \frac{32\mu L\bar{v}}{\rho g d^2}$$

This is called the Hagen–Poiseuille formula. Considering the velocity distribution equation (6.3) further

$$u = -\frac{a^2 - r^2}{4\mu}\frac{\partial p}{\partial x}$$

the velocity distribution is paraboloidal.

The maximum velocity V is given when $r = 0$, so

$$V = -\frac{a^2}{4\mu}\frac{\partial p}{\partial x}$$

The ratio of the maximum to the mean velocity is

$$\frac{V}{\bar{v}} = \frac{(-a^2/4\mu)(\partial p/\partial x)}{(-a^2/8\mu)(\partial p/\partial x)}$$

∴

$$V/\bar{v} = 2$$

The location of the point in the fluid at which the local velocity equals the mean velocity is specified by the radius r_m which is given by

$$\bar{v} = -\frac{a^2}{8\mu}\frac{\partial p}{\partial x} = -\left(\frac{a^2 - r_m^2}{4\mu}\right)\frac{\partial p}{\partial x}$$

so

$$r_m = a/\sqrt{2}$$

6.2.1 Kinetic energy correction

It is rarely necessary to consider the kinetic energy of a laminar flow through a pipeline as velocities are usually low but it may be necessary to include kinetic energy changes in the energy balances that are used to solve such problems. Because the velocity profile is paraboloidal and not even approximately constant the use of the term $\bar{v}^2/2g$ to describe the kinetic energy/unit weight of fluid is inaccurate. A term $\alpha\bar{v}^2/2g$ must be used instead and α can be evaluated for laminar flow from purely theoretical considerations.

The kinetic energy flowing through an annulus of radius r is

$$\delta KE = w2\pi r\,\delta r\,u\,\frac{u^2}{2g}$$

and the total kinetic energy flowing through the pipe per second is

$$KE = \frac{2\pi w}{2g}\int_0^a u^3 r\,dr$$

Substituting for u

$$KE = \frac{2\pi w}{2g}\left(\frac{-(\partial p/\partial x)^3}{64\mu^3}\right)\int_0^a (a^2 - r^2)^3 r\,dr$$

$$= -\frac{\pi w}{64g\mu^3}\left(\frac{\partial p}{\partial x}\right)^3\left[\frac{a^8}{2} - \frac{3}{4}a^8 + \frac{3}{6}a^8 - \frac{a^8}{8}\right]$$

$\therefore \qquad KE = -\dfrac{\pi w(\partial p/\partial x)^3 a^8}{512g\mu^3}$

If we assume that the mean velocity is a satisfactory basis for calculating the kinetic energy of the flow

$$KE_m = w\pi a^2\,\frac{\bar{v}^3}{2g}$$

As

$$\bar{v} = -\frac{a^2}{8\mu}\frac{\partial p}{\partial x}$$

$$KE_m = -\frac{w\pi a^8}{512\mu^3 2g}\left(\frac{\partial p}{\partial x}\right)^3$$

so

$$KE_m = -\frac{w\pi a^8}{1024 g\mu^3}\left(\frac{\partial p}{\partial x}\right)^3$$

$\therefore$

$$\frac{KE_{true}}{KE_{mean\ velocity}} = 2$$

In a problem in which laminar flow occurs through a simple pipeline as illustrated in Fig. 6.9 this value of α can be used to give a more accurate result.

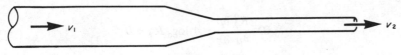

Fig. 6.9

Writing Bernoulli's equation allowing for energy losses but ignoring the losses at the area contraction gives

$$\frac{p_1}{w} + \frac{2v_1^2}{2g} + z_1 = \frac{p_2}{w} + \frac{2v_2^2}{2g} + z_2 + h_{f1} + h_{f2}$$

where h_{f1} is the energy loss in the large pipe and h_{f2} is the loss in the small pipe.

Equation (6.2) can be used to deal with the problem of flow through the annulus between concentric pipes (see Fig. 6.10).

$$u_r = \frac{r^2}{4\mu}\frac{\partial p}{\partial x} + A\log_e r + B$$

so when

$$r = R_1, \qquad u_r = 0$$

and

$$r = R_2, \qquad u_r = 0$$

$\therefore$

$$0 = \frac{R_1^2}{4\mu}\frac{\partial p}{\partial x} + A\log_e R_1 + B$$

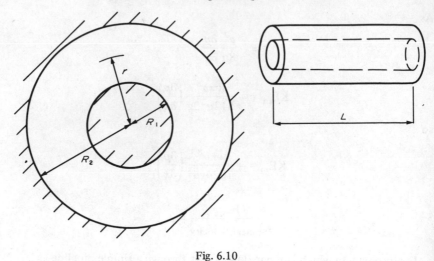

Fig. 6.10

and

$$0 = \frac{R_2^2}{4\mu}\frac{\partial p}{\partial x} + A \log_e R_2 + B$$

Subtracting

$$\frac{R_2^2 - R_1^2}{4\mu}\frac{\partial p}{\partial x} = -A \log_e \frac{R_2}{R_1}$$

so

$$A = -\frac{R_2^2 - R_1^2}{4\mu \log_e (R_2/R_1)}\frac{\partial p}{\partial x}$$

$$\therefore \quad B = \frac{1}{4\mu}\frac{\partial p}{\partial x}\left\{\frac{R_2^2 - R_1^2}{\log_e (R_2/R_1)}\log_e R_2 - R_2^2\right\}$$

$$\therefore \quad u_r = \frac{r^2}{4\mu}\frac{\partial p}{\partial x} - \frac{R_2^2 - R_1^2}{4\mu \log_e (R_2/R_1)}\frac{\partial p}{\partial x}\log_e r$$

$$- \frac{1}{4\mu}\frac{\partial p}{\partial x}\left\{R_2^2 - (R_2^2 - R_1^2)\frac{\log_e R_2}{\log_e (R_2/R_1)}\right\}$$

$$\therefore \quad u_r = -\frac{1}{4\mu}\frac{\partial p}{\partial x}\left\{(R_2^2 - r^2) - (R_2^2 - R_1^2)\frac{\log_e (R_2/r)}{\log_e (R_2/R_1)}\right\}$$

At the maximum velocity filament $\partial u/\partial r = 0$, so from

$$\frac{\partial u}{\partial r} = \frac{r}{2\mu}\frac{\partial p}{\partial x} + \frac{A}{r}$$

$$r_{max}^2 = -\frac{2\mu A}{\partial p/\partial x} \quad \text{and substituting for } A$$

gives

$$r_{max} = \sqrt{\left(\frac{R_2^2 - R_1^2}{2\log_e(R_2/R_1)}\right)}$$

6.3 Turbulent flow

In Section 6.1 Reynolds' experiment was described in which it was shown that the pressure drop in a pipe was linked to the velocity raised to a power of approximately 1·85. This suggests that the viscous shear stress at the wall of the pipe must also depend upon the mean velocity raised to a power of approximately 1·85.

If it is assumed that the relationship between the wall viscous shear and the mean velocity is $\tau = k\bar{v}^n$ where k and n are constants it becomes possible to deduce a law relating pressure drop and mean velocity. This relationship was first stated by Froude who performed experiments to determine the force required to tow long planks through water. He found that k and n were constants that depended to a certain degree upon the nature of the surface of the plank. He found n to be 1·83 for many types of surfaces.

The force acting upon the fluid contained within a length of pipe L of cross sectional area A and wetted perimeter P is caused by a pressure drop which in turn is balanced by the previously mentioned viscous shear acting over the surface of contact of the fluid with the pipe wall (Fig. 6.11).

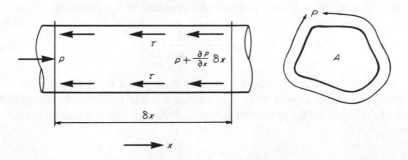

Fig. 6.11

The force causing flow is $-(\partial p/\partial x)\,\delta x A$ and the force opposing flow is $P\tau\,\delta x$. (The pressure gradient must be negative if flow is to occur in the direction of x increasing.) Equating these forces as the flow is to be steady

$$-\frac{\partial p}{\partial x}\,\delta x\,A = P\tau\,\delta x$$

and substituting for τ

$$-\frac{\partial p}{\partial x} = \frac{kv^n}{m}$$

where $m = A/P$ that is m is the mean thickness of flow.

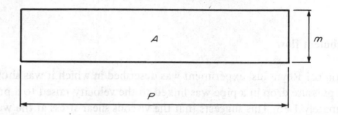

Fig. 6.12

To understand this last statement imagine the wetted perimeter laid out straight and construct a rectangle of area A as illustrated in Fig. 6.12. Its height will then be m that is A/P. This term is called the hydraulic mean radius. In American practice it is often denoted by R.

$$\therefore \qquad p = -kv^n x/m + C$$

When $x = 0$, $p = p_1$ and when $x = L$, $p = p_2$, so

$$p_1 - p_2 = \Delta p = kv^n L/m$$

Dividing through by w to obtain the head equivalent to the Δp value

$$h_f = \frac{\Delta p}{w} = \frac{kLv^n}{\rho g m} \qquad (6.4)$$

As n is nearly 2 (and in any case will later be made equal to 2 because of dimensional considerations) we may introduce a 2 into the denominator as this form of the equation facilitates the addition of energy loss and kinetic energy terms:

$$h_f = \frac{k'L}{\rho m}\frac{v^2}{2g}$$

For reasons that emerge from the dimensional considerations mentioned above k'/ρ is usually denoted by f. So the equation becomes

$$h_f = \frac{fL}{m}\frac{v^2}{2g} \qquad (6.5)$$

This result is ascribed to Darcy and Weisbach. For circular pipes

$$m = \frac{A}{P} = \frac{\pi}{4}d^2/(\pi d) = \frac{d}{4}$$

so

$$h_f = \frac{4fL}{d}\frac{v^2}{2g}$$

This form is used in the UK but in America and Europe the form used is

$$h_f = \frac{f'L}{d}\frac{v^2}{2g}$$

and

$$f' = 4f$$

It is true that far more pipes of circular cross section are made than any other type and this may be considered a sufficient justification for dropping the 4, but the author feels that British practice has much to commend it in that the original form of the fundamental equation is preserved. Throughout this book, wherever f values are quoted, the British value is intended.

6.4 The f number

In Chapter 3 dimensional analysis is used to obtain the same equation as has just been derived and f is shown to be a function of Reynolds number, the roughness number k/d (k is the mean height of the roughnesses), the ratio L/d and the Mach number v/c.

In long pipes the effect of L/d is insignificant. Strictly the effect of this ratio is in defining the entry length. Within the entry length the boundary layers have not become sufficiently thick to fill the pipe and there is a core of potential flow (see Fig. 6.13). At a distance greater than l_e the boundary layers adjacent to the opposite pipe walls meet and the entire flow is affected by friction. Clearly if the length of the pipe L is of the same order of magnitude as the entry length the nature of the flow is not completely controlled by frictional effects as it would be in a long pipe in which the length is much greater than the entry length. As the entry length is a function of the pipe diameter (and the

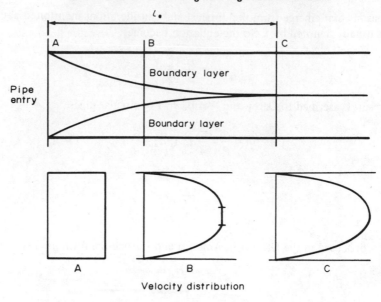

Fig. 6.13

Reynolds number) a large value of L/d indicates that the entry-length effect need not be considered. The entry length l_e is given by $l_e = 0 \cdot 058 \, Re \, d$.

The Mach number is rarely of significance in steady liquid flow problems because c is very large relative to v and the Mach number is negligibly small. For instance, a large value for v is 20 ft/s and c in a water pipeline may be 4000 ft/s so the Mach number is $0 \cdot 005$.

6.4.1 The variation of f with Reynolds number and roughness The roughness number cannot however be ignored as it is of more significance than the Reynolds number in deciding the value of f if its value is above a certain small limit. Thus the equation for f given in Chapter 3

$$f = \varphi \left(Re, \frac{k}{d}, \frac{L}{d}, \frac{v}{c} \right)$$

reduces to

$$f = \varphi \left(Re, \frac{k}{d} \right)$$

with only trivial limitations on its range of validity.

Using this result f could be plotted against Re and a family of curves should result, each with a different k/d value. In the dimensional analysis no distinction was made between laminar and turbulent flow so it should be possible to insert laminar flow results on the same graph. Now the Hagen–Poiseuille equation can be obtained from the Darcy equation by making the function of f equal to $16/Re$. With this in mind Nikuradse performed a large number of experiments upon artificially roughened pipes and presented the results he obtained by using the method described above. (See Fig. 6.14.)

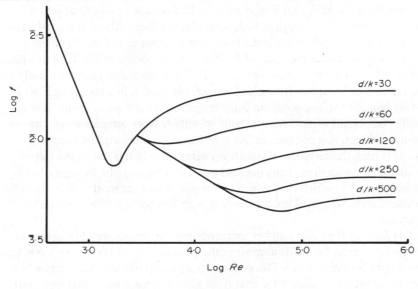

Fig. 6.14

This type of experiment must confirm the laminar flow result of course and for this zone f is indeed given by $f = 16/Re$.

At a Reynolds number of about 2000 (the lower critical Reynolds number) a break occurs in the graph and no satisfactory trends can be found until a Reynolds number of about 2800 has been exceeded (the upper critical Reynolds number). Above this value the results for various pipes can be arranged in curves with increasing values of f corresponding to increasing k/d values.

If the simple experiment described on page 177 in which the h and v values were plotted against one another is now reconsidered it will be realised that the upper and lower critical velocities are really manifestations of upper and lower critical Reynolds numbers and whereas the results obtained are *only* applicable to the one pipe transporting the particular liquid used (water), the log f against log Re plot applies to *all* pipes having large L/d and low v/c values transporting any newtonian fluid.

The surface of a pipe can be considered as a flat boundary rolled up into a pipe so what has been said about a flat surface can, with only minor modifications, be applied to a pipe. In a turbulent flow through a pipe there will be a laminar sub-layer which if thick enough will protect the turbulent core from the roughnesses on the boundary. It may be thinner than the roughnesses however and then these will generate additional turbulence in the main flow. From this it can be seen that a pipe is hydraulically smooth if its laminar sublayer is thicker but hydraulically rough if its laminar sub-layer is thinner than the height of the roughnesses.

As the laminar sub-layer decreases in thickness with increasing Reynolds number it can be seen that a pipe may be hydraulically smooth at low Reynolds numbers but rough at high Reynolds numbers. Also if the roughnesses are very large they will saturate the flow with turbulence and the Reynolds number will not affect the value of f. This is clearly shown in Fig. 6.14 in which at large k/d values f remains constant for all Reynolds numbers above a fairly low value. Clearly, at low Reynolds numbers the fluid is in a state which while not laminar cannot be considered fully turbulent. At any point in the flow the steady component of the instantaneous velocity is large compared with the turbulent fluctuating component. At high Reynolds numbers however the reverse is true, the turbulent fluctuations being large in relation to the steady translational component. Thus the onset of turbulence can be thought of as a gradual process in which fluid which is just not in laminar motion is in a weakly turbulent state, this turbulence increasing with Reynolds number and wall roughness.

The form of Reynolds number indicates that turbulence should increase as its value increases. Reynolds number is the ratio of inertial forces to viscous forces acting upon a fluid element. The viscous forces damp out any disturbance of uniform motion or dissipate the energy of any unstable motion that may start within the flow. If this viscous force is small in relation to the forces tending to magnify any initial disturbance or instability (such as the inertial forces) the fluid motion can be expected to degenerate into a random chaotic motion of fluid molecules, that is to become turbulent. Thus a large Reynolds number corresponds to unstable, that is turbulent flow whereas a small value corresponds to stable, that is laminar flow.

Turbulence has been described as a random chaotic motion in which particle motions are continuously varying in an unpredictable manner. This is certainly true of strongly developed turbulence, but in weak turbulence structure can exist, the fluid being filled with small or large vortices depending upon circumstances. A tranquil flow in a river moving over a rough bed which is deeply submerged shows the presence of these large vortices as patches of upwelling fluid which cover the surface. Flow behind an obstruction in a free surface flow may show a Karman vortex sheet in which the vortices are very large. Such coarse-grained turbulence can be regarded as one end of a spectrum of turbulence which grades down through finer-grained turbulence to a turbulence which involves fluid masses of microscopic size.

6.4.2 The value of f In the equation (6.5)

$$h_f = \frac{4fL}{d}\frac{v^2}{2g} \quad \text{or} \quad \frac{fL}{m}\frac{v^2}{2g}$$

f is a variable. In obtaining this expression from equation (6.4)

$$h_f = \frac{kLv^n}{\rho g m}$$

n was made 2 and $f/2$ replaced k/ρ. If k were a true constant then the alteration of n from 1·83 (or thereabouts) to 2 can only be compensated by making f a function of v which is such that $fv^2/2 = kv^n/\rho$. It is necessary to replace n by 2 not only to make the calculation easier by avoiding the awkward fractional index but also to get a dimensionally rational result. Inserting the dimensions into equation (6.4)

$$L = \frac{|k|LL^nT^{-n}}{ML^{-3}LT^{-2}L}$$

$\therefore$
$$|k| = ML^{-(n+1)}T^{n-2}$$

If n is non-integral, k cannot have rational dimensions so n *must* take the value of 2 and thus $|k| = M/L^3$. The value of f is related to τ. From Fig. 6.11

$$A \times \Delta p = P\Delta x \times \tau$$

$\therefore$
$$\Delta p/\Delta x = \tau/m$$

but

$$\Delta h_f = \frac{\Delta p}{w} = \frac{f\Delta x v^2}{2gm}$$

$\therefore$
$$\frac{\Delta p}{\Delta x} = \frac{\rho f v^2}{2m}$$

$\therefore$
$$\frac{\tau}{m} = \frac{\rho f v^2}{2m}$$

$\therefore$
$$\frac{\tau}{\rho} = (V^*)^2 = \frac{f v^2}{2}$$

and

$$f = \frac{2\tau}{\rho v^2} = 2\left(\frac{V^*}{v}\right)^2 \qquad (6.6)$$

In laminar flow, from purely theoretical considerations, it can be seen that $f = 16/Re$. In turbulent flow, the problem is far more complex and was originally tackled by purely empirical methods. The early experimental formulae are of doubtful value when applied in circumstances which are significantly different from those in which the formula was obtained. Examples of such early formulae are

Weisbach

$$f = 0.0036 + 0.00429/\sqrt{v}$$

Bazin

$$f = 0.00294(1 + 0.3736/d)$$

Kutter and Ganguillet

$$f = 64.4 \left\{ \frac{1 + (41.6 + 0.00281l/h)N/\sqrt{m}}{1.811/N + 41.6 + 0.00281l/h} \right\}^2$$

where N is a roughness number.

Thropp

$$h_f = C^n l v^n / m^x$$

where C, n and x are coefficients chosen from a list which specifies the values appropriate to the pipe surface condition.

Modern Formulae (Smooth bore pipes)

Blasius, 1913

$$f = 0.079/Re^{1/4}$$

This is applicable to pipes in which Re lies between 4000 and 100 000.

Lees, 1924

$$f = 0.0018 + 0.153/Re^{0.35}$$

Again, Reynolds number must lie between 4000 and 400 000

Hermann, 1930

$$f = 0.00135 + 0.099/Re^{0.3}$$

$$4000 < Re < 2\,000\,000$$

Nikuradse, 1932

$$f = 0.0008 + 0.05525/Re^{0.237}$$

6.5 The Prandtl mixing-length hypothesis applied to pipe flow

In Chapter 5 the theory of the boundary layer based on the Prandtl mixing-length hypothesis was developed. The velocity distribution equation is of use here

$$u = 2 \cdot 5 \sqrt{(\tau_0/\rho)} \log_e y + C$$

where u is the local velocity at a distance y from the pipe wall and τ_0 is the wall shear stress. On the centreline of the pipe $y = r$ and $u = V_{max}$ so

$$u = V_{max} + 2 \cdot 5 \sqrt{\left(\frac{\tau_0}{\rho}\right)} \log_e \frac{y}{r}$$

Denoting

$$\sqrt{(\tau_0/\rho)} \text{ by } V^*$$

($\sqrt{\tau_0/\rho}$ has the dimensions of velocity)

$$u = V_{max} + 2 \cdot 5 V^* \log_e (y/r) \tag{6.7}$$

A simpler result that is sufficiently accurate for all practical purposes can be used providing that the Reynolds number is less than 100 000

$$u/V_{max} = (y/r)^{1/7} \tag{6.8}$$

This one seventh power law is widely used in engineering.

The index $1/7$ can be progressively decreased as Reynolds number increases above 100 000.

Whereas Prandtl assumed that the shear stress from the wall to the pipe centre was constant, Van Karman assumed it to decrease linearly from its maximum value at the pipe wall to zero at the centre and obtained the following result

$$u = V_{max} + 2 \cdot 5 V^* \{ \sqrt{(1 - y/r)} + \log_e [1 - \sqrt{(1 - y/r)}] \}$$

The difference between the Van Karman and the Prandtl result is very small except near the walls.

6.6 The velocity distribution in smooth pipes

6.6.1 The laminar sub-layer Both the Prandtl and the Van Karman result give an infinite negative value for the local velocity at the pipe wall; an impossible result. At very small distances from a boundary it is hard to believe that fluid motions can be completely randomised as is required if the flow is to be turbulent. Fluid motion at such extremely small distances must be influenced by the presence of the wall and must be laminar. This suggests that in the case of a turbulent flow through a pipe there must be a peripheral annulus of fluid adjacent to the pipe wall which is in laminar motion, a laminar sub-layer. This layer is of course very thin. For such a layer the turbulent theory described before cannot apply and it is the presence of this layer that explains why the impossible negative infinite velocity does not occur at the boundary.

It is reasonable to assume that the velocity distribution in a very thin layer such as this is linear as non-linear effects cannot be very significant. If the layer is assumed to have a thickness δ' and the velocity at the edge of the layer is denoted by u_l

$$u_l/\delta' = \partial u/\partial y$$

so

$$\tau = \mu(\partial u/\partial y) = \mu u_l/\delta'$$

$\therefore$

$$\frac{\tau}{\rho} = V^{*2} = \frac{\mu u_l}{\rho \delta'}$$

$\therefore$

$$u_l/V^* = \rho \delta' V^*/\mu$$

The term on the right hand side can be seen to be a type of Reynolds number, and as would be expected it has a specific value for conditions at the edge of the laminar sub-layer.

6.6.2 The buffer zone

The turbulent core of flow in the pipe does not change suddenly into the laminar sub-layer at points close to the pipe wall. Instead there is a transition zone, called the buffer zone, across which the state of flow changes from fully turbulent at the radius at which it joins the turbulent core to fully laminar at the the laminar sub-layer.

At the edge of the laminar sub-layer the value of $\rho V^* \delta'/\mu$ is 5 and at the junction of the buffer zone with the turbulent core the value is 70. However a satisfactory theory can be developed if the buffer zone is ignored. Some authors consider that the thickness of the laminar sub-layer should be increased to include part of the buffer zone and that the rest of it should be ignored. Such authors use a value of $\rho V^* \delta'/\mu$ of between 10 and 11·6 for the laminar sub-layer thickness. Rewriting equation (6.7)

$$u = V_{max} + 2 \cdot 5 V^* \log_e (y/r)$$

At the edge of the turbulent core

$$y = \delta' \quad \text{and} \quad u = u_l$$

$\therefore$

$$u_l = V_{max} + 2 \cdot 5 V^* \log_e \delta'/r$$

$\therefore$

$$\frac{u_l}{V^*} = \frac{V_{max}}{V^*} - 2 \cdot 5 \log_e \frac{r}{\delta'} = \frac{\rho V^* \delta'}{\mu} = 11 \cdot 6 \tag{6.9}$$

At the edge of the laminar sub-layer

$$\frac{\rho V^* \delta'}{\mu} = 11 \cdot 6 \quad \text{so} \quad \delta' = \frac{11 \cdot 6 \mu}{\rho V^*}$$

Substituting for δ' in equation (6.9)

$$11 \cdot 6 = \frac{V_{max}}{V^*} - 2 \cdot 5 \log_e \frac{\rho V^* r}{11 \cdot 6 \mu}$$

$$\therefore \qquad \frac{V_{max}}{V^*} = 11 \cdot 6 + 2 \cdot 5 \log_e \frac{1}{11 \cdot 6} + 2 \cdot 5 \log_e \frac{\rho V^* r}{\mu}$$

$$\therefore \qquad \frac{V_{max}}{V^*} = 5 \cdot 5 + 2 \cdot 5 \log_e \frac{\rho V^* r}{\mu}$$

$$\therefore \qquad \frac{u}{V^*} = 5 \cdot 5 + 5 \cdot 75 \log_{10} \frac{\rho V^* y}{\mu} \qquad (6.10)$$

This equation gives the velocity distribution in smooth circular pipes. However if the mean height of the roughnesses on the wall of a rough pipe do not penetrate the laminar sub-layer, it too may be treated as a smooth pipe.

6.7 The velocity distribution in rough pipes

Prandtl assumed that in a rough pipe the edge of the turbulent core was located at a distance y_c from the wall and that y_c was given by

$$y_c = ck$$

where c is a constant and k is the mean height of the wall roughnesses. From equation (6.7)

$$\frac{u}{V^*} = \frac{V_{max}}{V^*} - 2 \cdot 5 \log_e \frac{r}{y}$$

To get u at the edge of the turbulent core substitute $y = ck$

$$\frac{u_c}{V^*} = \frac{V_{max}}{V^*} - 2 \cdot 5 \log_e \frac{r}{ck}$$

From this equation

$$\frac{V_{max}}{V^*} = \frac{u_c}{V^*} + 2 \cdot 5 \log_e \frac{r}{ck}$$

Substituting into (6.7) gives

$$\frac{u}{V*} = \frac{u_c}{V*} + 2 \cdot 5 \left(\log_e \frac{r}{ck} - \log_e \frac{r}{y} \right)$$

so

$$\frac{u}{V*} = \frac{u_c}{V*} - 2 \cdot 5 \log_e c + 2 \cdot 5 \log_e \frac{y}{k}$$

For any given roughness $u_c/V* - 2 \cdot 5 \log_e c$ is approximately constant and in the case of completely rough pipes this constant is $3 \cdot 7$. In common logarithms

$$\frac{u}{V*} = 8 \cdot 5 + 5 \cdot 75 \log_{10} \left(\frac{y}{k} \right)$$

When k is less than δ' the smooth pipe result should hold so replacing the 8.5 value by

$$\beta = 5 \cdot 50 + 5 \cdot 75 \log_{10} \frac{\rho V*k}{\mu}$$

will transform the rough pipe equation into the smooth pipe equation.

Thus if a pipe is hydraulically smooth (the mean height of roughnesses are less than the laminar sub-layer thickness and the turbulent core is protected from their influence) then $\rho V*k/\mu \leqslant 5$ and

$$\beta = 5 \cdot 50 + 5 \cdot 75 \log_{10} (\rho V*k/\mu)$$

When the pipe is completely rough (the roughnesses protrude through the laminar sub-layer)

$$\rho V*k/\mu \geqslant 70 \quad \text{and} \quad \beta = 8 \cdot 50$$

Transitionally rough pipes are defined by $5 < \rho V*k/\mu < 70$ (see Fig. 6.15).

Commercial pipes often lie in the transition zone but tend to come into the rough zone as they tuberculate with age. However the same pipe may operate anywhere on this curve depending upon the value of $V*$ which depends upon τ and finally upon the pressure difference applied to the ends of the pipe. Thus a pipe operating under a small head difference will carry a small flow and its laminar sub-layer thickness will be large, possibly greater than the roughness height, so that the smooth pipe formula is applicable. However under a large head difference, the laminar sub-layer may be thinner than the roughness height and the rough equation will then apply. The same pipe may be physically rough yet hydraulically rough or smooth according to the value of $\rho V*k/\mu$.

In the theory of rough pipes we have assumed that the roughnesses all have the same height. In commercial pipes there is a spectrum of roughness and not all of the roughnesses pierce the laminar sub-layer at the same time. As a result the theoretical discontinuity shown in Fig. 6.15 does not in fact occur; instead the smooth curve shown describes the real situation.

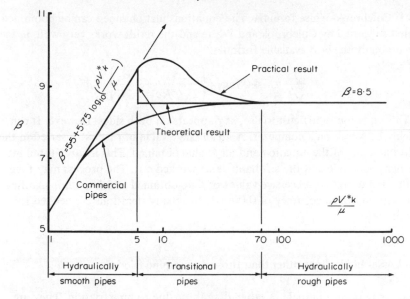

Fig. 6.15

6.8 The universal pipe friction laws

For smooth pipes, from equation (6.10)

$$u/V^* = 5.5 + 5.75 \log_{10} (\rho V^* y/\mu)$$

Considering the flow through an annulus in a circular pipe, integrating to give the total flow, and dividing by the cross sectional area of the pipe gives an expression for the mean velocity of flow:

$$v/V^* = 1.75 + 5.75 \log_{10} (\rho V^* r/\mu)$$

From equation (6.6)

$$V^* = v \sqrt{\frac{f}{2}}$$

∴ for a smooth pipe

$$1/\sqrt{f} = -1.81 + 4.07 \log_{10}(2\, Re \sqrt{f})$$

where $Re = \rho v D/\mu$ and $D = 2r$.

These results are usually adjusted slightly to fit experimentally obtained values a little better:

$$1/\sqrt{f} = 4.0 \log_{10}(Re\, 2\sqrt{f}) - 1.6$$

Following a similar procedure the corresponding result for rough pipes can be obtained:

$$1/\sqrt{f} = 4.0 \log_{10} (D/2k) + 3.48$$

6.8.1 Colebrook–White formula The equations just obtained can be combined as first suggested by Colebrook and White and the result works very well, in fact it is probably the best available formula.

$$\frac{1}{\sqrt{f}} = -4 \cdot 0 \log_{10} \left\{ \frac{2 \cdot 51}{2\sqrt{f} Re} + \frac{k}{3 \cdot 7D} \right\}$$

This equation is difficult to use, as f appears on both sides. However it can quickly be solved on a computer. A mean range value of f is first inserted in the right hand side of the equation and an f value obtained. This is substituted into the right hand side and the left hand value worked out. The process may then be iterated until two successive values of f are obtained which are not significantly different, say to an accuracy of 0·00001. Not many iterations are needed for this process.

6.9 Losses in pipelines other than those due to pipe friction

Losses may occur in pipelines other than those due to pipe friction. They are localised and are said to be local losses. All of them are caused by increase in turbulence due to boundary layer separation.

These losses occur at the positions listed below:

(1) At entry to a pipe.
(2) At a sudden increase in pipe diameter.
(3) At a sudden decrease in pipe diameter.
(4) At bends, elbows, tees and divergent taper sections.
(5) At partially closed valves.
(6) At exit from the pipe.

In all these cases, the most significant feature is flow expansion. An example that will best illustrate such a sudden expansion of a flow is that of a sudden increase in pipe diameter (see Fig. 6.16). The metal annulus that joins the two pipes exerts a pressure on the fluid p_3. The magnitude of this pressure must be close to p_1

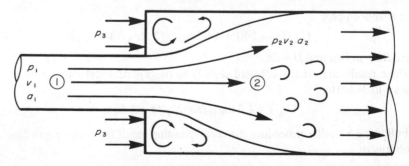

Fig. 6.16

because it is the pressure in the fluid just outside the parallel sided jet in which the pressure is p_1. For the jet to be parallel sided p_3 must equal p_1. This is one of the main assumptions underlying this analysis. The second assumption is that the pressure p_3 acts uniformly over the annulus. If this pressure distribution is significantly non-uniform the flow is curvilinear at high speeds and in the eddies in the corners of the section these conditions are not present. The assumption of uniform pressure distribution on the annulus is effectively true in practice.

Applying Bernoulli's equation from ① to ②

$$\frac{p_1}{w} + \frac{v_1^2}{2g} + z_1 = \frac{p_2}{w} + \frac{v_2^2}{2g} + z_2 + h_L$$

where h_L is the energy loss/unit weight of fluid

$$\therefore \qquad \frac{p_1 - p_2}{w} = \frac{v_2^2 - v_1^2}{2g} + h_L$$

assuming that changes in z values are negligible.

Applying the total force equation

$$\left(\frac{p_1}{w} + \frac{v_1^2}{g}\right) a_1 - \left(\frac{p_2}{w} + \frac{v_2^2}{g}\right) a_2 + \frac{F_x}{w} = 0$$

Now

$$F_x = p_3(a_2 - a_1) = p_1(a_2 - a_1)$$

$$\therefore \qquad \frac{p_1 a_1 - p_2 a_2 + p_1 a_2 - p_1 a_1}{w} = \frac{a_2 v_2^2 - a_1 v_1^2}{g}$$

$$\therefore \qquad \frac{(p_1 - p_2) a_2}{w} = \frac{a_2 v_2^2 - a_1 v_1^2}{g}$$

$$\therefore \qquad \frac{v_2^2 - v_1^2}{2g} + h_L = \frac{v_2^2 - a_1 v_1^2 / a_2}{g}$$

but

$$a_1 v_1 = a_2 v_2$$

$$\therefore \qquad h_L = \frac{v_2^2 - v_2 v_1}{g} - \frac{v_2^2 - v_1^2}{2g}$$

$$= \frac{v_1^2 - 2v_1 v_2 + v_2^2}{2g}$$

$$= \frac{(v_1 - v_2)^2}{2g}$$

Also

$$\frac{p_1 - p_2}{w} = \frac{v_2^2 - v_1^2}{2g} + \frac{v_1^2 - 2v_1v_2 + v_2^2}{2g}$$

∴
$$\frac{p_2 - p_1}{w} = \frac{(v_1 - v_2)v_2}{g}$$

Thus although there is an energy loss the pressure increases across a sudden flow expansion.

6.9.1 The sudden flow contraction The convergent flow from the larger pipe causes a vena contracta to form (Fig. 6.17). This sort of convergent flow is stable

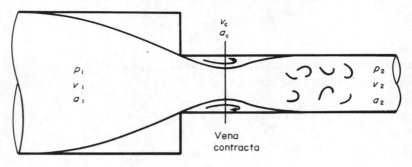

Fig. 6.17

and very little energy is lost in it. Downstream of the vena contracta a flow divergence occurs and energy is consequently lost. This loss is given by the sudden expansion expression

$$h_L = \frac{(v_c - v_2)^2}{2g}$$

Now $a_c v_c = a_2 v_2$ and $a_c / a_2 = C_c$, the coefficient of contraction.

∴
$$h_L = \left(\frac{1}{C_c} - 1\right)^2 \frac{v_2^2}{2g}$$

If C_c is given the extreme value of 0·582 the expression becomes

$$h_L = 0·5 \frac{v_2^2}{2g}$$

and therefore

$$\frac{p_1 - p_2}{w} = \frac{1·5v_2^2 - v_1^2}{2g}$$

This gives an upper estimate of the energy loss in a sudden flow contraction.

6.9.2 Pipe Entrance Flow at the entrance of a pipe (Fig. 6.18) is an example of a flow contraction so

$$h_L = 0.5 \frac{v_\rho^2}{2g}$$

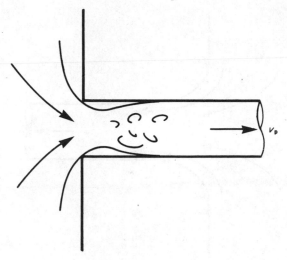

Sharp edged mouthpiece

Fig. 6.18

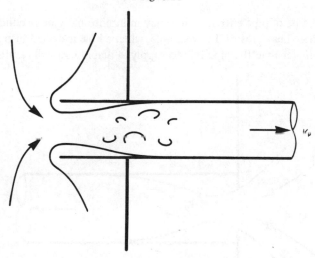

Re-entrant or Borda mouthpiece

Fig. 6.19

The mouthpiece shown in Fig. 6.19 can be analysed using the energy and force equations. The result is

$$h_L = v_\rho^2/2g \qquad\qquad (6.11)$$

Figure 6.20 illustrates the best mouthpiece, but this is obviously far more expensive to make. For this mouthpiece the loss is given by

$$h_L = 0.05\,(v_p^2/2g)$$

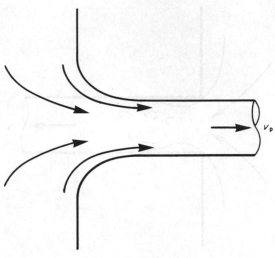

Bell mouth entrance

Fig. 6.20

A certain type of pipe entrance that may unintentionally be produced can cause extremely heavy losses. For example, when a hole is drilled into a tank of a smaller diameter than the pipe it is to supply, a diaphragm will result (Fig. 6.21).

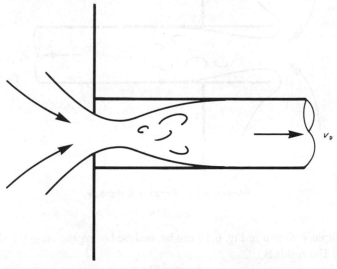

Fig. 6.21

This gives a very heavy loss due to excessive flow expansion. This type of loss can explain the failure of many domestic water supplies to deliver an adequate flow.

6.9.3 Losses at bends, elbows and tees In Chapter 5 the loss mechanism that occurs in bends and tees is dealt with in detail. Both boundary layer separations that occur in these circumstances cause losses that are described by a formula like

$$h_L = F(v_\rho^2/2g) \qquad\qquad (6.12)$$

where F is a number that depends upon the parameters defining the geometry of the section and the flow. Manufacturer's tables may be referred to for precise values of F for various flow circumstances.

To refresh the reader's memory sketches of the various boundary layer separations that can occur in these cases is given in Figs. 6.22a, b and c.

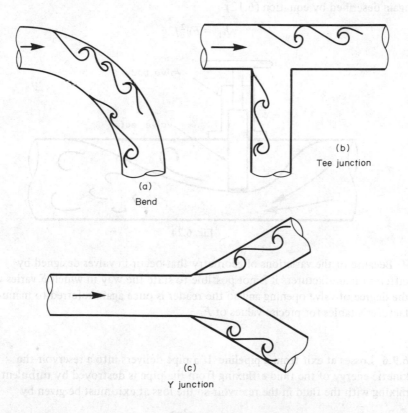

(a)
Bend

(b)
Tee junction

(c)
Y junction

Fig. 6.22

6.9.4 Losses at divergent taper sections Loss at a divergent taper section (Fig. 6.23) is a further example of boundary layer separation. The head loss caused by this mechanism depends upon the change of velocity caused by the taper and upon its angle of divergence. Again for precise details of the losses involved a manufacturer's tables should be consulted. However it is worth noting that if the angle of divergence is greater than $60°$ it is more economic to use a sudden flow expansion than a taper.

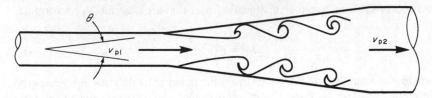

Fig. 6.23

6.9.5 Losses at partially closed valves Figure 6.24 illustrates losses at a partially closed valve. Clearly the loss mechanism is again that of flow expansion. This is again described by equation (6.12)

$$h_L = F v_\rho^2 / 2g$$

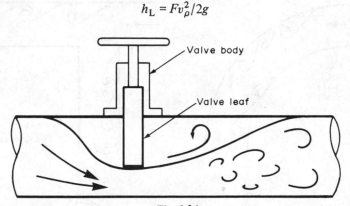

Fig. 6.24

Because of the variations of geometry that occur in valves designed by different manufacturers it is not possible to state the way in which F varies with the degree of valve opening and so the reader is once again referred to manufacturer's tables for precise values of F.

6.9.6 Losses at exit from a pipeline If a pipe delivers into a reservoir the kinetic energy of the fluid effluxing from the pipe is destroyed by turbulent mixing with the fluid in the reservoir so the loss at exit must be given by

$$h_L = v_\rho^2 / 2g$$

This loss may also be seen as a flow expansion in which the final flow velocity is zero: $h_L = (v_\rho - 0)^2/2g$.

If the pipe discharges straight to atmosphere it will deliver fluid which possesses a kinetic energy of $v_\rho^2/2g$ per unit weight of fluid delivered. This constitutes a flow of energy that is being rejected from the system. Although this energy is not being destroyed it is effectively unavailable. It is accounted for in an energy balance however and so it is essential to differentiate between rejected kinetic energy and exit loss.

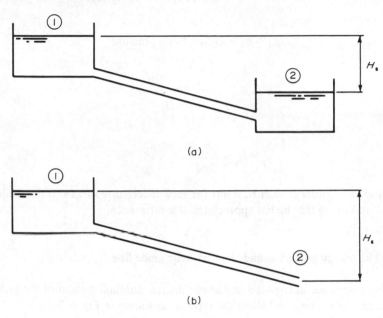

(a)

(b)

Fig. 6.25

Consider two simple pipelines (Figs. 6.25a and b) to illustrate this point and write the energy equations. From Fig. 6.25a

$$\frac{p_1}{w} + \frac{v_1^2}{2g} + z_1 = \frac{p_2}{w} + \frac{v_2^2}{2g} + z_2 + \frac{4fLv_\rho^2}{2gd}$$

$$+ \underbrace{\frac{1}{2}\frac{v_\rho^2}{2g}}_{\text{entry loss}} + \underbrace{\frac{v_\rho^2}{2g}}_{\text{exit loss}}$$

In this case

$$p_1 = p_2 = \text{atmospheric presssure}$$

$$v_1 = v_2 \approx 0$$

$$z_1 - z_2 = H_s$$

so

$$H_s = \frac{4fLv_\rho^2}{2gd} + 1.5\frac{v_\rho^2}{2g}$$

From Fig. 6.25b

$$\frac{p_1}{w} + \frac{v_1^2}{2g} + z_1 = \frac{p_2}{w} + \frac{v_2^2}{2g} + \underbrace{\frac{4fLv_\rho^2}{2gd}}_{\text{rejected KE}} + \underbrace{\frac{1}{2}\frac{v_\rho^2}{2g}}_{\text{entry loss}} + z_2$$

Here

$$p_1 = p_2 = \text{atmospheric pressure}$$
$$v_1 \approx 0$$
$$v_2 = v_\rho$$
$$z_1 - z_2 = H_s$$

so

$$H_s = \frac{4fLv_\rho^2}{2gd} + 1.5\frac{v_\rho^2}{2g}$$

These two results are identical but for very different reasons and the reader should make sure that he has appreciated this difference.

6.10 The energy grade line and the hydraulic grade line

It is sometimes useful to plot a line above the longitudinal section of the pipe that represents the energy head along the pipeline as shown in Fig. 6.26.

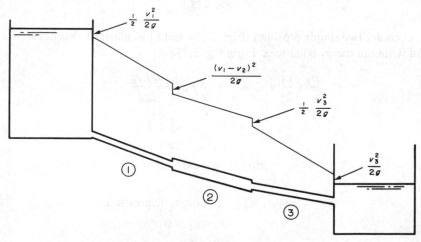

Fig. 6.26

In pipe 1 the slope of the energy grade line is $4f_1v_1^2/(2gd_1)$. In pipe 2 the slope is $4f_2v_2^2/(2gd_2)$ and in pipe 3 the slope is $4f_3v_3^2/(2gd_3)$.

In other cases, for example when the absolute pressure head is of interest because gas is being released or vapourous cavitation occurring, it may be more helpful to plot the pressure head line–this line is called the hydraulic grade line and its slope is called the hydraulic gradient. This line is similar to the energy grade line but not identical (see Fig. 6.27).

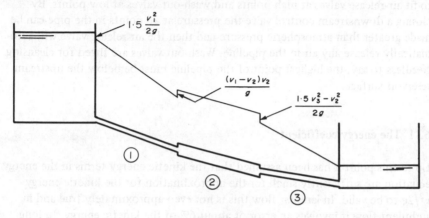

Fig. 6.27

This illustration demonstrates the difference between the energy grade line and the hydraulic grade line but it does not show its application in practice.

At the point x from the upstream reservoir in Fig. 6.28 the pressure head is p_x/w above atmospheric pressure. When the pipe line rises above the hydraulic grade line the pressure head becomes negative and if it rises to a sufficient height above the grade line the head may fall so low that gas may be released from the water. If this gas collects at the high point the flow area will be altered and a weir flow will occur over the high point. This will cause a flow reduction which in turn will reduce the friction loss in the upstream portion of the pipe. This will

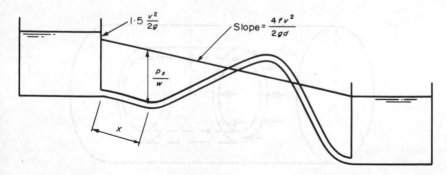

Fig. 6.28

cause the grade line to rise so preventing the development of the low pressure and so stopping the release of gas. The flow will thus fluctuate and this may cause water hammer to develop.

This situation must be avoided and the hydraulic grade line can be used to assess the risk of its occurrence. Gas release occurs at about 8 ft (2·4 m) head absolute and as the height of the water barometer is 34 ft (10·2 m) the hydraulic grade line should never be lower than 26 ft (7·9 m) below the pipe. It is usual to fit air-release valves at high points and wash-out valves at low points. By closing a downstream control valve the pressure at all points in the pipe can be made greater than atmospheric pressure and then the air release valve will automatically release any air in the pipeline. Wash-out valves are fitted for cleansing. Needless to say, the highest point of the pipeline must be below the upstream reservoir surface.

6.11 The energy coefficient

Up to this point it has been assumed that the kinetic energy terms in the energy equation are sufficiently small for the approximation for the kinetic energy $v^2/2g$ to be valid. In laminar flow this is not even approximately true and in turbulent flow it involves an error of about 6% of the kinetic energy. In long pipelines where kinetic energies represent a very small fraction of the total head this error is trivial and may actually represent an error of less than 0·5% of the total head. In short pipelines this is not necessarily the case and then a correction must be made to the value of the kinetic energy.

The kinetic energy of a flow is uniformly distributed across the flow only if the velocity distribution is uniform. At the walls the fluid, having a very small velocity has even less kinetic energy and conversely in the main core of the flow, the velocity and kinetic energy are large. The kinetic energy is dependent on the square of the velocity and the square of the mean velocity cannot equal the mean of the squares of local velocities so the term $v^2/2g$ cannot correctly represent the true average kinetic energy per unit weight.

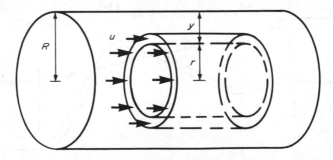

Fig. 6.29

Consider the kinetic energy of flow through an annulus (see Fig. 6.29). The kinetic energy flowing through this annulus per second is given by

$$w \times 2\pi ru \times \frac{u^2}{2g} \delta r$$

and therefore the total KE per second flowing through the pipe is

$$\frac{\pi w}{g} \int_0^R u^3 r\, dr$$

Assuming that the smooth-pipe velocity distribution equation (6.8) applies

$$u/V = (y/R)^{1/n}$$

where V is the centreline velocity. But $r = R - y$ so $dr = -dy$

$$\therefore \quad \text{KE/second} = -\frac{\pi w}{g} V^3 \int_R^0 \left(\frac{y}{R}\right)^{3/n} (R - y)\, dy$$

$$= -\frac{\pi w V^3}{R^{3/n} g} \left[\frac{R y^{1+3/n}}{1 + 3/n} - \frac{1}{2 + 3/n} y^{2+3/n} \right]_R^0$$

$$= \frac{\pi R^2 w V^3}{g} \left(\frac{n}{n+3} - \frac{n}{2n+3} \right) = \frac{\pi R^2 w V^3}{g} \frac{n^2}{(2n+3)(n+3)}$$

∴ The ratio α of the true kinetic energy flowing per second to the kinetic energy based on the mean velocity is

$$\frac{w\pi R^2 \dfrac{V^3}{2g} \left[\dfrac{2n^2}{(2n+3)(n+3)} \right]}{w\pi R^2 (v^3/2g)} = \frac{2n^2}{(2n+3)(n+3)} \left(\frac{V}{v} \right)^3 \qquad (6.13)$$

It is now necessary to evaluate V/v. The elemental flow through the pipe annulus is given by

$$\delta Q = 2\pi ru\, \delta r$$

$$= -2\pi V (y/R)^{1/n} (R - y)\, dy$$

$$\therefore \quad Q = -\frac{2\pi V}{R^{1/n}} \int_R^0 (R y^{1/n} - y^{1+1/n})\, dy$$

$$= -\frac{2\pi V}{R^{1/n}} \left[\frac{1}{1 + 1/n} R y^{1+1/n} - \frac{1}{2 + 1/n} y^{2+1/n} \right]_R^0$$

$$= 2\pi R^2 V \left(\frac{n}{n+1} - \frac{n}{2n+1} \right) = \pi R^2 V \frac{2n^2}{(n+1)(2n+1)}$$

∴

$$v = \frac{Q}{\pi R^2} = V \frac{2n^2}{(2n+1)(n+1)}$$

so

$$\frac{v}{V} = \frac{2n^2}{(2n+1)(n+1)}$$

Substituting back into equation (6.13)

$$\alpha = \frac{2n^2}{(2n+3)(n+3)} \left[\frac{(n+1)(2n+1)}{2n^2} \right]^3$$

$$= \frac{(n+1)^3 (2n+1)^3}{4n^4 (2n+3)(n+3)}$$

α is known as the Coriolis coefficient. n is usually 7 so

$$\alpha = \frac{8^3 \times 15^3}{4 \times 49 \times 49 \times 17 \times 10} = 1\cdot06$$

and

$$\frac{V}{v} = \frac{8 \times 15}{2 \times 49} = 1\cdot225$$

In laminar flow the kinetic energy ratio is 2 as demonstrated on page 183.

6.12 Momentum coefficient

Owing to the non-uniform velocity distribution the usual statement that the momentum, M, of the flow is given by

$$M = \rho a v^2$$

where a is the area and v is the mean velocity is not strictly true and a momentum coefficient must be employed if momentum effects are significant. By a precisely similar method as was used to derive the Coriolis coefficient the momentum coefficient can be obtained.

The total momentum flow rate is

$$M = \int_0^R \rho \times 2\pi r u^2 \, dr$$

substituting $u = V(y/R)^{1/n}$ gives

$$M = 2\pi R^2 \rho V^2 \frac{n^2}{(2n+2)(n+2)}$$

$\therefore$ Momentum/unit weight of flow $= \dfrac{\dfrac{\pi R^2 V^2 w}{g} \dfrac{2n^2}{(2n+2)(n+2)}}{\pi R^2 w v}$

$$= \left(\frac{V^2}{gv}\right) \frac{2n^2}{(2n+2)(n+2)}$$

The momentum of unit weight of fluid based on the mean velocity $= v/g$. Therefore the ratio β of the true momentum per unit weight of the flow to the momentum per unit weight of flow based on the mean velocity is given by

$$\beta = \frac{\dfrac{V^2}{gv} \dfrac{2n^2}{(2n+2)(n+2)}}{v/g} = \left(\frac{V}{v}\right)^2 \frac{2n^2}{(2n+2)(n+2)}$$

substituting for $(V/v)^2$ gives

$$\beta = \frac{(2n+1)^2(n+1)^2}{2n^2(2n+2)(n+2)}$$

When $n = 7$

$$\beta = \frac{15^2 \times 8^2}{2 \times 49 \times 16 \times 9} = \frac{50}{49}$$

$$= 1 \cdot 02$$

Inserting these two correcting coefficients enables more accurate expressions for energy and momentum to be written when velocity distributions are not uniform, for example in pipe flow. The energy equation becomes

$$\frac{p_1}{w} + \alpha \frac{v_1^2}{2g} + z_1 = \frac{p_2}{w} + \alpha \frac{v_2^2}{2g} + z_2 + h_f$$

and the momentum equation

$$\left(\frac{p_1}{w} + \beta \frac{v_1^2}{g}\right) a_1 - \left(\frac{p_2}{w} + \beta \frac{v_2^2}{g}\right) a_2 + \frac{F_x}{w} = 0$$

6.13 Flow in pipe networks

Provided that the network is not too complex relatively simple methods of analysis can be developed based on the continuity and dynamic equations. If the networks are made up of relatively long pipes and local turbulence losses are not significant a simple modification of the head loss equation yields a result of great value. In the equation

$$h_f = \frac{4fLv^2}{2gd}$$

replace v by $Q/[(\pi/4)d^2]$. Then

$$h_f = \frac{4fLQ^2}{(\pi/4)^2 2gd^5} = \frac{32fLQ^2}{\pi^2 gd^5}$$

In fps units this expression is closely given by

$$h_f = \frac{fLQ^2}{10d^5}$$

(Note that the number 10 in this expression is *not* a dimensionless number.) In metric units to an adequate accuracy

$$h_f = \frac{fLQ^2}{3d^5}$$

Consider a simple pipe network of the type illustrated in Fig. 6.30. A unit weight of water as it leaves reservoir A en route to reservoir C will lose energy

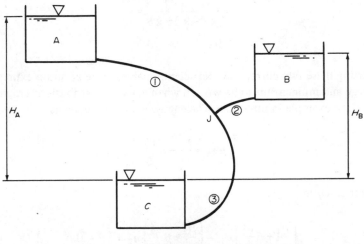

Fig. 6.30

in pipe 1 and more in pipe 3. It may also lose energy in passing through local turbulences at pipe entry, at the pipe junction and at the pipe exit.

$$\therefore \qquad \frac{p_A}{w} + \frac{v_A^2}{2g} + z_A = \frac{1}{2}\frac{v_1^2}{2g} + \frac{4f_1 L_1 v_1^2}{2g d_1} + \frac{k v_3^2}{2g}$$

$$+ \frac{4f_3 L_3 v_3^2}{2g d_3} + \frac{v_3^2}{2g} + \frac{p_C}{w} + \frac{v_C^2}{2g} z_C$$

where $k v_3^2 / 2g$ is the loss at the pipe junction.

Now in reservoirs surface velocities must be nearly zero so $v_A^2/2g$ and $v_C^2/2g$ can be neglected. $p_A = p_C = $ atmospheric pressure so

$$z_A - z_C = H_A = \left(\frac{1}{2} + \frac{4f_1 L_1}{d_1}\right)\frac{v_1^2}{2g} + \left(k + 1 + \frac{4f_3 L_3}{d_3}\right)\frac{v_3^2}{2g}$$

The same process can be applied to fluid leaving reservoir B en route to reservoir C and the result will be

$$H_B = \left(\frac{1}{2} + \frac{4f_2 L_2}{d_2}\right)\frac{v_2^2}{2g} + \left(k + 1 + \frac{4f_3 L_3}{d_3}\right)\frac{v_3^2}{2g}$$

Also

$$a_1 v_1 + a_2 v_2 = a_3 v_3$$

that is

$$Q_1 + Q_2 = Q_3$$

By substituting for v_1 and v_2 in terms of v_3 these three equations can be solved for v_1, v_2 and v_3.

If local turbulences are negligible the head–quantity equations can be used directly with some saving in effort. Then

$$H_A = \frac{f_1 L_1 Q_1^2}{3 d_1^5} + \frac{f_3 L_3 Q_3^2}{3 d_3^5}$$

$$H_B = \frac{f_2 L_2 Q_2^2}{3 d_2^5} + \frac{f_3 L_3 Q_3^2}{3 d_3^5}$$

and $Q_3 = Q_1 + Q_2$ in metric units.

The easiest way of solving these equations is to let $Q_1 = \alpha Q_3$. Then

$$Q_2 = (1 - \alpha)Q_3$$

$$H_A = \left(\frac{f_1 L_1}{3d_1^5}\alpha^2 + \frac{f_3 L_3}{3d_3^5}\right)Q_3^2 \qquad (6.14)$$

$$H_B = \left(\frac{f_2 L_2}{3d_2^5}(1 - \alpha)^2 + \frac{f_3 L_3}{3d_3^5}\right)Q_3^2 \qquad (6.15)$$

$$\therefore \qquad \frac{H_A}{H_B} = \left(\frac{f_1 L_1 \alpha^2/3d_1^5 + f_3 L_3/3d_3^5}{f_2 L_2(1 - \alpha)^2/3d_2^5 + f_3 L_3/3d_3^5}\right)$$

When values are substituted this reduces to a simple quadratic in α which when solved can be substituted back into either equation (6.14) or equation (6.15) to obtain Q_3. Q_1 and Q_2 can then be simply obtained.

The statement that the available head, H_A or H_B in Fig. 6.30 is used to overcome friction is satisfactory providing that it is remembered that it derives from the energy equation and therefore friction in this sense means all energy losses including local turbulence losses and rejected kinetic energy.

If pipes are arranged to operate in parallel, for example AJ and BJ in Fig. 6.30, the head loss from A to C is the head loss in AJ plus that in JC and the head loss from B to C is the head loss in BJ plus that in JC (ignoring local losses). It is a common error to say that the total loss in the system is the sum of the head losses in all three pipes. This statement is true *if* all the pipes are arranged in series so that every fluid element has to pass through all three but it is obviously untrue when some of the pipes are arranged to operate in parallel so that not all of the fluid passes through all of the pipes. This mistake is so common that it is considered necessary to draw the reader's attention to it.

6.13.1 Pipeline equipped with a nozzle A pipeline with a nozzle is illustrated in Fig. 6.31. Applying the energy equation to points 1 and 2

$$\frac{p_1}{w} + \frac{v_1^2}{2g} + z_1 = \frac{p_2}{w} + \frac{v_n^2}{2g} + z_2 + \frac{4fL}{d}\frac{v_p^2}{2g}$$

$$+ \left(\frac{1}{C_d^2} - 1\right)\left(\frac{v_n^2 - v_p^2}{2g}\right)$$

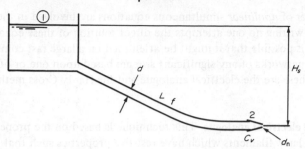

Fig. 6.31

Now $p_1 = p_2$ = atmospheric pressure. $v_1 = 0$ effectively and $z_1 - z_2 = H_s$

$$\therefore \quad H_s = \frac{v_n^2}{2g} + \frac{4fL}{d}\frac{v_p^2}{2g} + \left(\frac{1}{C_d^2} - 1\right)\left(\frac{v_n^2 - v_p^2}{2g}\right)$$

and

$$v_n = \frac{a_p}{a_n} v_p$$

$$H_s = \left[\left(\frac{a_p}{a_n}\right)^2 + \left(\frac{1}{C_d^2} - 1\right)\left\{\left(\frac{a_p}{a_n}\right)^2 - 1\right\} + \frac{4fL}{d}\right]\frac{v_p^2}{2g}$$

$$\therefore \quad H_s = \left[\frac{1}{C_d^2}\left\{\left(\frac{a_p}{a_n}\right)^2 - 1\right\} + \frac{4fL}{d} + \left(\frac{a_p}{a_n}\right)^2\right]\frac{v_p^2}{2g}$$

$$Q = a_p \sqrt{\left\{\frac{2gH_s}{(1/C_d^2)((a_p/a_n)^2 - 1) + 4fL/d + (a_p/a_n)^2}\right\}}$$

6.14 Analysis of pipe networks

For every pipe in a network the head loss equation can be written, an equation can be written for every junction which specifies that the flow into the junction must equal the flow out of the junction (Kirchoff's law—really a statement of continuity) and finally equations can be written which specify that head differences between any two points in the network must be the same irrespective of the route taken through the network. This gives enough equations completely to specify the problem. Hence a solution must be possible—at least theoretically. If the number of pipes in the network is greater than four or five this solution becomes extremely laborious and when the number is large, impossible. It must be remembered that a large number of *nonlinear* simultaneous equations are involved in the solution. At the time of writing no one attempts the direct solution of these equations although it is possible that it might be attempted on a large fast computer. All analyses of networks of any significant size are based upon one or other of two methods. These are the electrical analogue and the Hardy Cross methods.

6.14.1 The electrical analogue This technique is based on the property of either (1) glowing lamp filaments which have resistive properties such that $V = ki^n$ where V is the voltage across the lamp and i is the current through it (n has the value 1·83) or (2) ceramic resistance elements, for which a similar resistance law exists. Clearly a perfect analogy exists between the electrical and the hydraulic equations, that is

$$V = k_e i^n$$

$$h = k_h Q^n$$

Voltage is the analogue of head, current is the analogue of flow and n may be made the same by careful choice of the resistive elements. k_h and k_e can be made to bear the correct relation to one another by linking numbers of glowing bulbs together appropriately arranged in series or parallel to give a value of k_e that will be in appropriate scale to the value of k_h for which it is desired to model.

Once resistance elements of suitable properties have been manufactured they can be wired together and the currents and voltages measured which can then be directly scaled to give flows and heads as required. This technique is worth performing for medium sized networks. A model of the network is built, new resistance elements modelling new pipes that are being fitted into the pipe network and then added as required. Trials can be made of the effect upon the network of various sizes of pipes and an optimum chosen.

This approach is not economic for small networks and is too complex and expensive for very large networks. In such large networks the effects of installing a new pipe are estimated empirically. Pressure contours are plotted over the network, the information being obtained from data collected by direct measurement. These contours make it possible to estimate heads anywhere in the network and the effect of installing a new pipe can be assessed by analysing a limited portion of the surrounding network taking in sufficient to ensure that the alterations caused by fitting the proposed new pipe are negligible in the periphery of the section of network under detailed examination. It is also important to ensure that the section considered is not so large as to give rise to an excessively laborious analysis. After fitting the new pipe the pressure contours will need alteration, of course.

Thus, for small networks and for portions of a large network it is necessary to use an analytic technique to solve the network.

6.14.2 The Hardy Cross method This method is one in which a trial guess is made for the values of the flow in every circuit in the network. These trial guesses are improved by employing corrections as demonstrated later, the process being repeated until successive corrections become insignificant. If the first guesses are reasonably accurate convergence to the solution is quick, if the first trials are significantly wrong convergence is slower.

Two approaches are available and are used in circumstances to which they are best suited. They are head balancing and quantity balancing.

6.14.3 Head balancing Consider a loop in a complex pipe network (Fig. 6.32).

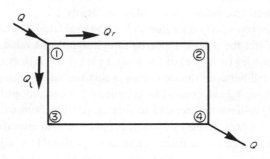

Fig. 6.32

Let Q_r and Q_l be the first values assumed and let Q_r' and Q_l' be the correct values. Then

$$Q_r' = Q_r + \delta Q$$

and

$$Q_l' = Q_l - \delta Q$$

where δQ is the error in the assumed values. Note that both $Q_r + Q_l$ and $Q_r' + Q_l'$ equal Q as indeed they must if they are to obey continuity.

Now the head lost along either route from junction 1 to junction 4 must be the same

$$\underset{1 \to 2 \to 4}{\Sigma h_f} = \underset{1 \to 3 \to 4}{\Sigma h_f}$$

∴
$$\Sigma k (Q_r')^n = \Sigma k (Q_l')^n$$

∴
$$\Sigma k (Q_r + \delta Q)^n = \Sigma k (Q_e - \delta Q)^n$$

∴
$$\Sigma (k [Q_r^n + n Q_r^{n-1} \delta Q + \cdots]) = \Sigma (k [Q_l^n - n Q_l^{n-1} \delta Q + \cdots])$$

Ignoring second and higher powers of δQ gives

$$\delta Q = -\frac{\Sigma kQ_r^n - \Sigma kQ_l^n}{n(\Sigma kQ_r^{n-1} + \Sigma kQ_l^{n-1})}$$

$$= -\frac{\Sigma h_r - \Sigma h_l}{n(\Sigma h_r/Q_r + \Sigma h_l/Q_l)}$$

or

$$\delta Q = -\frac{\Sigma h \text{ (paying attention to sign)}}{n\Sigma h/Q \text{ (ignoring signs)}}$$

The sign of flow around the network loop in a clockwise direction is taken as positive and in an anti-clockwise direction is negative.

Thus for any given loop the value δQ can be evaluated and then added to the original values and the process repeated until insignificant values of δQ are produced. If one leg (or more) of the loop is part of an adjacent loop, that is the pipe is shared between the two loops, a slight modification must be made to this procedure. δQ values must be calculated for both loops from the assumed values. The δQ values when applied to their respective loops will cause changes in the flow in the pipe that they have in common and an alteration made in one loop is communicated to the other. Thus all loops should be adjusted, corrections made throughout and then all loops corrected again. It is wrong to correct one loop fully and then start on the second which upon correction immediately throws the head balance out in the first.

The head-quantity equation

The Darcy or Fanning equation can of course be used but as f is a variable in the equation many people do not care to undertake the additional work necessary to evaluate it and then find h_f from

$$h_f = \frac{fLQ^2}{Nd^5}$$

(where N is either 10 or 3 according to the system of units in use). Instead, the Hazen–Williams equation is preferred

$$v = 1 \cdot 318 \, Cm^{0 \cdot 63} i^{0 \cdot 54} \text{ ft/s}$$

(C depends upon the roughness of the pipe). This leads to

$$h_f = \left(\frac{1 \cdot 594}{C}\right)^{1 \cdot 85} \frac{L}{d^{4 \cdot 87}} Q^{1 \cdot 85}$$

As C is a true constant this equation is exactly of the type assumed in the Hardy Cross analysis: $h_f = kQ^n$ where $k = (1 \cdot 594/C)^{1 \cdot 85} L/d^{4 \cdot 87}$ and $n = 1 \cdot 85$. The expression for k involves awkward indices so it is usual to use a nomogram for solving this equation.

The Hazen-Williams equation is very widely used in engineering practice and it is easy to understand why, but it should be realised that the convenience of the equation is no measure of its accuracy; and also it only applies to the flow of water at one specific temperature. The Colebrook-White formula for f is applicable to steady, non-spatially varied, flow of *all* newtonian fluids in pipes.

Example of head balancing

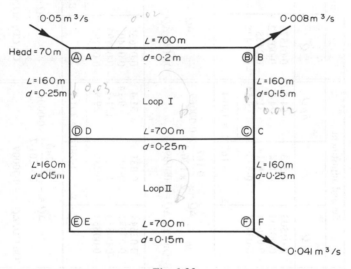

Fig. 6.33

For the pipes making up the network in Fig. 6.33, $f = 0 \cdot 005$, so

$$h_f = 0 \cdot 005 \, LQ^2 / 3d^5 = kQ^2$$

Thus $k_{AB} = 3650$, $k_{BC} = 3510$, $k_{CD} = 1190$, $k_{DA} = 273$

$$k_{CE} = 273, \quad k_{EF} = 15400, \quad k_{FD} = 3510$$

Table 6.1 shows how the network can be solved. A fourth or maybe fifth adjustment could be performed with advantage. Note that flow around a loop in a positive direction is taken as positive.

6.14.3 Quantity balancing In head balancing an error in the flow around a circuit of a network was assumed.

Table 6.1

Pipe	Dia. m	Length m	First adjustment			Second adjustment			Third adjustment			Head m
			Q assumed	h_f	h_f/Q	Q	h_f	h_f/Q	Q	h_f	h_f/Q	
AB	0·2	700	+0·02	1·46	73	0·01503	0·823	54·8	0·01572	0·903	57·4	A: 70
BC	0·15	160	+0·012	0·505	42·1	0·00703	0·174	24·7	0·00772	0·209	27·1	B: 69·107
CD	0·25	700	−0·016	−0·304	19	−0·02644	−0·831	31·4	−0·2435	−0·825	31·3	C: 68·9
DA	0·25	160	−0·03	−0·246	8·2	−0·03497	−0·333	9·54	−0·03428	−0·321	9·36	D: 69·72
				$\Sigma = 1·415$	$\Sigma = 142·3$		$\Sigma = -0·167$	$\Sigma = 120·4$		$\Sigma = 0·078$	$\Sigma = 125·2$	Error = 0·044 Loop is not yet balanced

$$\Delta Q = \frac{-1·415}{284·6} = -0·00497$$

On CD, $\Delta Q = -0·00497 - 0·00547 = -0·001044$

$$\Delta Q = \frac{-0·167}{2 \times 120·4} = 0·00069$$

On CD, $\Delta Q = 0·00009$

$$\Delta Q = \frac{-0·078}{250·4} = -0·000312$$

On CD, $\Delta Q = -0·000435$

Pipe	Dia. m	Length m	Q assumed	h_f	h_f/Q	Q	h_f	h_f/Q	Q	h_f	h_f/Q	Head m
CD	0·25	700	0·016	0·304	19	0·02644	0·831	31·4	0·02635	0·825	31·3	F: 68·583
DF	0·25	160	0·028	0·214	7·64	0·03347	0·306	9·14	0·03407	0·317	9·3	
FE	0·15	700	−0·014	−3·02	21·6	−0·00853	−1·121	131·3	−0·00793	−0·968	122·1	E: 69·55
ED	0·15	160	−0·014	−0·688	49·2	−0·00853	−0·256	29·9	−0·00793	−0·221	27·8	
				$\Sigma = -3·19$	$\Sigma = 292$		$\Sigma = -0·240$	$\Sigma = 201·7$		$\Sigma = -0·047$	$\Sigma = 190·5$	Error = 0·049 Loop is not yet balanced

$$\Delta Q = -\frac{-3·19}{2 \times 292} = 0·00547$$

On CD, $\Delta Q = 0·00547 + 0·001044$

$$\Delta Q = -\frac{-0·240}{2 \times 201·7} = 0·0006$$

On CD, $\Delta Q = -0·00009$

$$\Delta Q = \frac{-0·047}{381} = -0·000123$$

On CD, $\Delta Q = 0·000435$

Note that when a ΔQ value calculated for one circuit is applied to a pipe which is common to two circuits it changes sign when applied to the next circuit as its sense (clockwise or anticlockwise) changes.

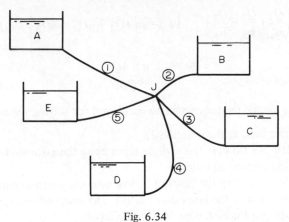

Fig. 6.34

In quantity balancing, an error in the head at a junction is assumed. Consider a junction such as illustrated in Fig. 6.34. A first trial value for the head at the junction H_J can be made. Assuming that this first trial is too large by an amount δh and H'_J is the correct head then

$$H'_J = H_J - \delta h$$

Calculating the flow in pipe 1 from the pipe formula

$$H_A - H_J = k_1 Q_1^n$$

the value Q_1 so obtained will be wrong by an amount of δQ. For the correct head and flow

$$H_A - H'_J = k_1 (Q_1 - \delta Q_1)^n$$

substituting for H_J in the equation gives $H_A - (H'_J + \delta h) = k_1 Q_1^n$. Subtracting gives

$$\delta h = k_1 Q_1^n - k_1 (Q - \delta Q_1)^n$$

$\therefore \qquad\qquad\qquad\qquad \delta h \approx nk Q_1^{n-1} \delta Q_1$

$\therefore \qquad\qquad\qquad\qquad \delta h = nh_1\, \delta Q_1 / Q_1$

This error can be written in terms of the flow errors in all the other pipes

$$\delta h = \frac{nh_1}{Q_1} \delta Q_1 = \frac{nh_2}{Q_2} \delta Q_2 = \frac{nh_3}{Q_3} \delta Q_3 = \ldots$$

$\therefore \qquad \dfrac{Q_1}{h_1} \delta h = n\, \delta Q_1, \dfrac{Q_2}{h_2} \delta h = n\, \delta Q_2, \dfrac{Q_3}{h_3} \delta h = n\, \delta Q_3 \ldots$

$\therefore \qquad \left(\dfrac{Q_1}{h_1} + \dfrac{Q_2}{h_2} + \dfrac{Q_3}{h_3} + \ldots\right)\delta h = n(\delta Q_1 + \delta Q_2 + \delta Q_3 + \ldots)$

$\therefore \qquad\qquad \delta h = \dfrac{n\Sigma\,\delta Q}{\Sigma Q/h}$

Thus a very simple procedure can be employed for solving the network.

(1) Assume a head at the junction J.

(2) Calculate the flows in the various pipes using the assumed head at the junction and the friction equation.

(3) Sum the flows into the junction; flow into the junction being assumed positive and flow out of the junction negative. The sum of these flows $\Sigma\,\delta Q$.

(4) Evaluate Q/h for each pipe and obtain $\Sigma Q/h$.

(5) Calculate δh.

Re-evaluate H_J', that is $H_J - \delta h$ and restart the calculation at step 1 again. Repeat the process until the value of δh becomes negligible.

Quantity balancing can be used to solve a network in a very similar way to head balancing. Estimates of heads at all the network junctions must first be made. Corrections of the heads at the junctions can then be calculated. The process can be repeated until successive values of junction heads are negligibly different. A tabular approach is desirable.

Worked examples

(1) A horizontal pipe of 7·5 cm diameter is joined by a sudden enlargement to a pipe of 15 cm diameter. Water is flowing through it at a rate of 0·0141 m³/s. If the pressure just before the junction is 6 m head, what will be the pressure head in the 15 cm pipe downstream of the junction? What will be the power loss at the junction?

Energy loss/unit wt of fluid $= (v_1 - v_2)^2/2g$

$$v_1 = \frac{0\cdot0141}{(0\cdot075)^2\pi/4} = 3\cdot192 \text{ m/s}$$

$$v_2 = \frac{d_1^2\pi/4}{d_2^2\pi/4}\,v_1 = \tfrac{1}{4}v_1 = 0\cdot7979 \text{ m/s}$$

$$h_L = \frac{(3\cdot192 - 0\cdot7979)^2}{19\cdot62} = 0\cdot292 \text{ m}$$

$$\frac{p_1}{w} + \frac{v_1^2}{2g} = \frac{p_2}{w} + \frac{v_2^2}{2g} + h_L$$

$$6 + 0.519 = \frac{p_2}{w} + 0.0324 + 0.292$$

$$\frac{p_2}{w} = 6.195 \text{ m}$$

Power loss $= wQh_L = 9810 \times 0.0141 \times 0.292$

$$= 40.39 \text{ watts (newton metres)}$$

(2) In rough pipes the Darcy f is given by:

$$1/f = 4.0 \log_{10}(r/k) + 3.48$$

In a certain district, the flow through a pipe of 10 cm diameter was found to be 0.007 m^3/s when the hydraulic gradient was 0.01. Fifteen years later this flow rate was shown to have decreased by 20%. A new 22.5 cm diameter pipe is to be installed to carry a flow of at least 0.056 m^3/s when operating under a hydraulic gradient of 0.015. For how long can this pipe be expected to carry this minimum flow. Assume that the roughness of the new pipe is the same as the initial roughness of the 10 cm pipe and that the roughness will increase linearly with time at the same rate as did that of the 10 cm diameter pipe.

Initially for the 10 cm dia. pipe

$$\frac{hf}{L} = \frac{4f}{d}\frac{v^2}{2g} = 0.01$$

$$v = \frac{0.007}{(\pi/4) \times 0.1^2} = 0.8913 \text{ m/s}$$

$$f = 6.175 \times 10^{-3}$$

$$1/\sqrt{f} = 12.73$$

After 15 years and with the same hydraulic gradient,

$$v = 0.8 \times 0.8913 = 0.713 \text{ m/s}$$

$$1/\sqrt{f} = 10.18$$

Initially $\qquad \log_{10}(r/k_0) = \dfrac{12.73 - 3.48}{4}$ (see p. 197)

$$= 2.312$$

$$r/k_0 = 204.9$$

$$k_0 = 0.05/204.9 = 2.44 \times 10^{-4} \text{ m}$$

After 15 years

$$\log_{10}(r/k_{15}) = \frac{10 \cdot 18 - 3 \cdot 48}{4} = 1 \cdot 675$$

$$k_{15} = 0 \cdot 05/47 \cdot 34 = 1 \cdot 056 \times 10^{-3}$$

$\therefore$ Rate of increase of k per year $= \dfrac{0 \cdot 001056 - 0 \cdot 0002441}{15} = 0 \cdot 000054$ m/year.

In the case of the 0·225 m pipe the flow will reduce to its minimum acceptable value of 0·056 m^3/s under a hydraulic gradient of 0·015 eventually.

$$\frac{h_f}{L} = 0 \cdot 015 = \frac{4f}{0 \cdot 225} \left. \frac{0 \cdot 056}{(\pi/4)0 \cdot 225^2} \right/ 19 \cdot 62$$

$\therefore$
$$f = 0 \cdot 008346$$

$\therefore$
$$1/\sqrt{f} = 10 \cdot 95$$

$\therefore$
$$\log_{10}\frac{r}{k} = \frac{10 \cdot 95 - 3 \cdot 48}{4} = 1 \cdot 867$$

$\therefore$
$$r/k = 73 \cdot 5 \quad \text{so} \quad k = 0 \cdot 1125/73 \cdot 54$$

$\therefore$
$$k = 0 \cdot 00153.$$

$\therefore$
$$t = \frac{0 \cdot 00153 - 0 \cdot 000244}{0 \cdot 000054}$$

$$= 24 \text{ years}$$

So the 22·5 cm diameter pipe will transport 0·056 m^3/s for at least 24 years.

(3) Calculate the horsepower that can be delivered to a factory 6·5 km distant from a hydraulic power house through three horizontal pipes each 150 mm in diameter laid in parallel, if the inlet pressure in maintained constant at 540 N/cm^2 and the efficiency of transmission is 94%. If one of the pipes becomes unusable, what increase of pressure at the power station would be required to transmit the same delivery pressure as before. What would be the efficiency of transmission under these conditions? $f = 0 \cdot 0075$ in both cases.

$$\text{Head at power station} = \frac{5 \ 400 \ 000}{9810} \text{ metres}$$

$$= 550 \cdot 5 \text{ m}$$

$$\text{Head loss due to friction} = 6\% \text{ of } 550 \cdot 5$$

$$= 33 \cdot 03 \text{ m}$$

$$\text{Flow/pipe: } 33\cdot03 = \frac{4 \times 0\cdot0075 \times 6500}{19\cdot62 \times 0\cdot15} v^2$$

$$v = 0\cdot706 \text{ m/s}$$

$$Q/\text{pipe} = \frac{\pi}{4} \times 0\cdot15^2 \times 0\cdot706$$

$$= 0\cdot01248 \text{ m}^3/\text{s}$$

Delivery head $= 517\cdot43$ m

$$\text{Power delivered/pipe} = \frac{WQH}{1000} \text{ kW}$$

$$= \frac{9810 \times 0\cdot01284 \times 517\cdot43}{1000}$$

$$= 63\cdot33 \text{ kW/pipe}$$

Total power delivered $= 3 \times 63\cdot33$

$$= 189\cdot99 \text{ kW}$$

When one pipe becomes unusable and the delivery pressure is to remain unaltered,

$$h_d = 517\cdot43 \text{ m}$$

The power to be delivered must remain the same as before of course, i.e. 190 kW, so power/pipe = 95·0 kW

$$95\cdot00 = \frac{WQh_d}{1000}$$

$$Qh_d = \frac{95\,000}{9810} = 9\cdot6845$$

Also if h_n is the new head required at the pumping station, then

$$h_n - h_d = \frac{4fL}{2gd} v_n^2$$

$$= \frac{4 \times 0\cdot0075 \times 6500}{19\cdot62 \times 0\cdot15} v_n^2$$

$$v_n = 0\cdot12285 \, (h_n - h_d)^{1/2}$$

and

$$Q_n/\text{pipe} = \frac{\pi}{4} d^2 v_n = 2 \cdot 171 (h_n - h_d)^{1/2} \times 10^{-3}$$

$$2 \cdot 171 \times 10^{-3} (h_n - h_d)^{1/2} h_d = 9 \cdot 6835$$

$$(h_n - h_d)^{1/2} = \frac{9 \cdot 6835}{2 \cdot 171 \times 10^{-3} \times 517 \cdot 43}$$

$$h_n = 74 \cdot 309 + 517 \cdot 43$$

$$= 591 \cdot 739 \text{ m}$$

$$\text{Pumping pressure} = 591 \cdot 74 \times 9810 \text{ N/m}^2$$

$$= 580 \cdot 496 \text{ N/cm}^2$$

$$\text{Transmission efficiency} = \frac{517 \cdot 43}{591 \cdot 739} \times 100$$

$$= 87 \cdot 4\%$$

(4) A reservoir, A, with its surface at an elevation of 26 m above datum supplies water to two other reservoirs, B and C. The surface level in B is 7·6 m above datum and that in C is at datum level. From A to a junction, J, a pipe of 15 cm dia., 244 m length is used. The branch, JB, is 7·5 cm dia. and 61 m length. The branch JC is 10 cm dia. and 91 m length. Taking f as 0·0075 throughout and neglecting local losses, calculate the flow through the net.

$$h_f = \frac{4fLv^2}{2gd} = \frac{4fL(4Q/\pi d^2)^2}{2gd} = \frac{64fLQ^2}{19 \cdot 62\pi^2 d^5}$$

$$= \frac{fLQ^2}{3 \cdot 026 d^5}$$

$$h_A - h_B = \frac{fL_A Q_A^2}{3 \cdot 026 d_A^5} + \frac{fL_B Q_B^2}{3 \cdot 026 d_B^5} = 26 - 7 \cdot 6 = 18 \cdot 4$$

$$\therefore \quad 18 \cdot 4 = \frac{0 \cdot 0075 \times 244}{3 \cdot 026 \times 0 \cdot 15^5} Q_A^2 + \frac{0 \cdot 0075 \times 61}{3 \cdot 026 \times 0 \cdot 075^5} Q_B^2$$

$$= 7963 \cdot 9 Q_A^2 + 63711 \cdot 21 Q_B^2$$

$$h_A - h_C = \frac{fL_A Q_A^2}{3 \cdot 026 d_A^5} + \frac{fL_C Q_C^2}{3 \cdot 026 d_C^5} = 26 - 0 = 26$$

$$\therefore \quad 26 = 7963 \cdot 9 Q_A^2 + \frac{0 \cdot 0075 \times 91}{3 \cdot 026 \times 0 \cdot 10^5} Q_C^2$$

$$7963 \cdot 9 Q_A^2 + 22554 \cdot 5 Q_C^2$$

Let

$$Q_B = \alpha Q_A \text{ then } Q_C = (1 - \alpha)Q_A$$

$$\therefore \quad 18\cdot4 = 7963\cdot9Q_A^2 + 63711\cdot21\alpha^2 Q_A^2 \quad (1)$$

$$26 = 30518\cdot4Q_A^2 - 45109\alpha Q_A^2 + 22554\cdot5\alpha^2 Q_A^2 \quad (2)$$

Dividing equation (1) by (2) and simplifying

$$\alpha^2 + 0\cdot6686\alpha - 0\cdot2855 = 0$$

$$\therefore \quad \alpha = 0\cdot29598$$

$$Q_A = \sqrt{\left(\frac{18\cdot4}{7963\cdot9 + 63711\cdot21 \times 0\cdot29598^2}\right)} = 0\cdot036853 \text{ m}^3/\text{s}$$

$$Q_B = 0\cdot0369 \times 0\cdot29598 = 0\cdot01091 \text{ m}^3/\text{s}$$

$$Q_C = 0\cdot0369(1 - \alpha) = 0\cdot025943 \text{ m}^3/\text{s}$$

Check

$$h_j = 26\cdot0 - 7963\cdot9Q_A^2$$

$$= 15\cdot18$$

$$h_j - 63711\cdot21Q_B^2 = 7\cdot596 \text{ (error} = 0\cdot004)$$

$$h_j - 22554\cdot5Q_C^2 = -0\cdot000\ 063$$

Note that these calculations were performed on a desk top calculator. High accuracy is needed to obtain satisfactory results.

Questions

(1) Two water tanks A and B having constant cross-sectional areas of 100 sq ft [10 m²] and 50 sq ft [4 m²] are connected by a 2" [5 cm] diameter pipe, 500' [150 m] long, for which $f = 0\cdot01$. Determine the time taken for 1250 gallons [5700 kg] to be transferred from tank A to tank B, if the initial difference in level is 7' [2·1 m].

Answer: 115·47 minutes [135·8 minutes].

(2) Two reservoirs A and B whose constant difference of level is 20' [6 m] A being the higher, are connected by a 6" [15 cm] diameter pipe, 5000' [1500 m] long, for which the coefficient of friction f is 0·005. Water can be drawn from the pipe through a valve at a point C, distant 2000' [600 m] from the reservoir A. Find, neglecting the velocity head and all losses other than pipe friction, the quantity of water entering reservoir B when the quantity drawn off through the valve at C is (a) zero, and (b) 0·4 ft³/s [0·011 m³/s]. Draw the hydraulic gradient for case (b).

Answers: (a) 0·5 cusec [0·0135 m³/s] (b) 0·3 cusec [0·00798 m³/s]
(c) $h_{AC} = 15\cdot7'$ [4·74 m] $h_{CB} = 4\cdot3'$ [1·26 m].

(3) A reservoir discharges into a second reservoir through a pipe 15" [37·5 cm] diameter and 6000' [1800 m] long. The difference of water levels in the two reservoirs is 25' [7·5 m]. Find the discharge, taking f = 0·008. If the level of the pipe midway between the reservoirs is 50' [15 m] below the water level in the higher reservoir, find the pressure head at this point.

Answer: 4ft^3/s [0·108 m^3/s], 37·50' [11·25 m].

(4) A pipeline conveying water from a reservoir to a hydro-electric plant comprises 19 000' [5800 m] of concrete lined tunnel followed by 1500' [460 m] of steel pipe. The ratio between the diameters of tunnel and steel pipe is 1·4/1. The total head from reservoir level to turbine tail race is 416' [128 m] and the turbines, which have an efficiency of 90%, are required to develop 32 000 HP [24 000 kW] when the loss of head in the pipe and tunnel is 32' [9·8 m]. The coefficient of friction f is 0·0030 for the tunnel and 0·0036 for the steel pipe. Determine (a) the quantity of water required, and (b) the diameter of the tunnel. Neglect the loss of head at entrance to tunnel and at the point where the diameter changes.

Answer: (a) 11·2' [3·44 m], (b) 816·1 ft^3/s [23 m^3 s].

(5) Two reservoirs whose difference of level is 45' [14 m] are connected by a pipe ABC whose highest point B is 5' [1·5 m] below the level in upper reservoir A. The portion AB has a diameter of 8" [20 cm] and the portion BC a diameter of 6" [15 cm] the friction coefficient for each being 0·005. The total length of the pipe is 10 000' [3050 m]. Find the maximum allowable length of the portion AB if the pressure head at B is not to be more than 10' [3·0 m] below atmospheric pressure. Neglect velocity head in the pipe, loss of head at pipe entry and loss of head at change of section.

Answer: 6781·5' [2032 m].

(6) Two sharp ended pipes of 2" [5 cm] and 4" [10 cm] diameter, each 100' [30 m] long, are connected in parallel between two reservoirs whose difference of level is 25' [7·5 m]. Find (a) the flow in cusec for each pipe and draw the corresponding hydraulic gradients; (b) the diameter of a single pipe 100' [30 m] long which will give the same flow as the two actual pipes. Assume that the entrance loss is 0·5 $v^2/2g$ and take f = 0·008 in each case.

Answers: 0·193 cusec [0·0052 m^3/s], 1·052 cusec [0·0286 m^3/s], 4·29" [0·107 m].

(7) Develop the standard formula for the energy loss per unit weight of flow at a sudden enlargement in a pipeline carrying a fluid, and state the assumptions made. At a sudden enlargement from 12" [0·30 m] to 24" [0·6 m] diameter carrying water, the hydraulic gradient line rises 0·4' [0·12 m]. Calculate the quantity of water flowing.

Answer: 6·5 ft^3/s [0·178 m^3/s].

(8) Determine the conditions for transmission of maximum power through a pipe assuming loss of head by friction only. Calculate the maximum power available at the far end of a hydraulic pipeline 3 miles long [5 km] and 8″ [20 cm] diameter when water at 1000 lb/sq. in [700 N/cm^2] pressure is fed in at the near end. Take the friction coefficient f as 0·007.

Answer: 527 hp [377 kW].

(9) A pipeline connects a reservoir to a nozzle. The first 800′ [240 m] of the pipe is 15″ [38 cm] dia. and the last 1200′ [360 m] is 12″ [30 cm] dia. If the water level in the reservoir is taken as datum, the two pipes join at − 180′ [−54 m] and the discharge is at −240′ [−72 m]. The coefficient of velocity of the nozzle which is 3″ [7·5 cm] dia. is to be taken as 0·96 and there is no contraction. If the friction coefficient for both pipes is taken as 0·008 calculate the discharge from the nozzle and the water horsepower of the jet. Find also the pressure in the 12″ [30 cm] pipe just after the change of section.

Answers: 5·4 ft^3/s [0·148 m^3/s], 116 hp [82 kW], 75 psi [50·13 N/cm^2].

(10) Calculate the diameter of a pipe to convey gas from a holder to a power station having the following particulars given: gas consumption 20 000 ft^3/hour [0·157 m^3/s], length of pipe 0·5 miles [800 m], delivery 50′ [15 m] above the entrance to the pipe, pressure at holder 4″ [10 cm] water and at the power station 2″ [5 cm], density of gas 0·045 [7] and of air 0·08 lb/ft^3 [12·5] [N/m^3], f = 0·005.

Answer: 8·22″ [20·9 cm].

(11) A tank feeds a pipe A which forks into two pipes B and C. At the ends of B and C are streamlined nozzles, each $1\frac{1}{2}$″ dia. (C_v = 0·97), which discharge to atmosphere and the heads in the tank over them are 50′ and 75′ respectively. Pipe A is 50′ long and 3″ dia. Pipe B is 50′ long and 3″ dia. Pipe C is 100′ long and 3″ dia. Taking f = 0·01 throughout and neglecting all losses except pipe friction, calculate the rates of discharge through the pipes B and C.

Answers: Q_C = 0·455 ft^3/s, Q_B = 0·34 ft^3/s.

7 Open Channel Hydraulics

The difference between bounded flows and flows in which free surfaces are present is that the flow areas in free-surface flows are not determined by their fixed boundaries, but by the need to maintain continuity of flow while obeying dynamic laws. Before examining the complex conditions that can arise in free surface flows, it is necessary to consider the simplest case, that is uniform or normal flow.

7.1 Uniform flow

Uniform flow is flow at constant depth. It can only occur in a prismoidal channel, that is, a channel in which the geometry of the cross section does not vary through the channel. If the depth is constant in an open prismoidal channel and the flow is steady, the mean velocity must be constant throughout the length of the channel. Frictional losses must therefore also be constant throughout the channel. These frictional energy losses must be exactly balanced by energy gains if the flow velocity is to be maintained constant and these can only be obtained from the potential energy that the fluid loses as it falls down the slope of the channel. Alternatively, uniform flow may be seen as a situation in which the weight component of the fluid acting down the channel slope causes the fluid to accelerate until a velocity is reached at which the fluid frictional shear stresses acting over the channel sides and bed exactly balance it. (See Fig. 7.1.)

Now $wA \, \delta x \sin i$ is the force component acting down the slope on the fluid element due to its weight and $\tau P \, \delta x$ is the frictional force exerted by the boundaries upon the fluid. (A is the flow area, τ the frictional shear stress at the boundary and P the length of the wetted boundary.) If the fluid velocity is to remain constant, as it must if the flow area is to remain constant,

$$wA \, \delta x \sin i = \tau P \, \delta x$$

so

$$\tau = wA \sin i/P$$

230

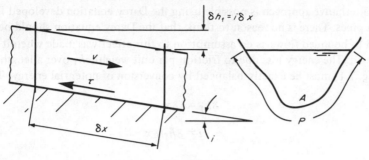

Fig. 7.1

For channels normally encountered in engineering practice the value of i is always small and the usual small-angle approximation applies: $i = \sin i = \tan i$, so

$$\tau = wAi/P$$

where i can be taken as either the channel bed slope in radians or its sine or tangent. This approximation may seem a little extreme, but it must be realised that uniform flow can only occur in channels of small slope, because when slopes are large the flow becomes aerated and also develops roll waves upon its surface, so the uniform flow approach is not then applicable anyway. The study of aerated flow and the development of roll waves is still in the area of research.

The shear stress τ is dependent upon v^n where n has a value between 1·75 and 2.

$$\tau = kv^n$$

This is Froude's law, but if n is assumed to be 2 the value of k will vary with a power of v. This assumption that $n = 2$ must be made if decimal indices are to be avoided.

$$kv^2 = w\frac{A}{P}i$$

so

$$v = \sqrt{\left(\frac{w}{k}\frac{A}{P}i\right)}$$

Usually A/P is denoted by m which is called the hydraulic mean depth.

$$\therefore \qquad v = C\sqrt{(mi)}$$

where

$$C = \sqrt{(w/k)}$$

C is itself a variable depending on the roughness of the surface and Reynolds number. This result was first developed by de Chezy and is known as the Chezy equation.

An alternative approach is possible using the Darcy equation developed for flow in pipes. There is no reason to think that the Darcy equation should be limited to bounded flows as no assumption to this effect was made when it was developed. The energy loss due to friction per unit weight δh_f over a length δx (see Fig. 7.1) must be exactly balanced by conversion of potential energy. Thus

$$\delta h_f = i\,\delta x$$

∴
$$i = \delta h_f / \delta x$$

but

$$\frac{dh_f}{dx} = \frac{fv^2}{2gm} = i$$

∴
$$v = \sqrt{(2g/f)}\sqrt{(mi)}$$

∴
$$v = C\sqrt{(mi)}$$

and

$$C = \sqrt{(2g/f)}$$

This approach links the Chezy C and the Darcy f. The flow in a channel is given by $Q = CA\sqrt{(mi)}$. If flow is uniform the quantity $CA\sqrt{m}$ is sometimes called the *conveyance* of the channel.

Much research has been performed to establish the variables upon which C depends. As usual the early results of such research are simple and easy to use, but inaccurate, while the more recently obtained results are complex, awkward to use but relatively accurate. Some of the various equations obtained are given below.

7.2 Formulae for the Chezy C

The Manning Formula, 1889

Manning expressed Strickler's formula as

$$C = 1 \cdot 0\, m^{1/6}/n \text{ (SI units)}$$

$$C = 1 \cdot 486\, m^{1/6}/n \text{ (fps units)}$$

This is a very popular formula and is still used. n is a dimensional quantity and therefore varies according to the system of units in use. n lies between $0 \cdot 009$ and $0 \cdot 045$ for the extreme ranges of roughness. The formula applies when turbulence is fully developed and as this is the case in most channel flows the Manning equation has a wide range of application.

The Bazin formula, 1897

Bazin gave

$$C = \frac{157 \cdot 6}{1 + n/\sqrt{m}} \text{ (fps units)}$$

$$C = \frac{87}{1 + n/\sqrt{m}} \text{ (SI units)}$$

The roughness number varies from 0·11 to 3·17 in the fps units, and from 0·0607 to 1·75 in the SI units. The Bazin formula was developed from work performed on model channels and is therefore not very accurate for large channels.

The Pavlovski formula, 1925

This formula is in common use in the Soviet Union.

$$C = m^y/n \text{ (SI units)}$$
$$y = 2 \cdot 5\sqrt{n} - 0 \cdot 13 - 0 \cdot 75\sqrt{m}(\sqrt{n} - 0 \cdot 10)$$

Approximate forms can be used:

$$\text{When } m < 1 \cdot 0 \text{ m} \quad y = 1 \cdot 5\sqrt{n}$$
$$\text{When } m > 1 \cdot 0 \text{ m} \quad y = 1 \cdot 3\sqrt{n}$$

The Powell Formula, 1950

$$C = -42 \log_{10}\left(\frac{C}{4Re} + \frac{\epsilon}{m}\right) \text{ (fps units)}$$

$$C = -23 \cdot 2 \log_{10}\left(\frac{C}{2 \cdot 2Re} + \frac{\epsilon}{m}\right) \text{ (SI units)}$$

$$(7.1)$$

where Re is the Reynolds number for the flow, m is the hydraulic mean depth and ϵ is the mean height of the channel roughnesses. In most channels Reynolds number is very large and the flow is highly turbulent hence

$$C = 42 \log_{10}(m/\epsilon)(\text{fps units})$$

$$C = 23 \cdot 2 \log_{10}(m/\epsilon)(\text{SI units})$$

For very smooth channels

$$C = 42 \log_{10}(4Re/C)(\text{fps units})$$

$$C = 23 \cdot 2 \log_{10}(2 \cdot 2Re/C)(\text{SI units})$$

As C appears on both sides an iterative method of solution must be used; this method is well suited to a computer approach but is rather tedious by hand. It is recommended that a trial value of $C = 100$ [55] is inserted into the right-hand side of equation (7.1) and C thus evaluated. This value can then be applied to the right hand side and a second value for C obtained. Continue iteration of this process until two successive values of C are insignificantly different. Approximately five iterations are necessary in my experience if C is required to 1% accuracy. (At certain high values of ϵ/m and at intermediate Reynolds numbers this type of solution can become unstable.)

Table 7.1

Type of channel	Chezy C^*		Manning n		Bazin n	
	fps	SI	fps	SI	fps	SI
Smooth cement	130 to 170	71 to 93	0·009	0·009	0·109	0·0607
Well laid brickwork	110 to 140	61 to 77	0·012	0·012	0·290	0·16
Brickwork in bad condition	70 to 100	39 to 55	0·017	0·017	0·833	0·46
Natural channels in good condition	60 to 80	33 to 44	0·0225	0·0225	1·54	0·85
Natural channels in poor condition	35 to 55	19 to 30	0·030	0·030	2·355	1·200
Natural channels in bad condition	25 to 45	14 to 25	0·035	0·035	3·170	1·75

* The lower values apply to low hydraulic mean depths and the higher to larger hydraulic mean depths. Chezy C values are only useful as giving the basis of a trial calculation which can later be improved by a more detailed calculation using one of the more sophisticated formulae.

7.3 The Prandtl mixing-length hypothesis applied to uniform free surface flows

In Chapter 5 the formation of a turbulent boundary layer according to the mixing length hypothesis was described and the equation

$$u = 2 \cdot 5 \sqrt{(\tau_0/\rho)} \log_e y/C$$

developed. Now C in a rough channel is equal to $k/33$

$$u = 2 \cdot 50 \sqrt{(\tau_0/\rho)} \log_e (33y/k) \tag{7.2}$$

where τ_0 is the shear stress at the boundary, the velocity is u at a distance y from the boundary, and k is the mean height of roughnesses.

A uniform free-surface flow over a rough bed is an example of such a boundary layer flow. At the entrance to such a channel the boundary layer will be thin and will constitute only a small fraction of the flow depth. At a considerable distance downstream from the entrance the boundary layer will thicken until eventually it occupies the entire depth of the flow. This is one of the main differences between flow in natural channels and flow in experimental channels of the type found in laboratories. In natural channels the boundary layer occupies the entire flow depth but in experimental channels the boundary layer may only occupy one fifth of the channel depth. Any work done in experimental channels must be carefully examined from this viewpoint to ensure that results obtained have significance in relation to flow in natural channels. Many experimentally obtained friction formulae do not work well when applied to full size rivers and channels precisely for this reason.

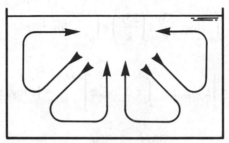

Secondary flow in straight narrow channels

Fig. 7.2

In broad channels the situation in the central regions of the flow can be closely approximated by a two dimensional approach and two dimensional boundary layer theory applies. In narrow channels and close to the sides of broad channels the boundary layers from the sides interact with those from the bed so that the problem becomes three dimensional and in addition secondary flows occur (Fig. 7.2). These two effects make the problem intractable.

For these reasons what follows applies only to broad channels. As $\sqrt{(\tau_0/\rho)}$ has the dimensions of velocity it will be denoted by V^* which is often called the shear velocity. This velocity does not necessarily describe the velocity of any particular fluid element.

Rewriting equation (7.2)

$$u = 2 \cdot 50 V^* \log_e(33y/k)$$

When $y = d$, $u =$ the fluid velocity at the surface, V. This is also the maximum velocity.

$\therefore$
$$V = 2 \cdot 50 V^* \log_e(33d/k)$$

and

$$\frac{u}{V} = \frac{\log_e(33y/k)}{\log_e(33d/k)}$$

$$\text{The flow/unit width} = q = \int_0^d u \, dy$$

$$= \frac{V}{\log_e (33d/k)} \int_0^d \log_e \frac{33y}{k} \, dy = \frac{V}{\log_e (33d/k)} \frac{k}{33} \int_0^d \log_e \frac{33y}{k} \, d\left(\frac{33y}{k}\right)$$

let $33y/k = z$

$$\therefore \qquad \int_0^d \log_e \frac{33y}{k} \, d\left(\frac{33y}{k}\right) = \int_0^{33d/k} \log z \, dz$$

$$\int_0^{33d/k} \log z \, dz = \left[z \log_e z - \int z \, d \log_e z \right]_0^{33d/k} = [z \log_e z - z]_0^{33d/k}$$

$$= \frac{33d}{k} \left(\log_e \frac{33d}{k} - 1 \right)$$

$$\therefore \qquad q = \frac{Vk}{33 \log_e (33d/k)} \frac{33d}{k} \left(\log_e \frac{33d}{k} - 1 \right) = Vd \left(1 - \frac{1}{\log(33d/k)} \right)$$

∴ Mean velocity

$$\bar{V} = \frac{q}{d} = V \left(1 - \frac{1}{\log_e (33d/k)} \right)$$

The height above the bed, y_m, at which this mean velocity will occur is given by

$$2 \cdot 5 V^* \log_e \frac{33 y_m}{k} = V \left(1 - \frac{1}{\log_e (33d/k)} \right)$$

$$= 2 \cdot 5 V^* \log_e \frac{33d}{k} \left(1 - \frac{1}{\log_e (33d/k)} \right)$$

$$\therefore \qquad \log_e (33 y_m/k) = \log_e (33d/k) - 1$$

$$\therefore \qquad \log_e (d/y_m) = 1$$

$$\therefore \qquad d/y_m = e$$

$$y_m/d = 1/e = 0 \cdot 368$$

This is usually approximated to 0·4 so the height of the the mean-velocity filament is 0·4 x depth. Hence, changing to common logarithms,

$$\bar{V} = 5.75\sqrt{(\tau_0/\rho)}\log_{10}(13.2d/k)$$

In a channel

$$wAi\,\delta x = \tau_0 P\,\delta x$$

considering forces acting on a fluid element of cross sectional area A, periphery
P and length δx.

$\therefore$
$$\tau_0 = wmi$$

and

$$\tau_0/\rho = gmi$$

For a broad channel $m = d$, so

$$\tau_0/\rho = gdi$$

$\therefore$ for a broad channel

$$\bar{V} = 5.75\sqrt{(gdi)}\log_{10}(13.2d/k)$$

Comparing this result with the Chezy formula

$$\bar{V} = C\sqrt{(di)}$$

(substituting d for m) it can be seen that

$$C = 5.75\sqrt{g}\log_{10}(13.2d/k)$$
$$C = 32.6\log_{10}(13.2d/k)\ \text{(fps units)}$$

and

$$C = 18.01\log_{10}(13.2d/k)\ \text{(SI units)}$$

From formulae such as this or the Powell formula, estimates of C at high
flow and large depth can be made by investigating the velocity distribution at
low flows and so finding k. If in high flow conditions this k value does not
alter, it can be used to obtain C at high depth values.

7.4 'Economic' channels

Up to this point no mention has been made of the effect of the shape of the
channel upon the flow. Quite obviously, it is important when designing a channel
to pick the cross section which will give the largest flow for the least cost. It is
not possible to develop a theory that will determine the most economic shape of
channel that will be applicable in all situations so the best or 'economic' shape
is defined as being that shape that gives the maximum discharge for a given cross

sectional area. As much of the cost of building a channel is incurred in its excavation this approximates to the condition of largest obtainable discharge for a given expenditure on the channel.

The flow in a channel is given by

$$Q = AC\sqrt{(mi)} = CA^{3/2}P^{-1/2}i^{1/2}$$

$$\therefore \qquad dQ = (\tfrac{3}{2}A^{1/2}P^{-1/2}\,dA - \tfrac{1}{2}P^{-3/2}A^{3/2}\,dP)Ci^{1/2}$$

For maximum discharge $dQ = 0$

$$\therefore \qquad\qquad 3P\,dA - A\,dP = 0$$

The velocity in a channel is

$$v = CA^{1/2}P^{-1/2}i^{1/2}$$

$$dv = (\tfrac{1}{2}A^{-1/2}P^{1/2}\,dA - \tfrac{1}{2}P^{-1/2}A^{1/2}\,dP)Ci^{1/2}$$

For maximum velocity $dv = 0$

$$\therefore \qquad\qquad P\,dA - A\,dP = 0$$

These two results have been obtained making the tacit assumption that the Chezy C is constant. If this assumption cannot be made and C depends upon other parameters such as the hydraulic mean depth then the expression for C must be substituted so that the differentiation process can include the variation of C. If the channel shape that gives maximum discharge for a given cross sectional area is required then δA can be specified as zero. So both these equations reduce to $\delta P = 0$. Now if the cross section consists of three straight boundaries, that is a trapezoidal cross section, the shape that obeys the equation can be completely defined. Most man-made channels are of this type. Consider a trapezoidal channel for which the side slopes are 1 in s and the bottom width is $2b$ (Fig. 7.3). Then

$$A = 2bd + sd^2$$
$$P = 2b + 2d\sqrt{(1 + s^2)}$$
$$\therefore \qquad P = A/d - sd + 2d\sqrt{(1 + s^2)}$$
$$dP/dd = 0 = -(A/d^2) - s + 2\sqrt{(1 + s^2)}$$
$$\therefore \qquad (A/d) + sd = 2d\sqrt{(1 + s^2)}$$

and from before

$$A/d = 2b + sd$$
$$\therefore \qquad b + sd = d\sqrt{(1 + s^2)}$$

This condition must be satisfied if discharge is to be maximum. It is now necessar to examine what this condition implies as regards the shape of the cross section. Referring to Fig. 7.3, $b + sd = $ AB and $d\sqrt{(1 + s^2)} = $ BC so the condition is that AB = BC.

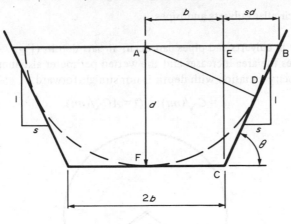

Fig. 7.3

If this is so then $\triangle ABD \equiv \triangle EBC$ as $AB = BC$, $\angle ADB = \angle BEC = 1$ right angle and $\angle EBC$ is common to both.

$\therefore \qquad\qquad\qquad AD = EC \quad \text{and} \quad EC = AF$

If $AD = AF$ it must be possible to inscribe a semicircle within the channel cross section which will be tangential to all three sides and have its centre in the surface of the flow. An economic channel is defined by this condition.

The value of the hydraulic mean depth for an economic channel is always half the depth.

$$m = \frac{A}{P} = \frac{2bd + sd^2}{2b + 2d(1 + s^2)^{1/2}}$$

$$m = \frac{2bd + sd^2}{2b + 2(b + sd)} = \frac{2bd + sd^2}{2(2b + sd)} = \frac{d}{2}$$

In the case of a rectangular channel the economic channel breadth is equal to twice the depth.

The statement that a channel in which a semicircle can be inscribed is economic, defines a class of trapezoidal cross sections all of which are economic and have the same cross sectional area. However, the one for which the side slope is 60° is better than all the others. This means that the best channel is half a regular hexagon. From before

$$P = (A/d) - sd + 2d \sqrt{(1 + s^2)}$$
$$dP/ds = -d + (2d/2)(1 + s^2)^{-1/2} 2s = 0$$
$\therefore \qquad\qquad\qquad 1 = 2s/(1 + s^2)^{-1/2}$
$\therefore \qquad\qquad\qquad s = 1/\sqrt{3}$
$\therefore \qquad\qquad\qquad \theta = 60°$

7.5 Flow in circular culverts and pipes

Uniform flow in culverts and pipes can occur at part depths (Fig. 7.4). As the depth increases the area increases and the wetted perimeter also increases so that the mean velocity variation with depth is not straightforward.

$$v = C\sqrt{(mi)}; \quad Q = AC\sqrt{(mi)}$$

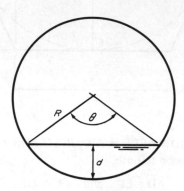

Fig. 7.4

If the Manning expression is used,

$$C = 1\cdot49 \, m^{1/6}/n$$

so

$$v = 1\cdot49 \, m^{2/3} i^{1/2}/n \quad \text{and} \quad Q = 1\cdot49 \, A m^{2/3} i^{1/2}/n$$

Therefore

$$v = \frac{1\cdot49}{n}\left(\frac{A}{P}\right)^{2/3} i^{1/2} \quad \text{and} \quad Q = \frac{1\cdot49}{n}\frac{A^{5/3}}{P^{2/3}} i^{1/2}$$

$$d = R - R\cos(\theta/2) = D/2[1 - \cos(\theta/2)]$$

$$A = R^2\theta/2 - R^2\cos(\theta/2)\sin(\theta/2)$$

$$\therefore \qquad A = R^2(\theta - \sin\theta)/2$$

$$P = R\theta \text{ so } m = \frac{R}{2}\left(1 - \frac{\sin\theta}{\theta}\right)$$

$$\therefore \qquad v = \frac{1\cdot49}{n} i^{1/2}\left[\frac{R}{2}\left(1 - \frac{\sin\theta}{\theta}\right)\right]^{2/3}$$

$$Q = \frac{1\cdot49}{n} i^{1/2}\frac{\theta R^{8/3}}{2^{5/3}}\left(1 - \frac{\sin\theta}{\theta}\right)^{5/3}$$

$$\frac{v}{v_{\text{full}}} = \left(1 - \frac{\sin \theta}{\theta}\right)^{2/3}$$

and

$$\frac{Q}{Q_{\text{full}}} = \frac{\theta}{2\pi} \left(1 - \frac{\sin \theta}{\theta}\right)^{5/3}$$

7.6 Gradually varied, non-uniform flow in channels

All flows approximate to uniform flow if the channel is sufficiently long. Flow may on the other hand be non-uniform. If depth variations cause surface slopes which are not large the flow is non-uniform and gradually varied but if slopes are large the flow is non-uniform and rapidly varied. (See Fig. 7.5.)

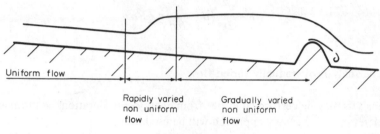

Uniform flow

Rapidly varied non uniform flow

Gradually varied non uniform flow

Fig. 7.5

Before analysing gradually varied non-uniform flow it is necessary to consider the three types of open channel flow: super-critical, critical and sub-critical flow.

A device that affects the flow depths upstream of it must signal the depth changes that it causes to the upstream fluid. This it does by initiating waves that travel upstream altering depths and flow velocities as they traverse the fluid. Once these velocities and depths have been adjusted to fit the depth imposed by the hydraulic control downstream the flow state becomes steady. If at an upstream point the flow velocity in the channel is greater than or equal to the velocity of the wave trains coming upstream from the control, the waves will be brought to rest and a surface discontinuity, an hydraulic jump, will form. If a free surface flow is occurring in which the flow velocity is greater than the velocity of transmission of a small wave upon its surface (wave celerity) conditions downstream cannot affect conditions upstream, but conditions upstream can affect conditions downstream. Conversely if flow velocities are less than the wave celerity, conditions downstream control conditions upstream. Clearly the ratio flow velocity/ wave celerity determines which of these situations occurs. (This is similar to the conditions that occur in compressible flow and the ratio flow velocity/wave celerity is strictly analogous to the Mach number: flow velocity/local velocity of sound). It will later be shown that the wave celerity c is given by $c = \sqrt{(gd)}$ (p. 272) so the ratio v/c becomes $v/\sqrt{(gd)}$. This is called the Froude number by some American and British writers, and the Boussinesq number by other British writers. Note that in this book the square of this number v^2/gd has been called the Froude number and the reader should always carefully check which expression is being used when Froude numbers are used. The critical value of this number is unity as this occurs when the local flow velocity equals the local wave celerity. A flow which has a velocity greater than the wave celerity is called supercritical, shooting or torrential and its Froude number is greater than 1. A flow in which the flow velocity is less than the wave celerity is called subcritical, streaming or tranquil and its Froude number is less than unity. These three classes of flow are very different. At this point it is worth noting the existence of another number, the *kineticity*. Kineticity λ is the ratio of the kinetic energy of a flow per unit weight to the depth energy.

$$\lambda = \frac{v^2}{2g} / d = \frac{v^2}{2gd} = \tfrac{1}{2}Fn$$

7.7 The analysis of gradually varied flow

The analysis may be performed either from energy considerations or momentum considerations. The energy approach will be used here.

In Fig. 7.6 τ denotes the shear stress exerted on the fluid by the bed. The energy equation is

$$\frac{p}{w} + \frac{v^2}{2g} + y = H$$

Now for any point C, $p/w = s$ if the pressure distribution is hydrostatic so $s + (v^2/2g) + y = $ constant but $s + y = d + z$. Therefore $d + (v^2/2g) + z = $ constant $= H$ where H is the total energy for unit weight of the fluid.

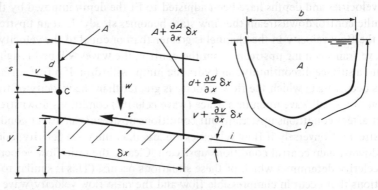

Fig. 7.6

The rate of change of H with x is

$$\frac{dH}{dx} = \frac{dd}{dx} + \frac{v}{g}\frac{dv}{dx} + \frac{dz}{dx}$$

but $dz/dx = \sin i$ and if i is small $\sin i \approx \tan i \approx i$. Applying the continuity equation

$$\frac{d(Av)}{dx} = 0$$

∴

$$v\frac{dA}{dx} + A\frac{dv}{dx} = 0$$

∴

$$\frac{dv}{dx} = -\frac{v}{A}\frac{dA}{dx}$$

But

$$\frac{dA}{dx} = \frac{dA}{dd} + \frac{dd}{dx} \quad *$$

and

$$\delta A = b\,\delta d \text{ (see Fig. 7.7)}$$

∴ $dA/dd = b$ where b is the surface breadth

∴

$$\frac{dv}{dx} = -\frac{vb}{A}\frac{dd}{dx}$$

∴

$$\frac{dH}{dx} = \frac{dd}{dx} - \frac{v^2 b}{Ag}\frac{dd}{dx} + i$$

Now dH/dx is the rate of change of energy/unit weight along the channel and this can only be caused by frictional losses. An estimate of the value of dH/dx can be made by considering the flow as if it were uniform at the depth and velocity occurring at a particular point. If the flow is uniform the energy lost

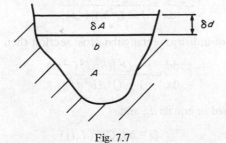

Fig. 7.7

per unit weight must equal the energy gained as the fluid moves down the slope and the surface slope is equal to the bed slope. Thus the energy loss per unit weight per unit distance is equal to the surface slope, j, which equals the bed slope. Also

$$v = C\sqrt{(mi)} = C\sqrt{(mj)}$$

$$\therefore \qquad j = v^2/(C^2 m)$$

So the energy loss rate $= j = v^2/C^2 m$.

$\mathrm{d}H/\mathrm{d}x$ must be negative as H must be decreasing due to friction, so

$$\mathrm{d}H/\mathrm{d}x = -j = -v^2/C^2 m$$

$$\therefore \qquad \frac{-v^2}{C^2 m} = \frac{\mathrm{d}d}{\mathrm{d}x} - \frac{v^2 b}{Ag}\frac{\mathrm{d}d}{\mathrm{d}x} + i$$

$$\therefore \qquad \frac{\mathrm{d}d}{\mathrm{d}x} = \frac{-i - v^2/(C^2 m)}{1 - (v^2 b/Ag)}$$

If the bed slope is regarded as positive when the slope is directed downwards in the direction of flow the sign of i in the above equation will have to be changed. This conversion is sensible as most people would consider a slope as positive if it causes flow. $v^2 b/Ag$ is the Froude number as can be seen from the following argument. Let A/b be denoted by δ where δ is the mean depth of the cross section. Then $v^2 b/Ag = v^2/g\delta$ which is the Froude number. Then

$$\frac{\mathrm{d}d}{\mathrm{d}x} = \frac{i - j}{1 - Fn}$$

where $j = v^2/C^2 m$ and is the energy loss per unit weight of fluid per unit length of the channel and the Froude number $Fn = v^2/g\delta$. (b is the surface breadth, not the base breadth nor the mean breadth).

This differential equation cannot be integrated even in its simplest form. Finite difference integration is the only possible way of dealing with the problem. As the velocity is variable as well as the depth, it is useful to express velocities in terms of the flow rate Q.

$$\frac{\mathrm{d}d}{\mathrm{d}x} = \frac{i - Q^2/(A^2 C^2 m)}{1 - Q^2 b/(A^3 g)} \qquad \checkmark \quad non.$$

If the channel is of uniform rectangular cross section then

$$\frac{\mathrm{d}d}{\mathrm{d}x} = \frac{i - Q^2/(b^2 d^2 C^2 m)}{1 - Q^2/(b^2 d^3 g)}$$

If the channel is broad m equals d_n and

$$Q = b d_n C\sqrt{(d_n i)}$$

so

$$Q^2/(b^2 d_n^3 C^2) = i$$

where d_n is the uniform flow depth.

When the depth of flow is critical the Froude number is 1 so

$$Q^2/(b^2 g) = d_c^3$$

∴
$$\frac{dd}{dx} = \frac{i - (d_n/d)^3 i}{1 - (d_c/d)^3} = i \left(\frac{d^3 - d_n^3}{d^3 - d_c^3} \right) \qquad (7.3)$$

This result is only applicable to a wide channel in which

$$m = \frac{bd}{b + 2d} = \frac{d}{1 + 2d/b}$$

when $b \to \infty$, $d/b \to 0$ and $m \to d$.

7.8 The specific force equation

When the momentum equation is applied to free surface flows the resulting equation is called the total force or specific force equation.

The form of the momentum equation that is applicable to a free surface flow is

$$\int_0^d p\, \delta A/w + \int_0^d v^2\, \delta A/g = \text{constant}$$

Now $\int_0^d p\delta A$ is the pressure force over a cross section of the flow of depth d and if the pressure distribution is hydrostatic the value of this integral is $wA\bar{z}$ where $\bar{z}$ is the depth below the surface of the centroid of the flow cross section.

The momentum equation thus becomes

$$A\bar{z} + Av^2/g = \text{constant}$$

Applying this result to two sections (subscripts 1 and 2) gives

$$A_1\bar{z}_1 + A_1 v_1^2/g = A_2\bar{z}_2 + A_2 v_2^2/g$$

This result is valid for any channel cross section but it is assumed that the two sections are sufficiently close for frictional forces due to the boundary to be negligible.

If the cross section is rectangular

$$A_1 = bd_1 \quad A_2 = bd_2 \quad \bar{z}_1 = d_1/2 \quad \text{and} \quad \bar{z}_2 = d_2/2$$

∴
$$d_1^2/2 + d_1 v_1^2/g = d_2^2/2 + d_2 v_2^2/g = S, \text{ the specific force}$$

7.9 The specific energy equation

Over a short length of a horizontal channel energy dissipation due to frictional forces from the boundary and turbulence can be ignored. Writing Bernoulli's equation for points 1 and 2 on a streamline

$$p_1/w + y_1 + v_1^2/2g = p_2/w + y_2 + v_2^2/2g$$

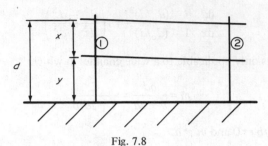

Fig. 7.8

If the streamlines are not curved the pressure distribution is hydrostatic so p_1/w and p_2/w are equal to x_1 and x_2 respectively (see Fig. 7.8).

∴
$$x_1 + y_1 + v_1^2/2g = x_2 + y_2 + v_2^2/2g$$

∴
$$d_1 + v_1^2/2g = d_2 + v_2^2/2g = E, \text{ the specific energy}$$

The specific energy is thus defined as the energy/unit weight of the fluid referred to the bed of the channel as datum. Note that the bed does not have to be horizontal and may even be discontinuous.
Summarising

$$d^2/2 + dv^2/g = S$$

where S is the specific force and

$$d + v^2/2g = E$$

v depends upon Q and d so

$$d^2/2 + Q^2/(b^2 dg) = S$$

and

$$d + Q^2/(2gb^2 d^2) = E$$

Plotting E against d for constant Q gives the curve shown in Fig. 7.9. A minimum value of E is obtained at a certain value of d and this can be obtained by differentiating with respect to d.

$$\frac{dE}{dd} = 1 - \frac{Q^2}{gA^3}\frac{dA}{dd} = 0$$

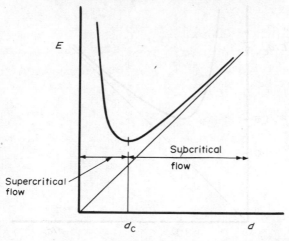

Fig. 7.9

but $dA/dd = b$, the surface breadth, so $Q^2b/A^3g = 1$; Now this is the Froude number, so for minimum specific energy the Froude number is unity and in a rectangular channel the critical depth d_c is obtained from

$$\frac{Q^2b}{b^3d^3g} = 1 \quad \therefore \quad d_c = (Q^2/b_g^2)^{1/3} = (q^2/g)^{1/3}$$

A suitable dimensionless form can be obtained by dividing the specific energy equation through by d_c

$$\frac{d}{d_c} + \frac{v^2/2g}{d_c} = \frac{E}{d_c}$$

In a rectangular channel

$$\frac{v^2}{2gd_c} = \frac{1}{2g}\left(\frac{Q^2}{b^2d^2}\right)\frac{1}{d_c}$$

Now

$$\frac{Q^2}{g(bd_c)^2d_c} = 1 \quad \text{that is, the Froude number is unity for minimum energy}$$

$$\therefore \quad Q^2/gb^2 = d_c^3$$

$$\therefore \quad v^2/2gd_c = \tfrac{1}{2}(d_c/d)^2$$

$$\therefore \quad d/d_c + \tfrac{1}{2}(d_c/d)^2 = E/d_c$$

This result is only applicable to a rectangular channel (see Fig. 7.10).
   Note that the critical depth is $\tfrac{2}{3}E$ as when $d = d_c$

$$1\cdot5 = E/d_c \quad \therefore \quad d_c = \tfrac{2}{3}E$$

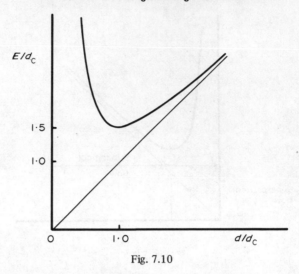

Fig. 7.10

Examining the specific force equation

$$d^2/2 + dQ^2/(gb^2d^2) = S$$

$$d^2/2 + Q^2/(gb^2d) = S$$

$dS/dd = d_c - Q^2/(bg^2d_c^2) = 0$ at the minimum

$\therefore$　　　　　　　　　$Q^2/(gb^2d_c^3) = 1$

That is, as for the specific energy case, minimum specific force occurs when the Froude number is unity.

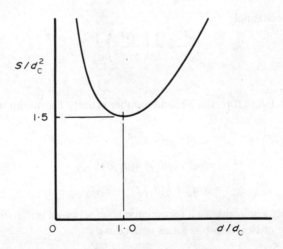

Fig. 7.11

Dividing the specific force equation by d_c^2 gives

$$\tfrac{1}{2}(d^2/d_c^2) + dv^2/d_c^2 g = S/d_2^2$$

∴

$$\tfrac{1}{2}(d/d_c)^2 + dQ^2/(d_c^2 b^2 d^2 g) = S/d_c^2$$

as before

$$Q^2/b^2 g = d_c^3$$

∴

$$\tfrac{1}{2}(d/d_c)^2 + d_c/d = S/d_c^2$$

This curve is sketched in Fig. 7.11.

As for the specific energy case the minimum value occurs when the depth is critical.

7.10 Flow profiles

If the depth of uniform or normal flow in a channel is greater than the critical depth (the depth at which the Froude number is unity) the bed slope is called mild. If it is less than the critical depth the bed slope is steep and if equal to the critical depth the bed slope is critical. Steep slopes as defined in this way are not steep in the usual sense of the word. Mild slopes are less than and steep slopes are greater than the critical slope.

Critical slopes, again, are slopes that have a normal flow depth which is equal to the critical depth, so $i = j$ and $Fn = 1$ simultaneously.

∴

$$i_c = v^2/(C^2 m)$$

and

$$v^2/(g\delta) = 1$$

∴

$$C^2 m i_c = g\delta$$

so

$$i_c = g\delta/(C^2 m)$$

In a broad channel $\delta = m$, so

$$i_c = g/C^2$$

Three zones can be defined (see Fig. 7.12).

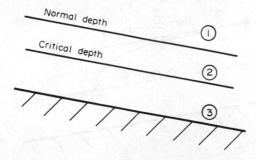

Fig. 7.12

(1) The zone of space above both the critical and normal depths.
(2) The zone of space between the normal depth and the critical depth.
(3) The lowest zone of space above the channel bed and below both the critical and the normal depth lines

As three types of slope exist (mild M, steep S and critical C) and three zones exist for M and S slopes and two zones for C slopes (the second zone does not exist for C slopes as $d_n = d_c$) eight different curves can be drawn. These are M1, M2 and M3; S1, S2 and S3; and C1 and C3 curves.

The M1 or backwater curve (Fig. 7.13)

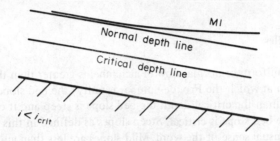

Fig. 7.13

$$\frac{dd}{dx} = \frac{i - v^2/(C^2 m)}{1 - v^2/(g\delta)}$$

As the depth is greater than the normal depth $v^2/C^2 m$ is less than i. Also $v^2/g\delta$ is less than 1 because the depth is greater than critical depth.

The numerator is positive and the denominator is also positive. Therefore dd/dx is positive so the depth must increase in the downstream direction. As the depth gets larger the $v^2/C^2 m$ and $v^2/g\delta$ terms become smaller so $dd/dx \to i$. This means that the water surface tends to the horizontal downstream and is asymptotic to the normal depth line far upstream.

The M2 or drawdown curve (Fig. 7.14)

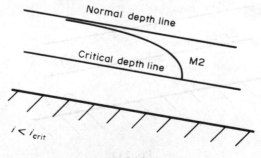

Fig. 7.14

If the depth is less than the normal depth the v^2/C^2m term is greater than i so the numerator is negative, but because the depth is greater than critical depth the Froude number is less than 1 so the denominator is positive. The value of the surface slope dd/dx is therefore negative, and the depth decreases in the downstream direction. When the depth reaches the critical depth the value of the Froude number becomes 1 and the denominator becomes zero. Thus the surface slope becomes $-\infty$. Far upstream the surface is asymptotic to the normal depth line.

The M3 curve (Fig. 7.15)

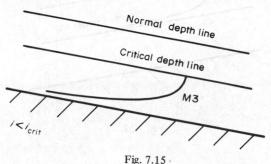
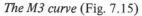

Fig. 7.15

In this case $v^2/C^2m \gg i$ and $v^2/g\delta > 1$ so dd/dx is positive. The depth therefore increases downstream. When it reaches the critical depth $v^2/g\delta = 1$ and $dd/dx = +\infty$. Upstream the surface slope approximates to

$$\frac{v^2/C^2m}{v^2/g\delta} = i_c$$

which, as i_c is greater than i, means that the surface slope is slightly above the horizontal, that is the slope, referred to the horizontal, is approximately $i_c - i$. This condition is hypothetical because when it applies the depth is zero, and the velocity is infinite, also the Chezy C cannot then be considered to have its normal value.

The S1 or ponding curve (Fig. 7.16)

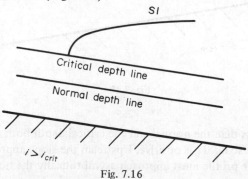

Fig. 7.16

In all S curves the critical depth line is above the normal depth line. In the S1 case $v^2/C^2m \ll i$ and $v^2/g\delta < 1$ so $\mathrm{d}d/\mathrm{d}x$ is positive. When $d = d_c$, $\mathrm{d}d/\mathrm{d}x = +\infty$. As the depth increases $v^2/C^2m \to 0$ and $v^2/g\delta \to 0$ so $\mathrm{d}d/\mathrm{d}x \to i$. The surface thus tends to the horizontal downstream.

The S2 curve (Fig. 7.17)

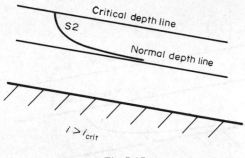

Fig. 7.17

$v^2/C^2m < i$ because the depth is greater than the normal flow depth but $v^2/g\delta > 1$ so $\mathrm{d}d/\mathrm{d}x$ is negative. The depth must then decrease downstream. When the depth reaches the normal depth v^2/C^2m equals i so $\mathrm{d}d/\mathrm{d}x \to 0$. The surface therefore approaches the normal depth line asymptotically downstream. Upstream the profile approaches the critical depth line which it must intersect perpendicularly.

The S3 curve (Fig. 7.18)

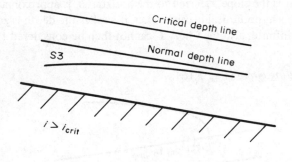

Fig. 7.18

The depth being less than the normal and the critical depth both v^2/C^2m and $v^2/g\delta$ must be large so $\mathrm{d}d/\mathrm{d}x$ is positive. Upstream the slope approximates to $i_c - i$ and downstream the profile must approach asymptotically the normal depth line.

The C curves (Fig. 7.19)

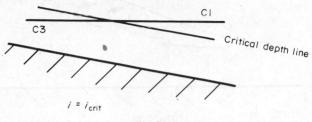

Fig. 7.19

In both C curves the normal depth and the critical depth lines are coincident.
For C curves

$$\frac{\mathrm{d}d}{\mathrm{d}x} = \frac{i_c - v^2/C^2m}{1 - v^2/g\delta} = \frac{g\delta/C^2m - v^2/C^2m}{(g\delta - v^2)/g\delta}$$

$$= \frac{g\delta}{C^2m}\left(\frac{g\delta - v^2}{g\delta - v^2}\right) = \frac{g\delta}{C^2m} = i_c$$

Thus both C curves are horizontal straight lines.

A general principle can be enunciated. When considering the shape of a
particular flow profile the curve must intersect the critical depth line perpendicularly and it must approach the normal depth line asymptotically.

Adverse slopes

Channels which have negative slopes, that is slopes which are directed upwards
in the direction of flow, are really special cases of mild slopes. In such channels
no normal flow can occur so the normal depth line disappears off the top of the
mild slope diagram and there can only be two cases, A2 and A3. These curves
are similar in shape to the M2 and M3 curves but extend over a shorter length
of the channel.

Before demonstrating the use of these curves in predicting the channel profiles
that occur in various circumstances it is necessary to examine a phenomenon that
is of frequent occurrence in open channels, namely, the formation of hydraulic
jumps.

7.11 The hydraulic jump

It is possible for very rapid changes in depth (surface discontinuities) to occur
in a channel in which the flow is non-uniform. These are called hydraulic jumps
(see Fig. 7.20).

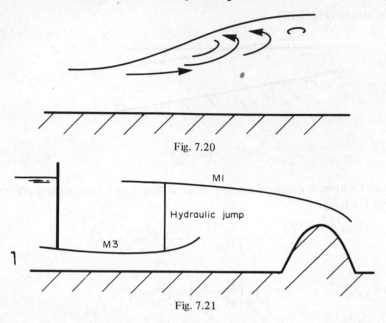

Fig. 7.20

Fig. 7.21

The conditions that determine whether a jump occurs are those at the upstream and downstream ends of the channel. If the flow upstream is constrained, by for example a sluice, to be supercritical while downstream, a weir compels it to be subcritical, the flow profiles cannot join smoothly (see Fig. 7.21) and consequently the profile is discontinuous. The jump must be located at a point in the channel at which the specific force just upstream is equal to the specific force just downstream of the jump. The specific energy of the flow changes across a jump because of the energy lost in the development of the strong turbulence in the roller that often develops on the front of the jump. For this reason the specific energy equation cannot be used to analyse the hydraulic jump.

Writing the specific force equation for a rectangular channel for points 1 and 2 (see Fig. 7.22).

$$d_1^2/2 + d_1 v_1^2/g = d_2^2/2 + d_2 v_2^2/g$$

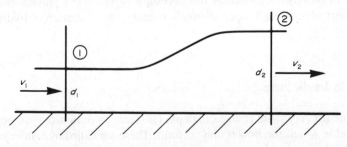

Fig. 7.22

also $bd_1v_1 = bd_2v_2$ (continuity equation)

$$\therefore \qquad v_1 = d_2v_2/d_1$$

$$\therefore \qquad (d_1^2 - d_2^2)/2 + [(d_2/d_1)^2d_1 - d_2]v_2^2/g = 0$$

Dividing through by d_2^2

$$(d_1^2 - d_2^2)/2d_2^2 + v_2^2(d_2/d_1 - 1)/gd_2 = 0$$

$$\therefore \qquad \tfrac{1}{2}[(d_1/d_2)^2 - 1] + Fn_2(1 - d_1/d_2)d_2/d_1 = 0$$

$$\therefore \qquad \tfrac{1}{2}(d_1/d_2 + 1) - Fn_2(d_2/d_1) = 0$$

$$\therefore \qquad \tfrac{1}{2}(d_1/d_2)^2 + \tfrac{1}{2}d_1/d_2 - Fn_2 = 0$$

$$\therefore \qquad d_1/d_2 = \frac{-\tfrac{1}{2} + \sqrt{(\tfrac{1}{4} + 2Fn_2)}}{1}$$

$$\therefore \qquad d_1/d_2 = -\tfrac{1}{2} + \tfrac{1}{2}\sqrt{(1 + 8Fn_2)}$$

If, instead of substituting for v_1, the substitution $v_2 = d_1v_1/d_2$ is used the following result is obtained

$$d_2/d_1 = -\tfrac{1}{2} + \tfrac{1}{2}\sqrt{(1 + 8Fn_1)}$$

Both results are useful; which is used depends upon the information available. If the upstream depth is known Fn_1 is known so the second result can be used. If the downstream depth is specified then Fn_2 is known so the first result can be used.

The specific force and specific energy diagram can conveniently be used when considering the hydraulic jump (see Figs. 7.23a and b). On the specific energy diagram two depths, for example d_1 and d_2', that have the same specific energy are called corresponding depths. On the specific force diagram two depths, for example d_1 and d_2 that have the same specific force are called conjugate depths.

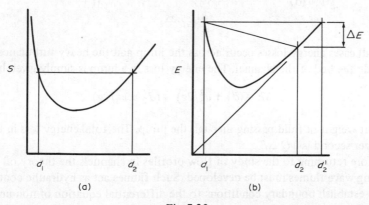

(a)   (b)

Fig. 7.23

Clearly, d_2^1 and d_2 are not the same and because of this an energy loss occurs. The fact that the line joining the two points on the specific force diagram that describe conditions upstream and downstream of a jump is nearly horizontal corresponds to the assumption that the specific force upstream equals the specific force downstream of the jump.

7.11.1 The nature of hydraulic jumps

For a hydraulic jump to occur the nature of the flow must change from supercritical to subcritical across the jump. The value of the Froude number of the approach flow decides the nature of the jump.

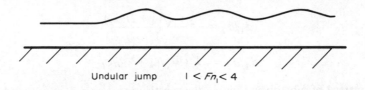

Undular jump $1 < Fn_1 < 4$

Fig. 7.24

If it lies between 1 and about 4 an undular jump occurs (see Fig. 7.24). Above 4 and below 100 the jump is weak and above 100 a roller forms on the front face of the jump (see Figs. 7.25a and b).

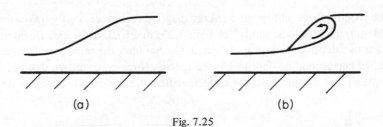

(a)   (b)

Fig. 7.25

In all cases energy losses occur across the jump and the heavy turbulence tends to erode the bed of the channel. The energy loss in a jump is simply given by

$$\Delta E = (d_1 + v_1^2/2g) - (d_2 + v_2^2/2g)$$

per unit weight of fluid passing through the jump. The total energy loss in the jump per second is $wQ\, \Delta E$.

Before returning to the study of flow profiles in channels the theory of standing wave flumes must be developed. Such flumes act as hydraulic controls and so establish boundary conditions to the differential equation of non-uniform steady flow.

7.12 The venturi flume

A venturi flume is similar to a venturimeter in that it is a flow metering device which operates by constricting the flow area. Most venturi flumes are built in channels of rectangular cross section and so are simply constrictions of the breadth of the channel. Sometimes a bed hump of streamlined cross section is fitted into such a constriction and arranged so that its highest point is located

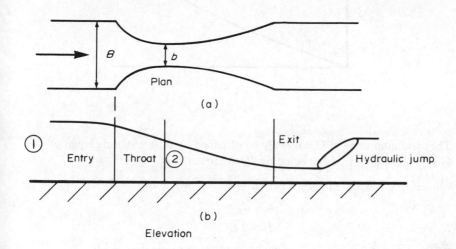

Plan

(a)

Entry   Throat ②   Exit   Hydraulic jump

(b)

Elevation

Fig. 7.26

at the throat of the constriction. The plan of a typical venturi flume is illustrated in Fig. 7.26a. Upstream of a venturi flume the constriction causes the depth to increase above the value it would have if the flume were not present. The surface within the flume itself falls as illustrated in Fig. 7.26b and reverts back to the downstream level via an hydraulic jump. The drop in the surface from the upstream entry of the flume to the throat is called a standing wave and all flumes causing such a phenomenon are called standing wave flumes. The hydraulic jump may occur inside or outside the flume depending upon the value of the downstream depth.

Applying the specific energy equation

$$E = d + v^2/2g$$

$$\therefore \qquad v_t = \sqrt{[2g(E - d)]}$$

$$\therefore \qquad Q = bd\sqrt{[2g(E - d)]} \qquad (7.4)$$

If Q is plotted against d a graph such as that illustrated in Fig. 7.27 is obtained.

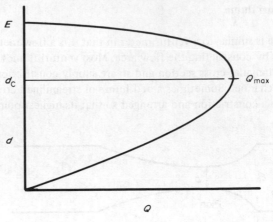

Fig. 7.27

The maximum value of Q can easily be obtained by differentiating equation (7.4) with respect to d and equating dQ/dd to zero.

$$dQ/dd = 0 = b\sqrt{(2g)}[(E - d)^{1/2} - \tfrac{1}{2}(E - d_c)^{-1/2}d_c]$$

$$\therefore \qquad E - d_c = d_c/2$$

$$\therefore \qquad d_c = 2E/3$$

So the value of d that produces maximum flow is the critical depth.

The maximum discharge is obtained by substituting this value of d into equation (7.4)

$$\therefore \qquad Q = b\tfrac{2}{3}E\sqrt{[2g(E - \tfrac{2}{3}E)]}$$

$$\therefore \qquad Q = 3{\cdot}09bE^{3/2} \quad \text{(in fps units)}$$

$$Q = 1{\cdot}705\,bE^{3/2} \quad \text{(in SI units)}$$

Introducing a coefficient of discharge

$$\left. \begin{aligned} Q &= 3{\cdot}09C_d bE^{3/2} \quad \text{(fps units)} \\ Q &= 1{\cdot}705C_d bE^{3/2} \quad \text{(SI units)} \end{aligned} \right\} \qquad (7.5)$$

Thus, if the flow profile passes through critical depth equations (7.5) are applicable.

The venturi flume is a device for which the specific energy is constant (if trivial frictional losses are ignored) but the breadth is variable. At the upstream part of the flume the breadth is B and it decreases gradually to b_t at the throat. At this point the specific energy curve defined by the equation $E = d + Q^2(2gb^2d^2)$ lies above curves applicable to larger b values as illustrated in Figs. 7.28a and b. The line of constant specific energy appropriate to the flow is a horizontal line starting at point 1, moving across the diagram from right to left and passing

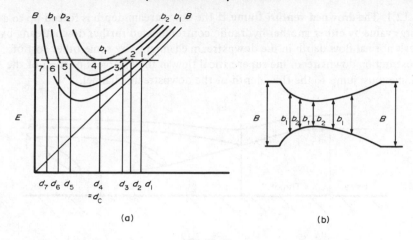

(a)   (b)

Fig. 7.28

through points 2 and 3 to 4 at the throat at which the depth becomes critical.
The breadth increases gradually through the divergent section, taking values b_2,
b_1 and finally B again, and correspondingly the specific energy line moves through
points 5, 6 and finally to 7. The depth has thus fallen progressively along the
flume from d_1 at entry to d_4 (=d_c) at the throat to d_7 at the exit. It should be
clear that the specific energy curve can never rise above the specific energy line
however small the throat breadth but, at the smallest flume breadth, the specific
energy curve must adjust itself so as to just touch the specific energy line, thus
giving critical depth at the smallest breadth. In fact the flow profile adjusts
itself by backing up in the upstream reach to a depth for which the corresponding
specific energy takes such a value that critical depth occurs at the narrowest
section.

Thus, as critical depth occurs at the throat, the critical depth equation applies:

$$Q = 3.09 C_d b E^{3/2} \quad \text{ft}^3/\text{s}$$

$$Q = 1.705 C_d b E^{3/2} \quad \text{m}^3/\text{s}$$

Now $E = d_u + v_u^2/2g$ where the subscript u denotes the upstream conditions.
In the upstream reach v_u is small so for a first approximation $E_{u1} = d_u$.
Therefore a first approximation for Q is $Q_1 = 1.705 C_d b d_u^{3/2}$ m^3/s. A first
approximation for v_u is $v_{u1} = Q_1/Bd_u$. A second approximation for E is
$E_{u2} = d_u + v_{u1}^2/2g$. Hence a second approximation for Q is
$Q_2 = 1.705 C_d b E_{u2}^{3/2}$ m^3/s. v_{u2} can now be calculated and a third approximation
for E_u can be obtained. This process should be repeated until two successive
values of Q are insignificantly different. This process rarely needs to be repeated
more than three times.

7.12.1 The drowned venturi flume

7.12.1 The drowned venturi flume If the downstream depth is forced up to a large value by either another hydraulic control located further downstream, by a high normal flow depth in the downstream channel or by some other sort of obstruction downstream the supercritical flow in the divergent portion of the flume must jump to the flow depth of the downstream channel.

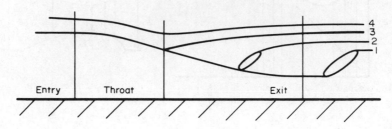

Fig. 7.29

As the flow depth downstream is increased this jump is forced upstream into the flume itself and finally up to the throat. It cannot pass the throat if the depth there is critical because the flow velocity there is equal to the wave velocity so any depth changes that occur downstream cannot be transmitted to the upstream reach as long as the depth at the throat remains critical. Once the downstream depth is increased to the point where the depth at the throat is forced above the critical value the depth upstream is compelled to increase also. The process described above can be followed in Fig. 7.29 by examining profiles 1, 2, 3 and 4. When the depth at the throat is greater than critical depth the flume is said to be drowned; the flume formula given before does not then apply.

If depths at the entry and the throat are considered, that is d_e and d_t then writing the specific energy equation

$$d_e + v_e^2/2g = d_t + v_t^2/2g$$

$$Bd_e v_e = b_t d_t v_t$$

$$\therefore \qquad v_e = \frac{b_t}{B}\frac{d_t}{d_e} v_t$$

$$\therefore \qquad d_e - d_t = \left[1 - \left(\frac{b_t}{B}\frac{d_t}{d_e}\right)^2\right]\frac{v_t^2}{2g}$$

$$\therefore \qquad v_t = \sqrt{\left\{\frac{2g(d_e - d_t)}{1 - (b_t d_t)^2/(Bd_e)^2}\right\}}$$

$$\therefore \qquad Q = C_d b_t d_t \sqrt{\left\{\frac{2g(d_e - d_t)}{1 - (b_t/B)^2(d_t/d_e)^2}\right\}}$$

This result is far more awkward to use than the free-discharge result as two depths must be measured which then require much more manipulation before the value of Q is obtained. For this reason it is usual to arrange the flume and the downstream channel in such a way that free discharge is obtained in the flume. At low flows it is probable that the flume will be drowned and if this is to be avoided a bed hump should be fitted but this has the disadvantage that bed humps cause the upstream depth to increase.

The ratio of the downstream depth to the upstream depth at which the depth at the throat is on the point of increasing is called the modular limit of the flume. It is usually possible to design the divergent portion of the flume to give a modular limit of at least 0·8. Values close to unity can be obtained by careful design.

7.13 Broad crested weirs and bed humps

The broad crested weir consists of an obstruction built across the flow which causes the upstream level to rise until flow over the weir occurs (Figs. 7.30 and 7.31). As the profile along the weir falls, it passes through the critical depth. The critical depth (in a rectangular channel) is two-thirds of the specific energy at the point in the channel at which the critical depth occurs.

$\therefore$
$$d_c = 2E/3 = (Q^2/b^2 g)^{1/3}$$

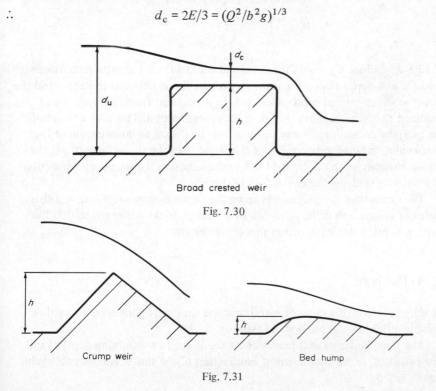

Broad crested weir

Fig. 7.30

Crump weir   Bed hump

Fig. 7.31

But as the critical depth occurs at the crest of the weir the specific energy there E is $E_u - h$ where E_u denotes the specific energy in the upstream channel. This is obtained as follows. From Bernoulli's equation

$$E_u = d_w + v_w^2/2g + h$$

Where d_w and v_w are the depth and velocity of flow at a point on the weir. As

$$d_w + v_w^2/2g = E$$

$$E_u = E + h$$

$\therefore$
$$E = E_u - h$$

$\therefore$
$$\tfrac{2}{3}(E_u - h) = (Q^2/b^2 g)^{1/3}$$

$\therefore$
$$Q = b\sqrt{g}\,\tfrac{2}{3}(E_u - h)^{3/2}$$

$$Q = 3\cdot09 b(E_u - h)^{3/2} \text{ ft}^3/\text{s}$$

$$Q = 1\cdot705 b(E_u - h)^{3/2} \text{ m}^3/\text{s}$$

Introducing a coefficient of discharge to allow for energy losses ignored in the analysis

$$Q = 3\cdot09 C_d b(E_u - h)^{3/2} \text{ ft}^3/\text{s}$$

$$Q = 1\cdot705 C_d b(E_u - h)^{3/2} \text{ m}^3/\text{s}$$

7.13.1 C_d values C_d values for weirs and venturi flumes usually lie between $0\cdot9$ and $1\cdot0$ and depend upon the geometry of the flume between the entry and the point at which critical depth is located. In the venturi flume analysis it was assumed that critical depth occurs at the throat section. This may not actually be true; the critical depth may occur a little upstream or downstream of the throat and the appropriate value of b is larger than the throat breadth. If the throat breadth is the value used for b in the equation the necessary correction will then be made by the coefficient of discharge.

The critical depth point moves along the flume as flow increases. Another point of design is thus the provision of an adequate throat length so that the critical depth point stays within this throat length.

7.14 The sluice

A sluice consists of a sheet of metal or wood that can be lowered into a flow leaving a flow area at the bottom (see Fig. 7.32).

The flow contracts as it passes under the sluice to a minimum depth d and the ratio d/d_s is the coefficient of contraction of the sluice. A reasonable value for C_c is $0\cdot6$.

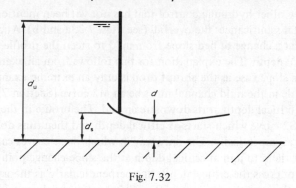

Fig. 7.32

Applying the specific energy equation

$$E = d_u + v_u^2/2g = d + v^2/2g$$

∴

$$v = C_v\sqrt{[2g(E - d)]}$$

∴

$$Q = C_v b C_c d_s \sqrt{[2g(E - C_c d_s)]}$$

$$Q = C_d b d_s \sqrt{[2g(E - C_c d_s)]}$$

Values of C_d lie between 0·45 and 0·6.

The various flow metering devices described are called hydraulic controls because they create a situation just upstream of themselves in which the flow depth is uniquely related to flow. In other words, they provide boundary conditions that can be applied to the differential equation $dd/dL = (i - j)/(1 - Fn)$ and so enable it to be integrated by finite difference methods as described earlier.

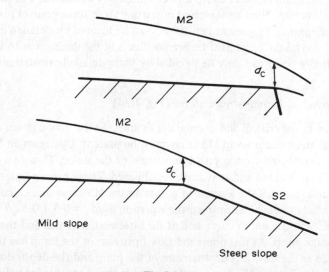

Fig. 7.33

There is one other hydraulic control that has not yet been mentioned, but which is of vital significance: the overfall (see Figs. 7.33a and b). At an abrupt bed drop and at a change of bed slope from mild to steep the profile must pass through critical depth. The explanation for this follows from an examination of the mild–steep slope case as the abrupt drop is only an extreme example of this. The flow profile in the mild channel must be an M2 curve (Section 7.10) which runs down to critical depth at its downstream end. The profile in the steep channel is an S2 curve which starts at critical depth and then runs down to the normal depth asymptotically. The only way in which these curves can be fitted together is for them to join at critical depth at the slope-change point. The profile does not cross the critical depth line perpendicularly as theory predicts, because when flow depths are near to the critical value curvilinear flow involves pressure distributions that are not hydrostatic. At the end of a channel, because of this effect, critical depth is usually found just a little upstream of the bed drop.

7.15 The prediction of flow profiles in channels

In Section 7.10 the various flow profiles were described, that is the M1, M2 and M3; S1, S2 and S3; and the C1 and C3 curves. In a real channel the breadth and bed slopes of different reaches may vary and although these profile elements are applicable providing that the breadth is constant in any particular reach, the fitting together of these profile elements to give the profile of the whole river or channel is not a simple process.

The first thing that must be done is to establish normal and critical depths throughout the channel. Next control points must be established. The possible profile elements must then be sketched in starting from these control points. Hydraulic jumps may be present and these must be located by detailed calculation. Often, several different profiles are possible and the decision as to which of these actually occur can only be decided by these detailed calculations.

A short channel with an upstream sluice (Fig. 7.34)

We first draw in the critical and normal depth lines. At the downstream end, critical depth must occur so an M2 curve must be present. Upstream an M3 curve starts at the contracted section just downstream of the sluice. Thus two control points have been located and two curves established. These cannot link except via an hydraulic jump. Take a series of points on the M3 curve and calculate the depths conjugate to each depth from the equation $d_2/d_1 = 0.5 + 0.5\sqrt{(1 + 8Fn)}$. This gives a conjugate depth curve and at the intersection of this and the M2 curve is the hydraulic jump. At this point the flow upstream of the jump has the same specific force as the flow just downstream of the jump and the depth downstream of the jump also fits the M2 curve at this point. If the channel is long the M2 curve may effectively attain the normal depth.

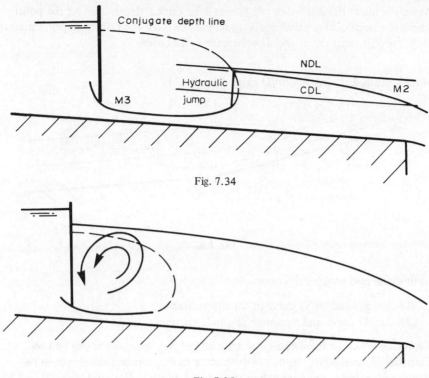

Fig. 7.34

Fig. 7.35

It may happen that the M2 curve, where it overlies the M3 curve, is everywhere higher than the conjugate depth curve, and in this case the sluice will run drowned as shown in Fig. 7.35. There will be a strong eddy just downstream of the sluice in this case. The drowning of the sluice will cause the level upstream of it to rise.

Slope change from mild to steep slope (Fig. 7.36)

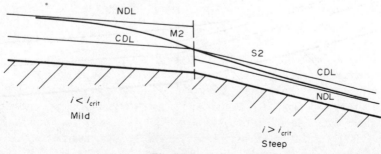

Fig. 7.36

As before insert the normal depth line and the critical depth line. At the point where the slope changes the profile must pass through the critical depth; upstream an M2 profile must occur and downstream an S2 curve.

Slope change from steep to mild slope (Fig. 7.37)

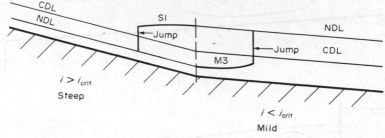

Fig. 7.37

In this case two possibilities exist.

    (1) A jump and an S1 curve in the steep channel.
    (2) An M3 curve and a jump in the mild channel.

To decide which of these cases will arise one of two approaches can be used. Calculate the conjugate depth corresponding to the normal flow depth in the steep section and if this is less than the normal depth in the mild channel an S1 curve must occur, and if it is greater than the the normal depth of the mild channel the M3 curve must occur. Alternatively, calculate the depth conjugate to the normal flow depth in the mild channel and if it is less than the normal flow depth in the steep channel the M3 curve cannot occur so the S1 must. If it is greater than the normal depth of the steep channel the M3 curve occurs.

Slope change from a large mild slope to a less mild slope

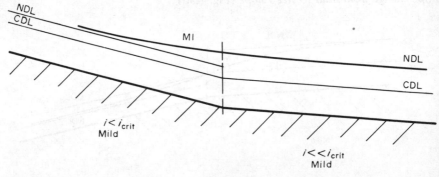

Fig. 7.38

Slope change from a small mild slope to a large mild slope

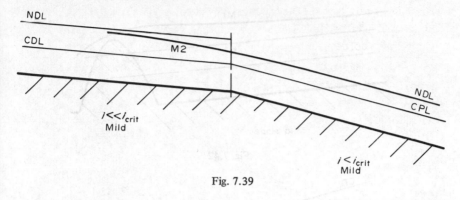

Fig. 7.39

Slope change from a large steep slope to a less steep slope

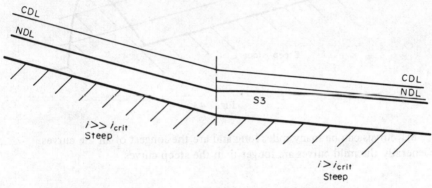

Fig. 7.40

Slope change from a small steep slope to a large steep slope

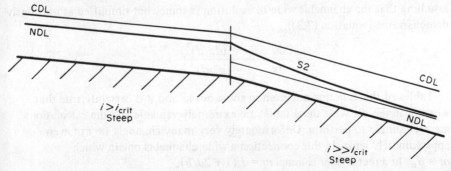

Fig. 7.41

The backwater curve

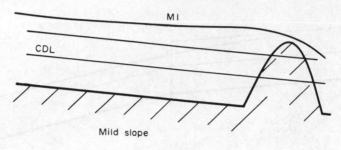

Mild slope

Fig. 7.42

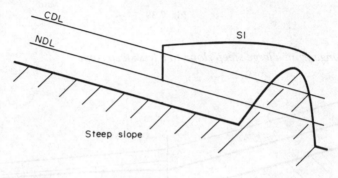

Steep slope

Fig. 7.43

M1 curves can be many miles long and are the longest of all the curves. Generally the mild curves are longer than the steep curves.

7.16 Method of integrating the gradually varied flow differential equation

Before the days of electronic computers it was desirable to find methods of simplifying the finite difference integration of this differential equation. By assuming that the channel is wide the solution is somewhat simplified as previously demonstrated (equation (7.3)).

$$\frac{\mathrm{d}d}{\mathrm{d}x} = \frac{i(1 - (d_{\mathrm{n}}/d)^3)}{1 - (d_{\mathrm{c}}/d)^3}$$

Tables of this function are given in some books and it is certainly true that when channels are wide these tables are extremely valuable and the calculations are very simple to perform. Unfortunately very many channels are not even approximately wide. In this connection a wide channel is one in which $m \approx d_{\mathrm{n}}$. In a rectangular channel $m = d/(1 + 2d/b)$.

Thus for $d/b = 1/10$; $m = 0.83d$ and for $d/b = 1/100$; $m = 0.98d$.

Thus a channel in which d is less than one hundredth of the width is effectively wide within two percent upon this basis of definition.

Many channels have varying widths and slopes in different reaches so any approach used must allow for this. Consider a short length of the channel over which it is reasonable to assume that the channel profile is approximately a straight line (see Fig. 7.44). d_m denotes the mean depth over the length of the

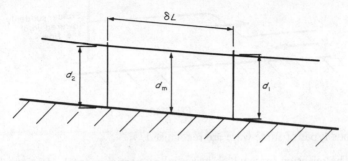

Fig. 7.44

element. d_1 can either be assumed to have been evaluated by a previous step of computation or to be located at an hydraulic control (such as a weir) which fixes its value by other considerations.

Choose a value of δd—the smaller the value chosen the more accurate the result but the more onerous the calculation will be. Then

$$d_2 = d_1 - \delta d \quad \text{so} \quad d_m = d_1 - \delta d/2$$

$v_m = Q/(bd_m)$ if the channel is rectangular.

$$m_m = bd_m/(b + 2d_m) \quad \text{so } i_m = v_m^2/(C^2 m_m)$$

$Fn_m = v_m^2/(gd_m)$. Hence the value of dd/dx can be evaluated for a short length of channel. But $dd/dx = \delta d/\delta L$

$\therefore$
$$\delta L = \left(\frac{1 - Fn_m}{i - j_m}\right) \delta d$$

The point at which the depth is $d_1 - \delta d$ has been located so a new point on the channel profile has been established. This process can be repeated until the entire curve has been developed. As the depth alters the value of C should be appropriately adjusted. Such adjustments can be made using this method for performing the finite difference integration. The method is well suited to programming on to a computer.

7.17 Surge waves in channels

An hydraulic jump is a surge wave that has moved to a point in a channel at which its celerity is exactly balanced by the flow velocity so causing it to become stationary. If a free surface flow is suddenly totally or partially arrested a surge wave travels upstream (see Fig. 7.45).

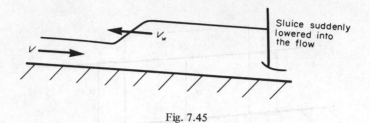

Fig. 7.45

Various types of surge wave are recognised.

(1) The rejection surge which is caused by partial or total rejection of the downstream flow, as illustrated in Fig. 7.45.

(2) The demand surge. This is caused by a sudden increase in flow at the downstream end shown in Fig. 7.46.

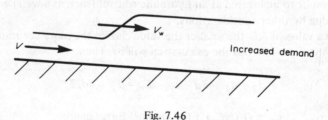

Fig. 7.46

(3) The flood surge which is caused by a sudden increase in supply upstream (such as a storm in a river catchment area or a dam burst)—see Fig. 7.47.

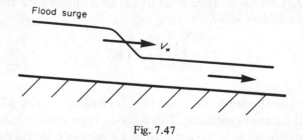

Fig. 7.47

(4) The ebb surge caused by a decrease of flow upstream (see Fig. 7.48).

The bore or eagre in the River Severn is a type of rejection surge (see Fig. 7.49).

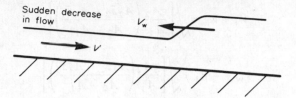

Fig. 7.48

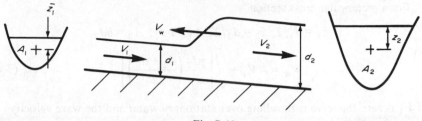

Fig. 7.49

The absolute velocity of the wave is V_w (this is not the wave celerity). By superimposing a velocity V_w in the left to right direction upon the entire flow there will be no alteration in the dynamics of the problem but the wave will be brought to rest and the flow boundaries will then be fixed in space (Fig. 7.50).

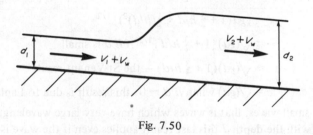

Fig. 7.50

Two equations can now be written; the continuity and the specific force equations.

$$A_1(V_1 + V_w) = A_2(V_2 + V_w)$$

$\bar{z}_1$ and $\bar{z}_2$ are the depth of the centroids of the cross sections below the surface.

$$wA_1\bar{z}_1 + wA_1(V_1 + V_w)^2/g = wA_2\bar{z}_2 + wA_2(V_2 + V_w)^2/g$$

$\therefore$
$$A_1(V_1 + V_w)^2/g - A_2(V_2 + V_w)^2/g = A_2\bar{z}_2 - A_1\bar{z}_1$$

$$A_1(V_1 + V_w)^2/g - A_1(V_1 + V_w)A_1(V_1 + V_w)/A_2g = A_2\bar{z}_2 - A_1\bar{z}_1$$

$\therefore$
$$A_1(1 - A_1/A_2)(V_1 + V_w)^2/g = A_2\bar{z}_2 - A_1\bar{z}_1$$

$$\therefore \qquad V_1 + V_w = \sqrt{\left[\frac{gA_2}{A_1}\left(\frac{A_2\bar{z}_2 - A_1\bar{z}_1}{A_2 - A_1}\right)\right]}$$

$$\therefore \qquad V_w = -V_1 + \sqrt{\left[\frac{gA_2}{A_1}\left(\frac{A_2\bar{z}_2 - A_1\bar{z}_1}{A_2 - A_1}\right)\right]}$$

This generalised result cannot be taken further as it stands unless the cross sectional shape is known.

For a rectangular cross section

$$\bar{z}_1 = d_1/2; \ \bar{z}_2 = d_2/2; \ A_1 = bd_1; \ A_2 = bd_2$$

so

$$V_w = -V_1 + \sqrt{\left[\frac{gd_2}{d_1}\left(\frac{d_1 + d_2}{2}\right)\right]} \qquad (7.6)$$

If V_1 is zero the wave is travelling over stationary water and the wave velocity V_w is then the celerity c so

$$c = \sqrt{\left[\frac{gd_2}{d_1}\left(\frac{d_1 + d_2}{2}\right)\right]}$$

$$\therefore \qquad c = \sqrt{\{gd_2[\tfrac{1}{2}(1 + d_2/d_1)]\}}$$

Now $d_2 = d_1 + h$ where h is the height of the wave. Dropping the subscript of d_1

$$c = \{gd(1 + \tfrac{3}{2} h/d + \tfrac{1}{2}(h/d)^2)\}^{1/2}$$

$$= \sqrt{(gd)}\{1 + \tfrac{3}{2} h/d\}^{1/2} \text{ if } h/d \text{ is small}$$

$$\approx \sqrt{(gd)}(1 + \tfrac{3}{4} h/d) - \text{the St Venant result}$$

$$c = \sqrt{(gd)} \text{ when } h/d \rightarrow 0 - \text{this result is due to Laplace.}$$

For very small waves, that is waves which have very large wavelengths in comparison with the depth d this last result applies even if the wave is of a translatory type.

Surge waves, however generated, can be solved using equation (7.6) for the wave velocity V_w

$$V_w = -V_1 + \sqrt{\left[\frac{gd_2}{d_1}\left(\frac{d_1 + d_2}{2}\right)\right]}$$

and the continuity equation

$$d_1(V_1 + V_w) = d_2(V_2 + V_w)$$

When solving equations such as these the value of d_2 must be available. Generally, the condition that initiates the surge will determine the value of d_2 but as the wave progresses d_2 changes as illustrated in Fig. 7.51. Usually the surge diminishes in height as it progresses upstream.

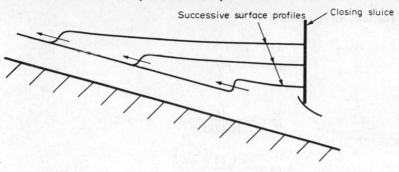

Fig. 7.51

Flow downstream of the jump can be analysed by normal methods for handling backwater curves. Finite difference methods must be used. Accurate solution of this problem is only possible on a fast computer and unsteady flow theory must be used.

Worked examples

(1) An open channel conveying water is of trapezoidal cross section. The base width is 1·5 m and the side slopes are at 60° to the horizontal. The channel bed slope is 1 in 400 and the depth is constant at 1 metre. Calculate the discharge in cubic metres per second if the Chezy C is calculated from the Bazin relationship $C = 87/(1 + 0·2/\sqrt{m})$ where m is the hydraulic mean depth.

$$m = A/P$$

$$A = 1 \times (3·0 + 2 \times \tan 30°)/2 = 2·077$$

$$P = 1·5 + 2 \times \sec 30° \times 1 = 3·809 \checkmark$$

$$m = 0·545 \text{ m}$$

$$C = 68·46$$

$$Q = CA\sqrt{(mi)}$$

$$= 68·46 \times 2·077\sqrt{0·545 \times \tfrac{1}{20}}$$

$$Q = 5·25 \text{ m}^3/\text{s}$$

(2) Water is flowing down a prismoidal channel of constant base width (15 m) and constant longitudinal slope (0·0005). The channel cross section is trapezoidal and the side slopes are 1:1. The Chezy C is 55. At a certain point in the channel the depth is decreasing at a rate 1 in 10 000 and at this point the depth is 5 m. Estimate the flow rate in the channel.

The relevant equation is:

$$\frac{\mathrm{d}d}{\mathrm{d}L} = \frac{i - j}{1 - Fn}$$

Now

$$\frac{\mathrm{d}d}{\mathrm{d}L} = - \frac{1}{10\,000}$$

$$\text{Area} = \frac{15 + 15 + 2 \times 5}{2} \times 5$$

$$= 100$$

$$P = 15 + 2 \times \sqrt{2} \times 5$$

$$= 29 \cdot 14$$

$$m = \frac{100}{29 \cdot 14}$$

$$= 3 \cdot 43$$

$$j = \frac{Q^2}{A^2 C^2 m} = \frac{Q^2}{A^2 \times 55^2 \times 3 \cdot 43}$$

$$Fn = \frac{v^2}{g\delta} \quad \text{and} \quad \delta = \frac{A}{15 + 2 \times 5} = 2 \cdot 5$$

$$\therefore \qquad Fn = 4 \cdot 077 \times 10^{-6} \times Q^2$$

$$\therefore \qquad -10^{-4} = \frac{5 \times 10^{-4} - Q^2 \times 9 \cdot 638 \times 10^{-9}}{1 - 4 \cdot 077 \times 10^{-6}}$$

$$\therefore \qquad Q = 244 \cdot 4 \ \mathrm{m^3/s}$$

(3) A venturi flume with a horizontal bed is fitted into a channel of 22·5 cm width. The throat width is 15 cm and the depth in the upstream section is 20 cm. Calculate the flow in the channel using a method of successive approximations to allow for the velocity of approach.

The relevant formula for a freely discharging venturi flume is

$$Q = 1 \cdot 705 b_t E^{3/2}$$

First approximation

Assume

$$E = d_1$$

$$Q_1 = 1 \cdot 705 \times 0 \cdot 15 \times 0 \cdot 20^{3/2}$$

$$= 0 \cdot 022875 \text{ m}^3/\text{s}$$

Upstream velocity

$$v_1 = \frac{0 \cdot 022875}{0 \cdot 225 \times 0 \cdot 2}$$

$$= 0 \cdot 508$$

$$E_1 = 0 \cdot 2 + \frac{0 \cdot 508^2}{19 \cdot 62}$$

$$= 0 \cdot 21317$$

Second approximation

$$E_1 = 0 \cdot 21317$$

$$Q_2 = 1 \cdot 705 \times 0 \cdot 15 \times 0 \cdot 21317^{3/2}$$

$$= 0 \cdot 02517 \text{ m}^3/\text{s}$$

$$v_2 = \frac{0 \cdot 02517}{0 \cdot 225 \times 0 \cdot 2}$$

$$= 0 \cdot 55936$$

$$E_2 = 0 \cdot 2 + \frac{0 \cdot 55936^2}{19 \cdot 62}$$

$$= 0 \cdot 21595$$

Third approximation

$$Q_3 = 1 \cdot 705 \times 0 \cdot 15 \times 0 \cdot 21595^{3/2}$$

$$= 0 \cdot 025665 \text{ m}^3/\text{s}$$

$$v_2 = \frac{0 \cdot 025665}{0 \cdot 225 \times 0 \cdot 2}$$

$$= 0 \cdot 57033$$

$$E_3 = 0 \cdot 2 + \frac{0 \cdot 57033^2}{19 \cdot 62}$$

$$= 0 \cdot 21658$$

$$Q_4 = 1 \cdot 705 \times 0 \cdot 15 \times 0 \cdot 21658^{3/2}$$

$$= 0 \cdot 025777 \text{ m}^3/\text{s}$$

This last value is insignificantly different from the previously calculated value so it can be accepted.

(4) An open channel having a trapezoidal cross section with side slopes of 1 vertical to 2 horizontal is required to discharge 15 m^3/s of water when running full. The longitudinal slope is 1 in 2000. Determine the depth d and the base width b to give minimum cross sectional area. $C = 70\ m^{1/6}$.

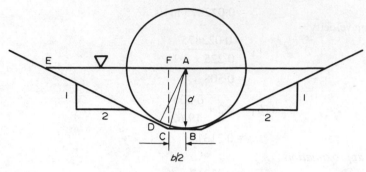

Fig. 7.52

For an economic section, a semicircle must be inscribable in the channel cross section. Under these circumstances, the hydraulic mean radius equals half the depth, d. The area $A = (2b + 4d)d/2$; $m = d/2$.

$$Q = AC\sqrt{(mi)} = A \times 70\ m^{2/3}i^{1/2}$$

$$15/(70i^{1/2}) = (b + 2d)d \times (d/2)^{2/3}$$

$$\triangle\ AED \equiv \triangle\ CFE$$

so

$$AE = CE$$

$$b/2 + 2d = \sqrt{[d^2 + (2d)^2]}$$

$$b = 2(\sqrt{5}d - 2d) = 0 \cdot 472\ d$$

$$9 \cdot 583 = 2 \cdot 472\ d^2(d/2)^{2/3}$$

$$d = 1 \cdot 977\ m$$

$$b = 0 \cdot 472 \times 1 \cdot 977 = 0 \cdot 933\ m$$

(5) The cross section of an irrigation canal is in the form of an equilateral triangle with its vertex downwards and equal side slopes. The flow depth just upstream of an hydraulic jump is 0·5 m when the flow is 1·0 m^3/s. Calculate the depth just downstream of the jump.

$$\text{Area of flow} = 2 \times (0 \cdot 5 \tan 30°) \times 0 \cdot 5/2$$

$$= 0 \cdot 1443\ m^2$$

$$\text{Velocity} = 0 \cdot 1/0 \cdot 1443 = 6 \cdot 928\ m/s$$

Relevant equation $\dfrac{v^2}{g} = \dfrac{gA_2}{A_1} \dfrac{A_1 z_1 - A_2 z_2}{A_1 - A_2}$

$$A_2 = 2 \times d_2 \tan 30° \times d_2/2 = 0.5774\, d_2^2$$

$$z_2 = d_2/3$$

$$z_1 = d_1/3$$

$\therefore \qquad v = \sqrt{\left[9.81\, \dfrac{d_2^2}{d_1} \left(\dfrac{\frac{1}{3} \times 0.5774 d_1^3 - \frac{1}{3} \times 0.5774 d_2^3}{0.5774 d_1^2 - 0.5774 d_2^2} \right) \right]}$

$\therefore \qquad v = 6.928 = \sqrt{\left[3.27 \left(\dfrac{d_2}{0.5} \right)^2 \left(\dfrac{0.125 - d_2^3}{0.25 - d_2^2} \right) \right]}$

This equation is most easily solved by successive approximations:

Try $d_2 = 2$	R.H.S. = 10·482—too high
Try $d_2 = 1$	R.H.S. = 3·906—too low
Try $d_2 = 1·5$	R.H.S. = 5·92—to low
Try $d_2 = 1·505$	R.H.S. = 6·949—too high
Try $d_2 = 1·502$	R.H.S. = 6·9287—adequate agreement

Depth downstream of hydraulic jump = 1·502 m

(6) A channel of rectangular cross section with a horizontal bed has a sluice gate. A flow of 3·6 m³/s per unit width is passing through the sluice with the gate so adjusted that the depth just upstream of it is 2·4 m. If the gate is suddenly partially closed so that only 1·0 m³/s per unit width is passing determine the velocity and depth with which the resultant surge starts to travel back up the channel.

$$\text{Wave speed} = V_w = -V + \sqrt{\left[\frac{g d_2}{d_1} \left(\frac{d_1 + d_2}{2} \right) \right]}$$

Flow down channel—flow through sluice

$$= V_w (d_2 - d_1)$$

$$3·6 - 1 = V_w (d_2 - 2·4)$$

$$V = 3·6/2·4 = 1·5 \text{ m/s}$$

$$\frac{2·6}{d_2 - 2·4} = V_w = -1·5 + \sqrt{\left[\frac{9·81 d_2}{2·4} \left(1·2 + \frac{d_2}{2} \right) \right]}$$

Solve by successive approximations:

Try

$$d_2 = 2 \cdot 8 \text{ m}$$

$$2 \cdot 6/0 \cdot 4 = 6 \cdot 5 \text{ m/s}; \ V_w = -1 \cdot 5 + \sqrt{\left[\frac{9 \cdot 81 \times 2 \cdot 8}{2 \cdot 4} (1 \cdot 2 + 1 \cdot 4)\right]} = 3 \cdot 955 \text{ m/s}$$

Increasing d_2 should give a better second approximation. Try

$$d_2 = 3 \cdot 0 \text{ m}$$

$$\frac{2 \cdot 6}{0 \cdot 6} = 4 \cdot 333; \ V_w = -1 \cdot 5 + \sqrt{\left[\frac{9 \cdot 81 \times 3}{2 \cdot 4}(1 \cdot 2 + 1 \cdot 5)\right]} = 4 \cdot 254 \text{ m/s}$$

Increase d_2 a little more

$$d_2 = 3 \cdot 01$$

$$\frac{2 \cdot 6}{0 \cdot 610} = 4 \cdot 2623; \ V_w = -1 \cdot 5 + \sqrt{\left[\frac{9 \cdot 81 \times 3 \cdot 01}{2 \cdot 4}(1 \cdot 2 + 1 \cdot 505)\right]} = 4 \cdot 269 \text{ m/}$$

So $d_2 = 3 \cdot 01$ m is acceptable.

$$\text{Wave speed} = 4 \cdot 26 \text{ m/s}$$
$$\text{Height of wave} = 3 \cdot 01 - 2 \cdot 4 = 0 \cdot 61 \text{ m}$$

Questions

(1) A semicircular channel is required to convey $1 \cdot 5 \text{ m}^3/\text{s}$ when flowing full, the gradient being 1 in 1000. Taking the Chezy C as 60, find the diameter of the channel.

Answer: $1 \cdot 74$ m.

(2) Working from first principles, derive Chezy's formula for uniform or normal flow in open channels, and show the location of the hydraulic gradient line. An open channel of rectangular section and breadth 1 m conveys water at a rate of $0 \cdot 3 \text{ m}^3/\text{s}$. C is 55 and the slope is 2 in 1000. Estimate the depth of normal flow.

Answer: $0 \cdot 286$ m.

(3) The breadth of a rectangular channel is twice the depth. Assuming the Chezy C to be 55, find the cross sectional dimensions of the channel and the slope to satisfy the following two conditions. (a) Discharge when flowing full = $0 \cdot 8 \text{ m}^3/\text{s}$. (b) Velocity when flowing half full = $0 \cdot 6 \text{ m/s}$.

Answers: $d = 0 \cdot 74$ m, $b = 1 \cdot 48$ m, $i = 1/2070$.

(4) An open channel of circular section 0·7 m diameter laid at a slope of 1 in 3600 conveys 0·16 m³/s when running three-quarters full; that is with $\frac{3}{4}$ of the vertical diameter immersed. If the Chezy $C = km^{1/6}$ where m is the hydraulic mean depth in metres, find the numerical value of k.

Answer: $k = 87\cdot435$.

(5) Obtain the ratio of depth to breadth for an open channel of rectangular cross section which requires minimum area for a given discharge. Calculate the dimensions of the section of such a channel which will convey 10 m³/s, the slope of the bed being 1 in 1500. Take C as 60 $m^{1/6}$.

Answer: $d = 1\cdot85$ m, $b = 3\cdot7$ m.

(6) A 1 m diameter conduit, 400 m long is laid at a uniform slope of 1 in 1500 and connects two reservoirs. When the levels of the reservoirs are low the conduit runs partially full and it is found that a normal depth of 0·671 m gives a rate of flow of 0·33 m³/s. The C value is given by km^n where k is a constant, m is the hydraulic mean depth and $n = 1/6$. Neglecting losses at exit and entry find (a) the value of k and (b) the discharge when the conduit is flowing full and the difference in level between the two reservoirs is 5 m.

Answers: $k = 51\cdot84$, $1\cdot807$ m³/s.

(7) The cross section of an open channel consists of a semicircular culvert with vertical sides, the overall depth being equal to the maximum width. The channel is required to convey 0·15 m³/s of water and in order to guard against overflow the depth of the stream is to be only 90% of the maximum possible depth of the channel. The slope of the bed is 1 in 1500 and $C = 70$. Calculate the dimensions of the cross section.

Answer: radius = 0·252 m; width = 0·504 m.

(8) The straight sides of an open channel slope outwards at 15° to the vertical and are tangential to the invert which is a circular arc of 0·5 m radius. Determine the gradient necessary to give a discharge of 0·6 m³/s when the maximum depth of the water is 1 m. $C = 55$.

Answer: 1 in 4026.

(9) Prove that the slope of the surface of the water in an open channel in which the flow is non-uniform is given by $dd/dD = (i - j)/(1 - Fn)$ the symbols having their usual meaning. A rectangular channel 15 ft [5 m] wide has a gradient of 0·00081 and the normal depth of flow is 3 ft [1 m]. The depth is increased by a weir which is placed across the channel and the same flow of water is maintained. Determine the slope of the surface at a point upstream where the depth is 3·5 ft [1·05 m]. Take C as 120 ft$^{1/2}$ [66 $m^{1/2}$].

Answer: $-4\cdot806 \times 10^{-4}$ [$-6\cdot81 \times 10^{-4}$].

(10) What do you understand by normal and critical flow conditions in a channel? A wide channel of slope $i = 1$ in 4000 is equipped with a sluice, just downstream of which the depth is reduced to 0·3 ft. The flow is 10 ft^3/s per ft width of channel. Find the distance from the sluice at which the hydraulic jump occurs in the downstream channel.

Answer: 51·06 ft using one step of integration.

(13) A horizontal rectangular channel having constant width of 5 ft [1·5 m] conveys water. Over a length of 500 ft [150 m] the depth decreases from 3 ft [0·9 m] to 2 ft [0·6 m]. Assuming $j = 0·01v^2/2gm$ [$j = v^2/44^2 m$] find the discharge. Use numerical integration.

Answer: With 1 step of integration 45·78 ft^3/s [1·24 m^3/s].

(14) A venturi flume is to be installed in a channel conveying water with the object of raising the level of the water upstream. The channel is rectangular in section and is 40 ft [12 m] wide, has a gradient of 1 in 6400 and a depth of water 5 ft [1·5 m]. The width of the throat section of the flume is 20 ft [6 m]. If the bed of the flume at the throat is a streamlined hump, find the necessary height of the hump in order that the depth of water on the upstream side shall be 6 ft [1·8 m]. Take $C = 140$ [77] for the channel, ignore hydraulic losses in the flume and assume that a standing wave is formed on the downstream side of the hump.

Answer: 1·09 ft [0·3465 m].

(15) Explain what is meant by critical depth of flow in a channel. If specific energy is defined by $h + v^2/2g$, where h is the depth measured from the bottom of a channel of rectangular section, show that the critical depth for a given flow is that satisfying the condition of minimum specific energy. A jump occurs downstream of a sluice spanning a channel of rectangular section which conveys water at normal depth of 3 ft [0·9 m] and velocity 4 ft/s [1·2 m/s]. The opening below the sluice is 9 in [0·23 m]. Estimate the height of the jump and the specific energy lost.

Answers: 2·213 ft [0·667 m], 1·15 ft [0·355 m].

(16) A venturi flume is installed in an open channel conveying water. Derive the formula for the discharge through the flume when a standing wave is formed on the downstream side, and explain why the depth of water at the throat section is then said to be the critical depth. The channel, rectangular in section, has a width of 30 ft [9 m] and the gradient of the bed is 1/3600. The bed of the flume at the throat is a streamlined hump 2 ft [0·6 m] high, and the width at the throat section is 20 ft [6 m]. If the depth of water in the channel at the entrance to the flume is 7 ft [2·1 m] estimate the quantity discharged through the flume. Also, find the

depth of water at the highest point of the hump, and the depth of water in the channel far upstream. Ignore hydraulic losses in the flume and take $V = 120\sqrt{(mi)}$ $[V = 66\sqrt{(mi)}]$ for the channel.

Answers: 730·2 cusec [19·861 m³/s], 3·45 ft [1·034 m], 5·91 ft [1·777 m].

(17) What is meant by an hydraulic jump in an open channel? Considering the case of a rectangular channel of constant width, determine from first principles the conditions required for the formation of such a jump and calculate the consequent loss of head in terms of the depth d_1 just before the jump and the depth d_2 just after it.

Answer: $(d_2 - d_1)^3/(4d_1d_2)$.

(18) The cross section of an irrigation canal is in the form of an equilateral triangle with vertex downwards and equal side slopes. The stillwater head upstream is H m. Calculate the critical depth and critical velocity of flow.

Answers: 0·8 H, 0·447 $\sqrt{(2gH)}$.

(19) The section of a horizontal channel conveying water is defined by the equation $b^2 = 16d$, where b is the width of the free surface and d is the maximum depth of flow. The section is symmetrical about the vertical centreline. At a certain section the velocity of the water is 15 ft/s [4·5 m/s] and the maximum depth is 2·5 ft [0·75/m]. Calculate the critical depth. If a hydraulic jump is formed, calculate the depth after the jump.

Answers: 3·58 ft [1·068 m], 4·92 ft [1·5 m].

(20) A river of depth H_1 is flowing with a velocity v towards the sea. Tide conditions cause a sudden increase in depth downstream to H_2 and this depth is maintained constant by the tide. Show that a bore will travel upstream with velocity V where

$$V = -v + \sqrt{\left[\frac{gH_2}{H_1}\left(\frac{H_1 + H_2}{2}\right)\right]}$$

(21) Explain what is meant by a hydraulic bore in an open channel. Describe the conditions of flow necessary for a bore to travel upstream, and show by a mathematical analysis for a rectangular cross-section channel, that a hydraulic jump may be considered as a bore with zero velocity. Discuss the use of a jump as an energy dissipator.

(22) A stream flows at its normal depth of 4 ft [1·2 m] through a rectangular channel 20 ft [6 m] wide. The bed slope is 1 in 2000 and the Chezy C may be taken as 100 [55]. A landslip into the stream causes a sudden increase of depth to 5 ft [1·5 m]. Show that this will cause a surge to travel upstream and, working from first principles, calculate how long it will take for the surge to reach a

village two miles [3 km] upstream from the landslip. Neglect storage effects and assume that the depth at the landslip does not increase with time. Show how your basic equations would be modified if the cross section of the stream were not rectangular. Discuss the error in the method caused by making the assumption mentioned above.

Answer: 18·2 minutes [17·06 minutes].

8 Pressure Transients

The development of large transient pressures can occur in pipelines whenever the velocity changes rapidly. Such transient pressures can cause major failure of the pipe and the consequent expense involved in repair can be large. It is probably true to say that the generation of pressure transients is the largest single hydraulic cause of pipeline failure. It is strange, therefore, to find that a large percentage of users are not aware of the danger to their pipelines caused by rapid velocity fluctuations, nor do they seem particularly interested in studying the considerable volume of literature that now exists on this subject.

It may be of interest to note at this point that a velocity change of 1 ft/s can generate a pressure head of approximately 125 ft if it occurs rapidly enough; if the pipeline is long, the required rate of change is not particularly large. Some catastrophic failures have occurred and it is now clear that the cost of a pipeline which will not fail under any circumstances is too high to be contemplated. Instead, a compromise must be sought which will reduce the risk of failure to an acceptable level and yet not be excessively expensive. The question remains of how to establish the amount of risk to which a pipeline may be exposed, and the theory that follows is the basis upon which it is assessed.

8.1 Rigid pipe theory of waterhammer

Pressure transients—or waterhammer, as the phenomenon is sometimes called—can be analysed by rigid pipe theory or elastic pipe theory. The method to be used depends upon circumstances and these will be specified later.

If a velocity fluctuation occurs over a relatively long period of time, and the accelerations of the fluid are small, the pressure generated will also be small. If the pipeline is of normal construction, these small pressure changes will cause very little deformation of the pipe, and it will be reasonable to treat it as a rigid pipe; furthermore, if the velocity of pressure wave propagation in the system is large, the time taken for a velocity fluctuation at one end of the pipe to be signalled to the rest of the fluid in the pipe will be very small in relation to the overall time of the velocity fluctuation. It will then be reasonable to analyse the motion of the fluid as if it were a solid body occupying the internal volume of the pipeline. If this were not true; if the wave velocity were low, or if the velocity fluctuation occurred in a very short time, it would be possible for a portion of the fluid in the downstream section of the pipeline to be at rest because of a valve closure at the downstream end, while fluid at the upstream end was still entering the pipe. This actually happens in very rapid valve closures (see Section 8.6) and under such circumstances it is not reasonable to write an equation that describes the fluid's motion and pressure changes for all points in the pipeline at the same instant in time. Thus, only if the velocity fluctuations occur over a relatively long time will it be correct to treat the fluid as if it were moving as a solid body. The assumptions that the pipe is rigid and that the fluid *can* be treated as a solid body are essential to the development of the rigid pipe theory.

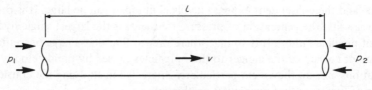

Fig. 8.1

Bearing these assumptions in mind, consider the case of decelerating flow in a pipeline. The fluid can reduce its velocity only if a decelerating force is applied to it and such a force is caused by a pressure difference between the ends of the pipe (See Fig. 8.1). In other words, p_2 must be greater than p_1 if the fluid is to be decelerated (neglecting frictional effects for the moment). The mass of fluid in the pipe is $\rho A l$ where A is the pipe cross sectional area. The force acting on the fluid from left to right is $(p_1 - p_2)A$. By Newton's second of motion

$$(p_1 - p_2)A = \rho A l \, dv/dt$$

Cancelling A, substituting $\rho = w/g$ and calling $(p_1 - p_2) \; \Delta p$ gives

$$\Delta p = +\frac{w}{g} l \frac{dv}{dt}$$

or

$$\frac{\Delta p}{w} = -h_i = +\frac{l}{g}\frac{dv}{dt} \tag{8.1}$$

where h_i is the head that must be created (perhaps by a valve closure) at the downstream end of the pipe in excess of the static head if the fluid is to be decelerated by the amount dv/dt.

Note that dv/dt is negative for a deceleration of the flow. The subscript i is used to show that this is a head caused by inertia changes, that is an inertia head.

Given this equation, a large range of problems can be solved.

8.2 Sudden valve opening at the end of a pipeline

The case of a sudden valve opening at the end of a pipeline is illustrated in Fig. 8.2. Immediately after the valve has opened, the fluid in the pipeline is at rest,

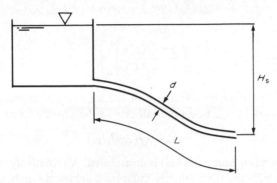

Fig. 8.2

but a net head of H_s is acting upon it and therefore it immediately experiences an acceleration. After a short time t its velocity will have increased to a value v and frictional forces will start to operate and will reduce the head available to produce acceleration by an amount $4fLv^2/(2gd)$ that is, h_f. Then the inertia head will be given by the equation

$$h_i = H_s - h_f$$

(Note that if local turbulence is occurring anywhere in the pipeline, this can be included in the friction loss term.)

$$\therefore \qquad +\frac{L}{g}\frac{dv}{dt} = H_s - \left(\frac{4fL}{d} + k\right)\frac{v^2}{2g}$$

where $kv^2/2g$ is the head loss due to local turbulence. Denote $4fL/d + k$ by K. Then

$$\frac{L}{g}\frac{dv}{dt} = \frac{2gH_s - Kv^2}{2g}$$

$$\therefore \qquad \frac{dv}{dt} = \frac{1}{2L}(2gH_s - Kv^2)$$

$$\therefore \qquad dt = \frac{2L}{2gH_s - Kv^2}\,dv \qquad\qquad (8.2)$$

This will be solved by partial fractions, so rewrite equation (8.2) as follows

$$dt = \frac{2L}{[\sqrt{(2gH_s)}]^2 - [\sqrt{K}v]^2} dv$$

$$= \frac{L}{\sqrt{(2gH_s)}} \left[\frac{dv}{\sqrt{(2gH_s)} - \sqrt{(K)}v} + \frac{dv}{\sqrt{(2gH_s)} + \sqrt{(K)}v} \right]$$

$$\therefore t = \frac{L}{\sqrt{(2gH_s)}\sqrt{(K)}} [-\log_e\{\sqrt{(2gH_s)} - \sqrt{(K)}v\} + \log_e\{\sqrt{(2gH_s)} + \sqrt{(K)}v\}]_0^v$$

$$= \frac{L}{\sqrt{(2gH_sK)}} \left[\log_e \left\{ \frac{\sqrt{(2gH_s)} + \sqrt{(K)}v}{\sqrt{(2gH_s)} - \sqrt{(K)}v} \right\}_0^v \right] \tag{8.3}$$

When time is infinite the demominator of the log expression must be zero. Therefore $\sqrt{2gH_s} = \sqrt{K}v_\infty$, v_∞ being the velocity in the pipeline at $t = \infty$.

$$\therefore \qquad V_\infty = \sqrt{(2gH_s/K)}$$

This is obvious when equation (8.3) is considered. Alternatively, when time approaches infinity the fluid velocity must have achieved steady state. Then, by the usual method of solving steady state flow in pipes,

$$h_f = H_s$$

$$\therefore \qquad Kv_\infty^2/2g = H_s \quad \therefore v_\infty = \sqrt{(2gH_s/K)}$$

Re-writing equation (8.3) in the following form

$$t = \frac{L}{\sqrt{(2gHK)}} \log_e \frac{\sqrt{(2gH_s/K)} + v}{\sqrt{(2gH_s/K)} - v}$$

and substituting v_∞ for $\sqrt{(2gH_s/K)}$

$$t = \frac{L}{\sqrt{(2gHK)}} \log_e \frac{v_\infty + v}{v_\infty - v}$$

$$\therefore \qquad \exp\left[\frac{\sqrt{(2gHK)}t}{L} \right] = \frac{v_\infty + v}{v_\infty - v}$$

Let $(\sqrt{(2gHK)}/L$ be denoted by c.

$$\therefore \qquad e^{ct} = \frac{v_\infty + v}{v_\infty - v}$$

$$\therefore \qquad v_\infty e^{ct} - v e^{ct} = v_\infty + v$$

$$\therefore \qquad v(1 + e^{ct}) = v_\infty(e^{ct} - 1)$$

$$\therefore \qquad v = v_\infty \left(\frac{e^{ct} - 1}{e^{ct} + 1} \right)$$

Since

$$c = \sqrt{[2gH_s(4fL/d + k)]}/L$$

substituting back for c gives

$$\frac{v}{v_\infty} = \frac{\exp\{\sqrt{[2gH_s(4fL/d + k)]}t/L\} - 1}{\exp\{\sqrt{[2gH_s(4fL/d + k)]}t/L\} + 1}$$

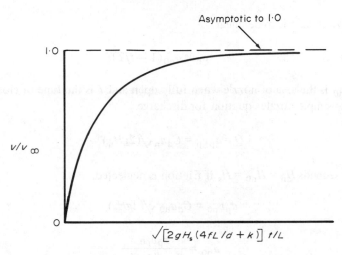

Fig. 8.3

Thus a solution for the pipeline velocity has been obtained which includes the effect of friction (see Fig. 8.3).

8.3 Slow uniform valve closure

Assume that a valve is closed in such a way that its area of opening at any instant t is being linearly decreased as t increases. A solution for the maximum pressure attained during the closure can be obtained. (The uniform closing of a spear valve closely approximates to this case.) This is illustrated in Fig. 8.4. Here a_p is the area of pipe, a_n the area of nozzle at time t seconds, H_n the head behind the nozzle at time t seconds, H_s the static head in the pipe, v_n the velocity of the jet at time t seconds and v_p the velocity in the pipe at time t seconds. In this case, as a nozzle is fitted on the end of the pipeline, the head lost due to friction in the pipe will be small and in the following analysis it will be neglected.

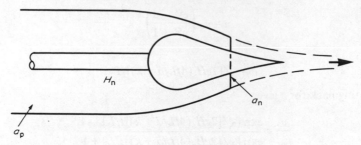

Fig. 8.4

If valve closure is to be linear

$$a_n = a_{n0}(1 - t/T) \tag{8.4}$$

where a_{n0} is the area of nozzle when fully open and T is the time of closure. Using the simple nozzle equation for discharge

$$Q = a_p v_p = C_d a_n \sqrt{(2gH_n)} \tag{8.5}$$

At $t = 0$ seconds $H_n = H_{n0} = H_s$ if friction is neglected.

$\therefore$
$$a_p v_{p0} = C_d a_{n0} \sqrt{(2gH_s)}$$

$\therefore$
$$a_{n0} = \frac{a_p v_{p0}}{C_d \sqrt{(2gH_s)}} \tag{8.6}$$

Substituting this value into equation (8.4) gives

$$a_n = \frac{a_p v_{p0}}{C_d \sqrt{(2gH_s)}} (1 - t/T)$$

then substituting this value of a_n into equation (8.5) gives

$$a_p v_p = C_d \frac{a_p v_{p0}}{C_d \sqrt{(2gH_s)}} \sqrt{(2gH_n)}(1 - t/T)$$

$\therefore$
$$v_p = v_{p0} \sqrt{(H_n/H_s)}(1 - t/T) \tag{8.7}$$

But

$$H_n - H_s = -\frac{L}{g} \frac{dv_p}{dt} \tag{8.8}$$

Differentiate equation (8.7) to obtain dv_p/dt.

$$\frac{dv_p}{dt} = \frac{v_{p0}}{\sqrt{(H_s)}} \left(\frac{1}{2} H_n^{-1/2} \left(1 - \frac{t}{T} \right) \frac{dH_n}{dt} - \frac{1}{T} H_n^{1/2} \right)$$

$\therefore$

$$\frac{dv_p}{dt} = v_{p0} \sqrt{(H_n/H_s)} \left(\frac{1 - t/T}{2H_n} \frac{dH_n}{dt} - \frac{1}{T} \right) \tag{8.9}$$

Substituting for dv_p/dt from equation (8.9) into (8.8)

$\therefore$

$$(H_s - H_n)\frac{g}{L} = v_{p0} \sqrt{(H_n/H_s)} \left(\frac{1 - t/T}{2H_n} \frac{dH_n}{dt} - \frac{1}{T} \right)$$

$\therefore$

$$\frac{dH_n}{dt} = \frac{2[(H_s - H_n)\sqrt{(H_s/H_n)}g/(v_{p0}L) + 1/T]H_n}{1 - t/T} \tag{8.10}$$

The method of integrating this equation will be discussed later. To estimate the maximum value of H_n, re-arrange equation (8.10) as follows

$$\frac{1}{2}\left(1 - \frac{t}{T} \right) \frac{dH_n}{dt} = (H_s - H_n)\sqrt{(H_sH_n)} \frac{g}{v_{p0}L} + \frac{H_n}{T} \tag{8.11}$$

when H_n is a maximum $dH_n/dt = 0$

$\therefore$

$$(H_s - H_n)\sqrt{(H_sH_n)} \frac{g}{v_{p0}L} + \frac{H_n}{T} = 0 \tag{8.12}$$

If the curve of H_n against t does not have a maximum in the normal algebraic sense the right hand side of the equation will still be zero when $t = T$ and H_n has its largest value here. If the maximum were to occur before t had reached its largest value, then dH_n/dt would be zero. In either case the right hand side of the equation (8.11) will be zero when H_n attains its largest value.

Rearranging equation (8.12) into dimensionless form

$$H_s \left(1 - \frac{H_n}{H_s} \right) \sqrt{(H_n/H_s)} \frac{gT}{v_pL} + \frac{H_n}{H_s} = 0$$

Then

$$\sqrt{(H_n/H_s)} = - \frac{v_pL/(gTH_s)}{1 - H_n/H_s} \frac{H_n}{H_s}$$

$\therefore$

$$\frac{H_n}{H_s} = \frac{[v_pL/(gTH_s)]^2[H_n/H_s]^2}{(1 - H_n/H_s)^2}$$

Let

$$v_p L/(gTH_s) = k$$

$$\frac{H_n}{H_s}\left[1 - \frac{2H_n}{H_s} + \left(\frac{H_n}{H_s}\right)^2\right] = k^2\left(\frac{H_n}{H_s}\right)^2$$

$$\therefore \qquad \left(\frac{H_n}{H_s}\right)^2 - (2 + k^2)\frac{H_n}{H_s} + 1 = 0$$

$$\therefore \qquad \frac{H_n}{H_s} = +\frac{(2 + k^2) \pm \sqrt{(4 + 4k^2 + k^4 - 4)}}{2}$$

$$\therefore \qquad \frac{H_n}{H_s} = +1 + \frac{k^2}{2} + \frac{k\sqrt{(4 + k^2)}}{2}$$

$$\therefore \qquad \frac{H_n}{H_s} = 1 + \frac{k^2}{2} + k^2\sqrt{\left(\frac{1}{4} + \frac{1}{k^2}\right)} \qquad (8.13)$$

k can be evaluated quite easily and H_n then solved.

The method of integrating to obtain the curve of H_n against t

Rewriting equation (8.10)

$$\frac{dH_n}{dt} = 2\frac{(H_s - H_n)\sqrt{(H_s/H_n)g/v_{po}L + 1/T}}{1 - t/T}H_n$$

When $t = 0$, $H_n = H_s$

$$\therefore \qquad \left(\frac{dH_n}{dt}\right)_{t=0} = \frac{2H_s}{T}$$

If the value of T is split up into n increments so that $dt = T/n$, the value of H_n at time $t_1 = dt$ is approximately

$$H_{nt_1} = H_s + 2H_s/T \, \delta t$$

Using this value of H_n, calculate $(dH_n/dt)_{t=t_1}$ and use it to get

$$H_{nt=t_2} = (H_n)_{t=t_1} + (dH_n/dt)_{t=t_1} \, \delta t$$

This process can be repeated as often as desired, or until $t = T$. The method used is a very simple finite difference integration method and will be accurate only if n is large. There are more refined methods of performing such finite difference integrations of course.

8.4 Elastic pipe theory

The rigid pipe theory is often applicable in circumstances that are of little interest to the engineer. It can be used only if valve closures are slow and the pressure surges so caused are small. Usually it is necessary to know the worst conditions under which the pipe may have to operate, and it is precisely these conditions that the rigid pipe theory cannot describe. It is thus essential to be able to predict the pressure rises generated when velocity fluctuations are rapid and the consequent pressure surges large.

It is helpful to consider an infinitely rapid, that is, instantaneous valve closure although in practice rapidly varying pipe velocities are produced by valve closures that are only moderately rapid.

When a valve at the end of a long pipeline is shut instantaneously, the fluid upstream of the valve will impinge upon the closed valve and its pressure will immediately rise. The fluid layer behind this now stationary fluid mass will in turn be brought to rest and its impact upon the first layer will cause its pressure to rise and also maintain the pressure of the first fluid layer. This process will then be repeated by the next layer and so on. Eventually all the fluid in the pipe will be brought to rest and its pressure will be raised throughout the pipe. The increase in pressure so created will cause the fluid to be compressed and the pipe itself to be distended. This process can be pictured as shown in Fig. 8.5.

In other words, a pressure surge runs up the pipe at a wave speed of magnitude c. The wave speed c is high, approximately 1200 m/s, but it is not infinite. It will now be realised that if valve closure speeds are slow, this value of c is relatively very large and it is this that justifies the application of rigid pipe theory to slow valve closures.

When the wave has reached the reservoir end of the pipe the situation is as

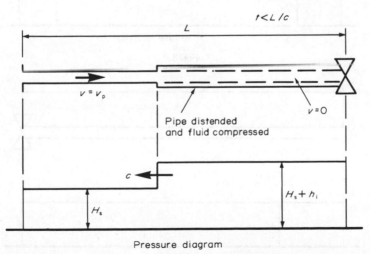

Pressure diagram

Fig. 8.5

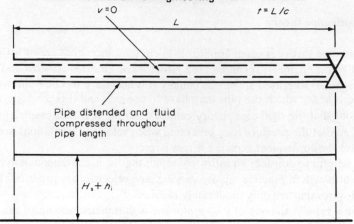

Pressure diagram

Fig. 8.6

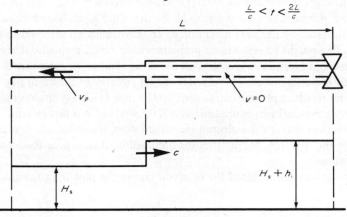

Pressure diagram

Fig. 8.7

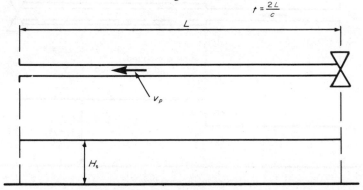

Pressure diagram

Fig. 8.8

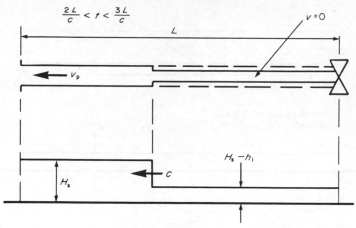

Pressure diagram

Fig. 8.9

shown in Fig. 8.6. The reservoir head is H_s but the head in the pipe is $H_s + h_i$. The fluid will now start to flow out of the pipe back into the reservoir. By considerations of energy this reversed flow will have a velocity of $-v_p$. This can be understood by remembering that the strain energy of the distended pipe walls and the compressed fluid must have been produced by conversion of the kinetic energy of the fluid in the pipe, and when the fluid flows back into the reservoir, this strain energy must be converted back into kinetic energy. As there is very little energy lost by friction in this process, the reverse flow must have the same energy as the original flow. The fluid layer at the reservoir end of the pipe will move towards the reservoir with a velocity v_p and its pressure will revert to H_s. Then the next layer will move in the upstream direction and its pressure will fall also. This process will continue, each layer moving upstream in turn, gaining velocity and losing pressure in the process. As the pressure drops back to its original value the pipe contracts back to its original diameter (Fig. 8.7).

This process continues and the pressure falls until eventually at a time $2L/c$ the entire pipe is at a pressure H_s and the fluid is moving at a velocity v_p towards the reservoir. This is illustrated in Fig. 8.8. This situation can exist only for an instant, because the fluid in the pipe will attempt to move away from the valve at velocity v_p. Immediately, the pressure of the downstream layer at the valve will drop and the fluid will come to rest. The next layer will then experience the same sequence of events and each layer in turn will suffer the pressure drop and be brought to rest. The drop in pressure will cause the pipe to contract. The situation will then be as shown in Fig. 8.9. At time $t = 3L/c$ the situation is as shown in Fig. 8.10. Now the pipe contains fluid at a pressure less than that in the reservoir, and it is in a contracted state. As soon as this state occurs, flow at velocity v_p into the pipe starts and the pressure returns to its original value H_s, the pipe returning to its original diameter (Fig. 8.11). At time $t = 4L/c$ the situation is as shown in Fig. 8.12.

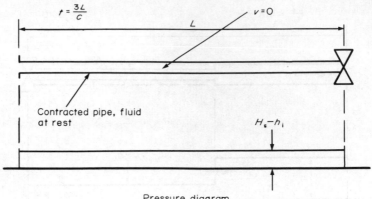

Pressure diagram

Fig. 8.10

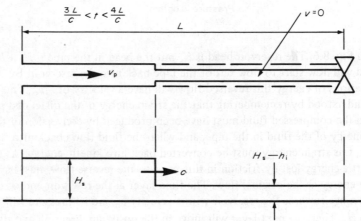

Pressure diagram

Fig. 8.11

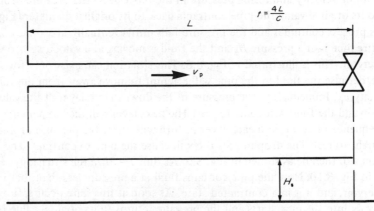

Pressure diagram

Fig. 8.12

This last state is exactly the same as the state just before valve closure occurred. Therefore the cycle will start again; a positive pressure wave running up the pipe causing a pressure $H_s + h_i$ then reflecting with equal magnitude but opposite sign, causing a negative pressure wave to run back to the valve and producing a pressure head of H_s. This negative wave then reflects at the closed valve with the same magnitude and the same sign, so a negative wave runs back up the pipe again, causing a pressure head $H_s - h_i$. This reflects at the reservoir with the same magnitude and opposite sign, so causing a positive wave to return to the valve end, giving a pressure head H_s.

A basic principle has been enunciated here: at an open end a pressure wave reflects negatively, but with the same magnitude as the incident wave, and at a closed end a pressure wave reflects positively and with the same magnitude as the incident wave.

Graphs of pressure head against time for an instantaneous valve closure are shown at the valve in Fig. 8.13 and at a point upstream from the valve in Fig. 8.14.

In all the foregoing, friction has been completely ignored. In some circumstances friction is almost negligible but it is not always so, and to include it in an analysis involves a very much greater effort.

If the valve closure is not instantaneous, pressure waves are generated sequentially by the closing movement of the valve. These waves follow one another up the pipe and are reflected negatively at the reservoir end. If the valve closure is completed before the first wave initiated has had time to travel up the pipe and return in its negative form, the sum of all the waves must be the same as would have occurred if the valve had been closed in zero time. The time taken by a wave to travel to the reservoir and back to the valve is $2L/c$. Thus all closures that occur in a time less than $2L/c$—the *pipe period*—will produce the same maximum pressure at the valve as would have been produced by an instantaneous valve closure, although the resulting wave form will be very different in shape. Such closures are called *sudden closures*.

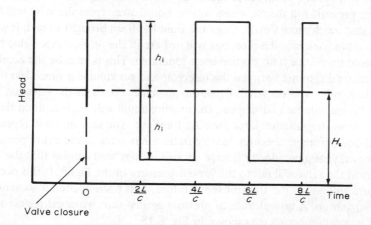

Fig. 8.13

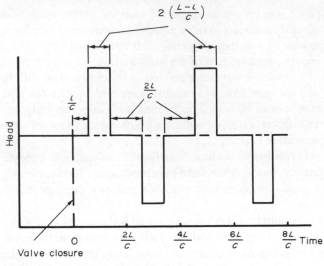

Fig. 8.14

A phenomenon that can cause gross modification of the events that have been described above occurs when the absolute head at the valve before closure, H_s, is less than the inertia head generated by the closure, h_i, minus the head at which gas release occurs in the liquid, about 2·4 m absolute for water. After the first wave reflection, that is after a time $2L/c$, the absolute head behind the valve is $H_s - h_i$. This value is less than the head at which gas bubbles evolve in the liquid so gas evolution will occur. The liquid will be moving away from the valve at this instant and the reduced pressure at the valve is necessitated by the requirement that the liquid must not leave the valve. As gas generates under these low pressures, the liquid *will* leave the valve and the movement of the fluid away from the valve will cause expansion of the gas filled space so generated. The pressure will fall at a much lower rate than would have occurred if the gas had not been present, and the movement of the liquid away from the valve will be decelerated much more slowly. Once the fluid has been brought to rest, it will start to move back into the pipe, but will not attain the original pipe velocity, as it would have done if no gas had been generated. This is because the accelerating pressure difference between the reservoir and gas volume is much less than would have otherwise occurred. When the movement towards the valve has caused the gas volume to disappear, the moving liquid will impinge upon the valve so generating another large pressure transient. The sequence will repeat as described before until energy loss by friction has damped the entire process.

If the original pipe velocity is large enough, it may well happen that the pressure at the valve will fall to the vapour pressure of the liquid. If this occurs, the liquid will boil at the ambient temperature. This phenomenon is sometimes called separation. A typical trace of pressure against time when gas release and/or vapour generation occurs is as shown in Fig. 8.15.

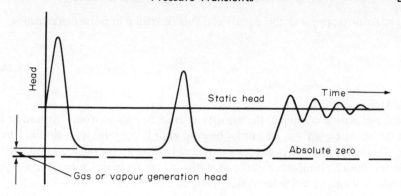

Fig. 8.15

Having described the phenomenon that occurs when a valve at the end of a pipeline is rapidly closed, it is necessary to show how the magnitude of the resulting pressure wave is calculated.

8.5 Pressure surge caused by instantaneous valve closure

At time t after a valve has closed, the situation will be as shown in Fig. 8.16. From equation (8.1)

$$h_i = -\frac{x}{g}\frac{dv}{dt}$$

where x is the length of pipeline over which the inertia head h_i is developed.

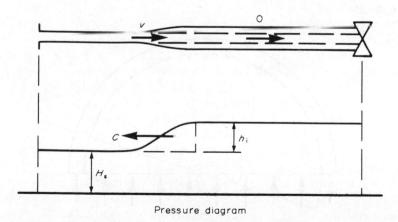

Pressure diagram

Fig. 8.16

Two unknowns appear in this equation if it is rewritten in finite difference form

$$h_i = -\frac{x}{g}\frac{\Delta v}{\Delta t} \qquad (8.14)$$

these being Δv and Δt.

If the closure is complete, the velocity change Δv will be from v to zero in a time Δt, so Δv equals $-v$. Δt cannot be zero even if the valve were shut in zero time. This is because the large pressure rise caused by a very rapid closure will cause the fluid to compress and the pipe distend, so creating a space in the pipe into which fluid can continue to flow even though the valve itself is shut. In other words, it is impossible to stop the fluid in zero time even though the valve is shut in zero time.

If the volume made available by fluid compression and pipe distension is calculated it becomes possible to calculate how long it will take for fluid travelling at the original pipe velocity to fill it and so obtain the time taken for the fluid to be brought to rest. Once this value has been obtained it is possible to calculate h_i from equation (8.14). It is now necessary to calculate the space produced by fluid compression and pipe distension.

The hoop stress f_h in the pipe wall is given by

$$f_h = p_i D_0 / 2T \qquad (8.15)$$

where T is the wall thickness. The longitudinal stress is given by

$$f_l = p_i D_0 / 4T \qquad (8.16)$$

These equations are readily derived by considering force compatibility. (See Figs. 8.17a and b.)

Various cases can be considered but only two will be dealt with here: (1) pipe restrained from extending longitudinally and (2) pipe permitted to extend longitudinally.

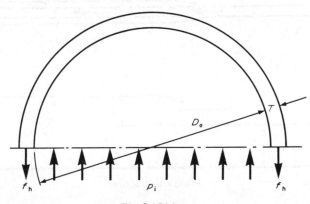

Fig. 8.17(a)

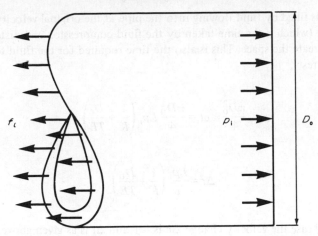

Fig. 8.17(b)

8.5.1. Restrained pipe A pipe is frequently restrained over its length by the fitting of anchor blocks which withstand the momentum and pressure forces acting at every bend

$$E = \frac{f_h}{\delta(\pi D)/(\pi D_0)}$$

$\therefore$
$$\frac{\delta D}{D_0} = \frac{f_h}{E}$$

so
$$\delta D = D_0 f_h / E$$

where E is Young's modulus of the pipe wall material. $\delta L = 0$ because of the restraining effect

$\therefore$
$$\delta D = \frac{p_i D_0^2}{2TE} \tag{8.17}$$

The total additional space generated in the pipe by fluid compression and pipe distension is

$$\frac{p_i}{K} \frac{\pi D_0^2 L}{4} + \frac{\pi D_0 L}{2} \delta D$$

(K is the bulk modulus of the fluid).

Substituting for δD from equation (8.17), this becomes

$$\frac{\pi D_0 L}{2} \left(\frac{p_i D_0}{2K} + \frac{p_i D_0^2}{2TE} \right) = \frac{\pi D_0^2 L}{4} p_i \left(\frac{1}{K} + \frac{D_0}{TE} \right)$$

This space is filled by fluid flowing into the pipe at the original velocity over the time Δt (which is the time taken by the fluid compression and distensibility effects to create the space. This is also the time required for the fluid to be brought to rest).

Then

$$\frac{\pi D_0^2}{4} v \, \delta t = \frac{\pi D_0^2}{4} L p_i \left(\frac{1}{K} + \frac{D_0}{TE} \right)$$

so

$$\Delta t = \frac{L p_i}{v} \left(\frac{1}{K} + \frac{D_0}{TE} \right)$$

Now in this case the velocity change Δv is $- v$ and Δt is as given above, so the equation

$$h_i = \frac{p_i}{w} = -\frac{L}{g} \frac{\Delta v}{\Delta t}$$

becomes

$$p_i = \frac{wL}{g} v \left[\frac{L p_i}{v} \left(\frac{1}{K} + \frac{D_0}{TE} \right) \right]^{-1}$$

$$\therefore \qquad p_i = v \left[\frac{g}{w} \left(\frac{1}{K} + \frac{D_0}{TE} \right) \right]^{-1/2} \tag{8.18}$$

8.5.2 Pipe unrestrained The additional space provided in such an unrestrained pipe by fluid compression and pipe distension is

$$\pi D_0 L \frac{\delta D}{2} + \frac{\pi D_0^2}{4} \delta L + \frac{\pi}{4} D_0^2 L \frac{p_i}{K}$$

Now

$$\delta D = \left(\frac{f_h}{E} - \frac{f_l}{E} \sigma \right) D_0$$

where σ denotes Poisson's ratio
and

$$\delta L = \left(\frac{f_l}{E} - \frac{f_h}{E} \sigma \right) L$$

substituting for f_h and f_l from equations (8.15) and (8.16)

$$\delta D = \frac{p_i D_0^2}{2TE}\left(1 - \frac{\sigma}{2}\right)$$

and

$$\delta L = \frac{p_i D_0 L}{2TE}\left(\tfrac{1}{2} - \sigma\right)$$

∴ Additional space

$$= \pi D_0 L \frac{p_i D_0^2}{4TE}\left(1 - \frac{\sigma}{2}\right) + \frac{\pi D_0^2}{4}\frac{p_i D_0 L}{2TE}\left(\tfrac{1}{2} - \sigma\right) + \frac{\pi}{4} D_0^2 L \frac{p_i}{K}$$

$$= \frac{\pi}{4} D_0^2 L p_i \left(\frac{1}{K} + \frac{D_0}{TE}\left(\tfrac{5}{4} - \sigma\right)\right)$$

Proceeding as for the restrained pipe

$$\Delta t = \frac{L p_i}{v}\left(\frac{1}{K} + \frac{D_0}{TE}\left(\tfrac{5}{4} - \sigma\right)\right) \tag{8.19}$$

and this can be developed as before to give

$$p_i = v \left/ \left[\frac{g}{w}\left\{\frac{1}{K} + \frac{D_0}{TE}\left(\tfrac{5}{4} - \sigma\right)\right\}\right]^{0.5} \right. \tag{8.20}$$

From the analysis above it is also possible to obtain an expression for the velocity of wave transmission. The fluid in the pipe is brought to rest in a time Δt by the action of closing the valve. During this interval the pressure surge will have travelled up the pipe from the valve end to the reservoir end, distance of L at a velocity c so that $c = L/\Delta t$.

Using equation (8.19) for Δt

$$c = L \left/ \left[L\frac{p_i}{v}\left\{\frac{1}{K} + \frac{D_0}{TE}\left(\tfrac{5}{4} - \sigma\right)\right\}\right] \right.$$

$$= v \left/ \left[p_i\left\{\frac{1}{K} + \frac{D}{TE}\left(\tfrac{5}{4} - \sigma\right)\right\}\right] \right.$$

Substituting for p_i from equation (8.20)

$$c = \frac{v}{v\left[\frac{g}{w}\left\{\frac{1}{K} + \frac{D}{TE}\left(\frac{5}{4} - \sigma\right)\right\}\right]^{-1/2}\left\{\frac{1}{K} + \frac{D}{TE}\left(\frac{5}{4} - \sigma\right)\right\}}$$

$\therefore$
$$c = \left[\frac{w}{g}\left\{\frac{1}{K} + \frac{D}{TE}\left(\frac{5}{4} - \sigma\right)\right\}\right]^{-1/2} \qquad (8.21)$$

(or

$$c = \left[\frac{w}{g}\left\{\frac{1}{K} + \frac{D}{TE}\right\}\right]^{-1/2}$$

in the case of the restrained pipe). The expression for p_i, equations (8.18) and (8.21), can be manipulated into a far more easily remembered form.

From equation (8.21)

$$\left\{\frac{1}{K} + \frac{D}{TE}\left(\frac{5}{4} - \sigma\right)\right\}^{1/2} = c^{-1}(w/g)^{-1/2}$$

Substituting the square-rooted expression into equation (8.20)

$$p_i = v(g/w)^{-1/2}c(w/g)^{1/2}$$
$$= wcv/g$$

or $h_i = p/w = cv/g$. This is the Allievi expression. This expression is also sometimes called the Joukowski equation. It can be very simply obtained if the value of c is already known.

Consider a wave traversing a length of pipe of length x (see Fig. 8.18). The fluid in the pipe length x will be decelerated by an amount Δv in a time Δt

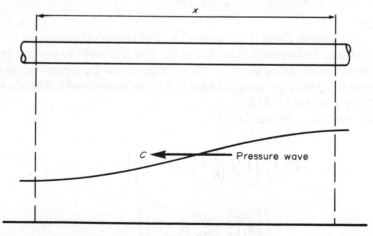

Fig. 8.18

where Δt is the time taken for the wave to traverse the length x that is $\Delta t = x/c$.

$$\Delta h_i = -\frac{x}{g} \times -\frac{\Delta v}{\Delta t} = \frac{x}{g} \frac{\Delta v}{x/c}$$

$\therefore$
$$\Delta h_i = c\Delta v/g$$

Integrating

$$h_i = cv/g \text{ as before.} \qquad (8.22)$$

The results obtained in equations (8.20) and (8.22) will accurately predict the magnitude of the pressure surge if the valve closure occurs in a time less than or equal to the pipe period $2L/c$.

Note that the expression given above for the wave speed c is a special form of the usual equation for the velocity of transmission of sound, $c = \sqrt{(K/\rho)}$. If $1/K'$ is written into the equation in place of $1/K + D/TE \left(\frac{5}{4} - \sigma\right)$ then from equation (8.21) for the unrestrained pipe

$$c = \left[\frac{w}{gK'}\right]^{-1/2}, \text{ but } w/g = \rho$$

$\therefore$
$$c = \sqrt{(K'/\rho)}$$

Thus the expression $\dfrac{1}{K'} = \dfrac{1}{K} + \dfrac{D}{TE}\left(\frac{5}{4} - \sigma\right)$ describes the way in which the effective bulk modulus of the fluid is modified by pipe distensibility. If the pipe has a very small distensibility, that is if $\dfrac{D}{TE}\left(\frac{5}{4} - \sigma\right)$ is very small, the pressure-wave speed approximates to that of sound and as K for water is 300 000 lbf/sq in and

$$\rho = 62 \cdot 4/32 \cdot 2 = 1 \cdot 93 \text{ slugs/ft}^3$$

$$c = 300\ 000 \times 144/1 \cdot 93 = 4700 \text{ ft/s } [1435 \text{ m/s}]$$

In a pipe of greater distensibility this figure can be as low as 2000 ft/s and in a rubber tube it can be as low as 100 ft/s.

The foregoing analysis applies only to simple unbranched pipelines in which valve closures are sudden, that is they occur in times less than $2L/c$. If $c = 4000$ ft/s and $L = 8000$ ft, $2L/c = 4$ s. If the pipe is of significant diameter, say 6 inches or more, the valve on it will also be quite large and impossible to close in such a short time. Although valve closures taking longer than $2L/c$ cannot develop as large pressure surges as those that close suddenly it is still necessary to predict the magnitudes of the pressure surges so caused. To do this, further techniques are available. The Schnyder-Bergeron graphical method is one such method but this has limitations. A better method is based on the solution of the full partial differential equations of waterhammer. This solution is a finite difference integration of the characteristic forms of the partial differential equations and this must be done on a computer. Lack of space precludes the presentation of these methods.

Worked examples

(1) At the downstream end of a pipeline of length 2000 m, a valve is fitted. When open, the pressure head upstream of the valve is 300 m and the flow velocity is 4 m/s. The valve closure time is 30 s and it closes in such a way that the flow area varies linearly with time. Estimate the pressure head at time 2 s and also calculate the maximum pressure head that occurs in the pipeline. Neglect all friction and elasticity effects.

The relevant equation is (8.11).

$$\frac{dH}{dt} = \frac{2[(H_s - H_n)\sqrt{(H_s/H_n)g/(v_0L)} + 1/T]H_n}{1 - t/T}$$

This equation must be integrated using finite difference methods.
Take time steps at 1 second intervals

$$t = 0: \quad H_n = H_s = 300; T = 30$$

$$\left(\frac{\Delta H}{\Delta t}\right)_{t=0} = 2 \times \frac{300}{30} = 20$$

$$\Delta H = 20 \times 1 = 20 \text{ m}$$

$$t = 1: \quad H_n = 320; H_s = 300; v_0 = 4 \text{ m/s}$$

$$\left(\frac{\Delta H}{\Delta t}\right)_{t=1} = \frac{2[(-20)\sqrt{(300/320)}9\cdot81/(4 \times 2000) + 1/30]320}{1 - 1/30}$$

$$\Delta H = 6\cdot35 \text{ m}$$

$$t = 2: \quad H_n = 326\cdot35$$

$$\left(\frac{\Delta H}{\Delta t}\right)_{t=2} = \frac{2[(-26\cdot35)\sqrt{(300/326\cdot35)}9\cdot81/8000 + 1/30]326\cdot35}{1 - 2/30}$$

$$\Delta H = 1\cdot65$$

so

$$H_2 = 328 \text{ m}$$

To calculate the maximum head generated during the closure, use the equation (8.13)

$$\frac{H_{max}}{H_s} = 1 + \frac{k^2}{2} + k^2 \sqrt{\left(\frac{1}{4} + \frac{1}{k^2}\right)}$$

where

$$k = \frac{v_0 L}{g H_s T} = \frac{4 \times 2000}{9 \cdot 81 \times 300 \times 30} = 0 \cdot 0906$$

$$\therefore \qquad k^2 = 0 \cdot 008208$$

$$\therefore \qquad H_{max}/H_s = 1 \cdot 0948$$

$$H_{max} = 328$$

Note that the calculation to obtain the pressure head at time 2 seconds is dependent upon the Δt interval chosen. A value as large as 1 second is likely to produce an inaccurate answer. However, it is interesting to note that a Δt value as large as 2 will produce a value for ΔH which is larger than H_{max}. This makes the point that a choice of an excessively large Δt value could produce a completely wrong result.

(2) A pipe fitted with a valve at its downstream end transports water at 3 m/s, the head upstream of the valve then being 150 m. The effective valve opening is suddenly reduced by one third. Calculate the instantaneous increase in pressure head behind the valve. Wave speed = 1300 m/s.

When the valve is full open the head upstream of it is H_{n0}. After the partial closure, it is H, the corresponding velocities in the pipeline then being v_0 and v.

The discharge $q = C_d a_v \sqrt{(2gH)}$

$$v = \frac{q}{a_p} = C_d \frac{a_v}{a_{v0}} \frac{a_{v0}}{a_p} \sqrt{(2gH)}$$

where a_v is the valve area and a_p is the pipe area.
Similarly,

$$v_0 = C_d a_{v0} \sqrt{(2gH_0)}/a_p$$

$$\therefore \qquad \frac{v}{v_0} = \eta \zeta \quad \text{where } \eta = a_v/a_{v0}$$

and

$$\zeta = \sqrt{(H/H_0)}$$

But after the closure

$$H = H_0 - \frac{c \Delta v}{g} \text{ (remember that } \Delta v \text{ is negative)}$$

$$\therefore \qquad H = H_0 - \frac{c}{g}(v - v_0)$$

$$= H_0 + \frac{c}{g}(v_0 - v) = H_0 + \frac{cv_0}{g}\left(1 - \frac{v}{v_0}\right)$$

$$\therefore \qquad \frac{H}{H_0} = 1 + \frac{cv_0}{gH_0}\left(1 - \frac{v}{v_0}\right) = 1 + \frac{cv_0}{gH_0}(1 - \eta\zeta)$$

$$\therefore \qquad \zeta^2 - 1 = \frac{cv_0}{gH_0}(1 - \eta\zeta)$$

Note that this is the first equation in the series known as Allievi's interlocking equations.

$$\eta = \tfrac{2}{3}; \quad c = 1300; \quad H_0 = 150; \quad v_0 = 3;$$

so

$$\frac{cv_0}{gH_0} = \frac{1300 \times 3}{9 \cdot 81 \times 150} = 2 \cdot 65$$

$$\therefore \qquad \zeta^2 - 1 = 2 \cdot 65(1 - 0 \cdot 667\zeta).$$

$$\therefore \qquad \zeta = 1 \cdot 222$$

$$\therefore \qquad H/H_0 = 1 \cdot 222^2 = 1 \cdot 492$$

$$\therefore \qquad H = 1 \cdot 492 \times 150$$

$$H = 223 \cdot 8 \text{ m.}$$

Pressure head rise = $223 \cdot 8 - 150 = 73 \cdot 8$ m.

(3) A pipeline consists of a 6 in diameter length connected at its upstream end to a length of 12 in diameter. At the downstream end of the 6 in diameter section, a spear valve is fitted. When the spear valve is fully open, the effective nozzle diameter is 1·5 in. The coefficient of discharge of this valve may be assumed constant for all nozzle settings at a value of 0·9. With the valve full open, the flow in the pipeline is 2·36 ft³/s. The wave speed may be assumed constant throughout the pipeline at a value of 4000 ft/s.

Calculate the rise in pressure head immediately behind the valve caused by a sudden partial valve closure from full open nozzle area to half open nozzle area. When the pressure wave reaches the junction of the 6 in and 12 in pipes, a partial transmission of the wave through the junction occurs and a negative reflected wave travels back down the 6 in pipe. Calculate the magnitude of the transmitted wave. Neglect friction.

The approach used to solve the first part of the question is the same as that used in example 2.

$$\zeta^2 - 1 = \frac{cv_0}{gH_0}(1 - \eta\zeta)$$

$$v_0 = \frac{2\cdot 36}{\pi \times 0\cdot 5^2/4} = 12\cdot 02 \text{ ft/s}$$

$$c = 4000, \quad g = 32\cdot 2, \quad v_0 = \frac{c_d a_{v0}}{a_0}\sqrt{(2gH_0)}$$

$$\therefore \quad H_0 = \left(\frac{a_0}{c_d a_{v0}}\right)^2 \frac{v_0^2}{2g} = \frac{1}{0\cdot 9^2} \times \left(\frac{6}{1\cdot 5}\right)^4 \times \frac{12\cdot 02^2}{64\cdot 4}$$

$$H_0 = 709\cdot 1 \text{ ft}$$

$$\frac{cv_0}{gH_0} = \frac{4000 \times 12\cdot 02}{32\cdot 2 \times 709\cdot 1} = 2\cdot 106$$

$$\eta = 0\cdot 5.$$

$$\zeta^2 - 1 = 2\cdot 106(1 - 0\cdot 5\zeta)$$

$$\therefore \quad \zeta = 1\cdot 313.$$

$$H/H_0 = \zeta^2$$

$$\therefore \quad H = 1\cdot 313^2 \times 709\cdot 1 = 1222 \text{ ft}$$

$$\therefore \quad \text{Wave magnitude} = H - H_0 = 513\cdot 3 \text{ ft}$$

The transmission of the wave through the junction is illustrated in Fig. 8.19.

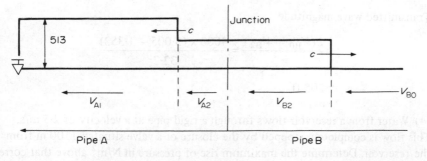

Fig. 8.19

Magnitude of wave transmitted into pipe B

$$= \frac{(V_{B0} - V_{B2})c}{g}$$

Value of V_{A1}

$$\frac{c(V_{A0} - V_{A1})}{g} = 513 \cdot 3. \qquad V_{A0} = 12 \cdot 02 \text{ from before}$$

so

$$V_{A1} = V_{A0} - \frac{g \Delta H_i}{c} = 12 \cdot 02 - \frac{32 \cdot 2 \times 513 \cdot 3}{4000} = 7 \cdot 888 \text{ ft/s}$$

Now reflected wave magnitude plus transmitted wave magnitude = initial wave magnitude,
so

$$\frac{c(V_{A1} - V_{A2})}{g} + \frac{c(V_{B0} - V_{B2})}{g} = 513 \cdot 3$$

$$\therefore \qquad V_{A1} - V_{A2} + V_{B0} - V_{B2} = \frac{513 \cdot 3 \times 32 \cdot 2}{4000} = 4 \cdot 132$$

By continuity

$$d_1^2 V_{A2} = d_2^2 V_{B2}$$

so

$$V_{A2} = (d_2/d_1)^2 V_{B2} = 4 V_{B2}$$

also

$$V_{B0} = (d_2/d_1)^2 V_{A0} = V_{A0}/4 = 3 \cdot 005$$

$$\therefore \qquad 7 \cdot 888 - 4 V_{B2} + 3 \cdot 005 - V_{B2} = 4 \cdot 132$$

$$5 V_{B2} = 6 \cdot 761 \quad \text{so } V_{B2} = 1 \cdot 352$$

Transmitted wave magnitude

$$= \frac{c(V_{B0} - V_{B2})}{g} = \frac{4000 \times (3 \cdot 005 - 1 \cdot 352)}{32 \cdot 2}$$

$$= 205 \text{ ft}$$

(4) Water from a reservoir flows through a rigid pipe at a velocity of 2·5 m/s. This flow is completely stopped by the closure of a valve situated 1100 m from the reservoir. Determine the maximum rise of pressure in N/m^2 above that corresponding to uniform flow when valve closure occurs in (a) 1 second, (b) 5 seconds.

$c = 1430$ m/s. In (b), assume that the pressure rises at a uniform rate with time and there is no damping of the pressure wave.

$$\text{The value of } 2L/c = 2200/1430$$

$$= 1\cdot538 \text{ s}$$

As this is greater than 1 second, (a) represents a sudden closure. The relevant equation is

$$\Delta H = \frac{c\Delta v}{g} = \frac{1430 \times 2\cdot5}{9\cdot81} = 364\cdot4 \text{ m}$$

$$\Delta p = \rho g \Delta H.$$

$$= 9810 \times 364\cdot4$$

$$= 3\cdot575 \times 10^6 \text{ N/m}^2$$

In case (b), the time of closure $>1\cdot538$ s so the sudden closure analysis will not be applicable. The question suggests that a rigid pipe–incompressible fluid theory could be used. The assumption that pressure rises at a uniform rate implies this.

Let

$$p = kt$$

but

$$p = \frac{-wL}{g}\frac{dv}{dt} = kt$$

so

$$\frac{dv}{dt} = -\frac{g}{wL}kt$$

$$\therefore \quad v = -\frac{gk}{wL}\frac{t^2}{2} + \text{constant}$$

When

$$t = 5, \quad v = 0$$

so

$$\text{constant} = \frac{25gk}{2wL}$$

hence

$$v = \frac{gk}{2wL}(25 - t^2)$$

when

$$t = 0, \quad v = 2\cdot5$$

so

$$2 \cdot 5 = \frac{25gk}{2wL}$$

$\therefore$
$$k = \frac{5wL}{25g} = \frac{5 \times 9810 \times 1100}{25 \times 9 \cdot 81}$$

$$= 220\ 000$$

so

$$p_{5\ \text{seconds}} = kt = 5 \times 220\ 000$$

$$= 1 \cdot 1\ \text{MN/m}^2$$

Questions

(1) Describe the pressure phenomena produced in a pipe by sudden valve closure when elasticity of the water is taken into account. The pressure change produced in a pipe by a sudden decrease of velocity dv is given by $dp = wa\ dv/g$ lbs/ft^2 where w is specific weight of water and c is the velocity of sound transmission in water. A pipe is 800 m long and the velocity of flow is 2·0 m/s. Determine the pressure produced by complete closure of a valve in 1 s and show how this pressure is not affected by the rate of valve closure. Also determine the pressure produced by valve closure in 5 s, assuming that the velocity decreases at a uniform rate as the valve closes. $c = 1430$ m/s.

Answers: $2 \cdot 86 \times 10^6$ N/m^2, $3 \cdot 2 \times 10^5$ N/m^2.

(2) Develop a formula for the rise of pressure in a pipe through which water is flowing at a constant rate due to the sudden closing of a valve, allowing for the expansion of the pipe and the compressibility of the water. A cast iron pipe is 0·15 m bore and 0·015 m thick. Calculate the maximum permissible flow in m^3/s if a sudden stoppage is not to stress the pipe to more than 5×10^7 N/m^2. Modulus of compressibility of water is 2×10^9 N/m^2. Young's modulus for cast iron is $1 \cdot 24 \times 10^{11}$ N/m^2.

Answer: $0 \cdot 135$ m^3/s.

(3) Describe the phenomenon of waterhammer which occurs in a pipeline when a valve in the pipeline is suddenly closed. Give a diagram against time for a point close to the valve and also for a point halfway along the pipe when a valve at the end of the pipe is suddenly closed. Prove that the rise of pressure at the valve $dp = dv\sqrt{(9810\ K/g)}$ in N/m^2 where dv is the change of velocity in the pipe if the valve is partially closed and K is the combined bulk modulus of water and pipe. Water flows in a pipe at 4 m/s and the head at the valve is 200 m. The valve is suddenly partially closed to 90% of its previous opening. Find dp if $K = 1 \cdot 45 \times 10^9$ N/m^2. Note that dv is *not* $= 0 \cdot 1 \times 12$.

Answer: $2 \cdot 32 \times 10^5$ N/m^2.

(4) Describe with diagrams of pressure against time, the phenomenon of water-hammer at the discharge valve end and halfway along a horizontal main, when the valve is suddenly closed. Show that if the speed of the water was originally V m/s, sudden and complete closure will cause a rise of pressure of $1\cdot18 \times 10^6 V$ N/m² approximately, given that the effective bulk modulus of elasticity of the water allowing for the stiffness of the pipe is $1\cdot4 \times 10^9$ N/m². A pipe carries water flowing at 4 m/s, the head at the valve being 200 m. The effective valve opening is suddenly reduced by one fifth. What is the instantaneous rise of head at the valve?

Answer: 50·5 m.

(5) Derive a formula for the pressure rise in a fluid flowing in a pipe when the valve at the end from which the fluid escapes is closed (a) slowly, and (b) very quickly. Water flows through a steel pipe 0·2 m internal dia. and 0·05 m thick with a velocity of 2·5 m/s. The length is 120 m. Calculate the maximum pressure rise if (a) the flow is reduced uniformly to zero in 5 seconds, (b) the valve is closed in a time which may be treated as instantaneous. Bulk modulus for water is 2×10^9 N/m², E for steel is 2×10^{11} N/m².

Answers: 6×10^4 N/m², $3\cdot47 \times 10^6$ N/m².

(6) A pipeline of length 4000 ft [1300 m] is fitted with a valve at its downstream end. When this valve is fully open the pressure head just upstream of the valve is 700 ft [200 m] and the flow velocity 10 ft/s [3 m/s]. The valve closes in 25 seconds in such a way that the area varies linearly with time. Neglecting friction and all elasticity effects estimate the maximum pressure head that occurs in the pipe line.

Answer: 751·3 ft [216·5 m].

(7) A pipeline of length 4000 ft [1300 m] conveys water at a velocity of 9 ft/s [3 m/s]. With this velocity in the pipe the pressure head behind a valve at the downstream end of the pipe is 100 ft [30 m]. This valve is suddenly partially closed to two thirds of its full open area and one pipe period later this valve is further partially closed to a flow area of one half of the full open area. The wave speed is 4000 ft/s [1300 m/s].

Neglecting friction calculate (a) the head behind the valve immediately after the first step of closure and (b) the head behind the valve immediately after the second step of closure.

Answer: 190·17 ft [58·5 m], 216·0 ft [69·2 m].

9 Surge Tanks

Any pipeline in which velocity fluctuations can occur can be subjected to large magnitude pressure transients. All pipelines are subjected to such velocity fluctuations when for example a valve is opened or closed or when an hydraulic turbine connected to it is started up or shut down. In Chapter 8 it was demonstrated that a rapid velocity fluctuation, that is one occurring in a time less than $2L/c$, causes a pressure change that can be as large as 130 feet head per foot per second of velocity change. It is therefore clear that a pipe must be protected against such rapid velocity fluctuations if it is not to be excessively costly. In other words, rapid velocity fluctuations must be converted into slow velocity fluctuations so that the pressure transients so generated will be diminished to insignificant values.

A device capable of achieving this result must also be capable of meeting a further requirement. When a turbine is started or a valve is suddenly opened the water in a long pipeline will take a significant time to accelerate under the influence of the reservoir head and in some cases this could be a serious hindrance to the efficient operation of the pipeline. To supply the water needed to give an approximation of steady flow during the starting-up phase of the pipeline operation it is necessary that an auxiliary source such as a small reservoir should be located near to the exit from the pipeline and connected to it from which water can be taken during the starting-up phase and which is replenished when the system is shut down.

This last requirement in effect defines the device needed. In its simplest form it consists of a vertical pipe of large diameter connected to the pipeline and located at a point as near to the exit as practical. The vertical pipe can act as a reservoir during start-up and can act as a point of relatively constant pressure so preventing the passage, almost completely, of any pressure transient into the upstream portion of the pipe. Such a device is called a surge tank. The simplest form of a surge tank is the vertical pipe described above and shown in Fig. 9.1.

If this surge tank is open to the atmosphere at the top it must be extended to an elevation higher than the reservoir static level if the tank is not to spill when the system is shut down. If it were located close to the exit and the pipe connected

312

Reservoir static water
level

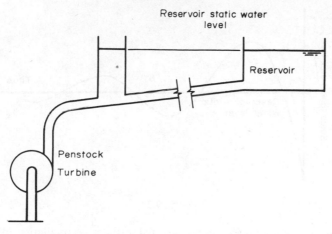

Fig. 9.1

to a turbine, the head on which was large, the surge tank would have to be impracticably high. Under such circumstances the surge tank must be so located that it will not have to be unreasonably high and the pipe between the turbine and the surge tank must then be capable of withstanding the pressure transients. This section of the pipe is called the penstock in American practice and is made of steel. The pipe connecting the surge tank to the reservoir is often made of concrete and usually consists of a concrete lined tunnel of larger dimensions than the penstock so the velocities in it are smaller than those in the penstock. Such concrete lined tunnels must be protected from elevated pressures as they can be cracked; this is the function of the surge tank.

When a pipeline fitted with a simple surge tank is operating in steady state, the level in the surge tank is lower than that in the reservoir by an amount equal to the friction head loss in the pipeline connecting the surge tank to the reservoir. When a shut-down occurs a pressure transient travels up the penstock reducing the velocity in it to zero. A reflection occurs at the surge tank and a negative wave travels back down the penstock. The usual waterhammer phenomenon occurs and this rapidly dies away due to the attenuating effects of friction. The penstock is relatively short so its pipe period ($2L/c$) is small. The fluid in the penstock can therefore be regarded as effectively stationary very soon after valve closure or turbine shut-down is complete. The water moving in the main pipeline is not brought to rest by the closure as it is able to flow into the surge tank. This flow causes the level in the surge tank to rise and this applies a decelerating head to the water in the pipeline. By the time that the surge tank water level has reached the reservoir water level the pipeline velocity will be reduced but will not be zero, consequently the level will continue to rise above reservoir static level and so a mass oscillation will be set up. This will be damped by friction (see Fig. 9.2).

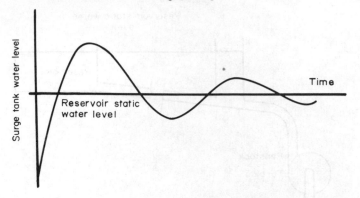

Fig. 9.2

In effect a pressure oscillation is converted into a mass oscillation in which the time scale of the oscillation is greatly increased and the pressure fluctuations very greatly reduced. Friction plays a great part in determining the nature of the oscillation; as friction increases, the oscillation alters as shown below in Fig. 9.3a, b and c. These types of oscillation do not occur in real systems because

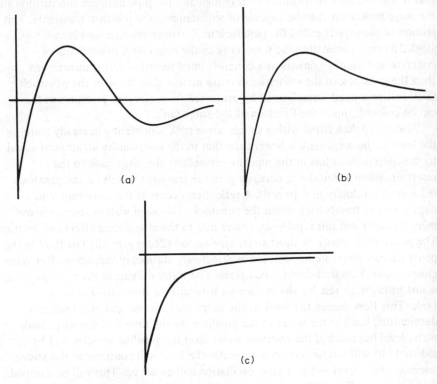

Fig. 9.3

such heavy frictional dissipation is not acceptable when the system is operating
in steady state.

   The surge tank is thus essentially a U tube in which the limbs of the U tube
are of greatly different cross sectional area and are connected by a pipe which
turbulently damps any oscillation that may occur. The parameters that must be
determined when designing a surge tank are therefore (1) the maximum amplitude
of the mass oscillation (2) the frequency of the oscillation and (3) the attenuation
of the oscillation. The basis of the analysis must therefore be the same as that of
a simple U tube with friction allowed.

9.1 The frictionless analysis

Before demonstrating the frictional analysis it is necessary to develop the analysis
ignoring friction. This gives results which are of varying accuracy; for example,
the frequency is given accurately but the value obtained for the amplitude is only
approximate. Even so it is useful in providing a preliminary basis for design.

   In steady state the water level in the surge tank will be the same as the
reservoir static water level because there is no friction (see Fig. 9.4). Let y be
measured positively upwards from the reservoir static water level (Fig. 9.5).

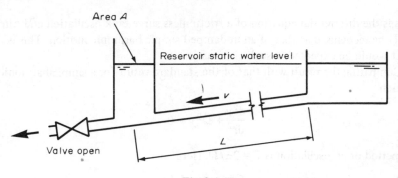

Fig. 9.4

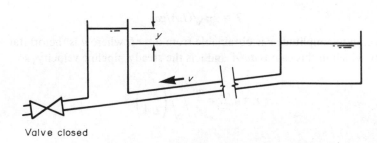

Fig. 9.5

Then two equations can be written, the continuity equation

$$A \, dy/dt = av \tag{9.1}$$

and the dynamic equation

$$y = -\frac{L}{g}\frac{dv}{dt} \tag{9.2}$$

the minus sign being introduced because dv/dt is a deceleration. Now

$$v = \frac{A}{a}\frac{dy}{dt} \quad \text{from equation (9.1)}$$

so

$$\frac{dv}{dt} = \frac{A}{a}\frac{d^2y}{dt^2} \tag{9.3}$$

$$\therefore \quad \frac{L}{g}\frac{A}{a}\frac{d^2y}{dt^2} + y = 0$$

so

$$\frac{d^2y}{dt^2} + \frac{ga}{LA}y = 0$$

This is the differential equation of a frictionless surge tank oscillation and can easily be recognised as that of an undamped simple harmonic motion. This is what would be expected in such circumstances.

Comparing the result with that of the standard result for a simple harmonic motion

$$\frac{d^2y}{dt^2} + \Omega^2 y = 0$$

the period of an oscillation is $T = 2\pi/\Omega$. Here

$$\Omega = \sqrt{(ga/LA)}$$

hence

$$T = 2\pi\sqrt{(LA/ga)}$$

The maximum amplitude r is obtainable from $u = r\Omega$ where u is the orbital velocity which in this case is av/A and v is the steady pipeline velocity, so

$$r = u/\Omega = \frac{av}{A}\Bigg/\sqrt{\left(\frac{ga}{LA}\right)}$$

that is

$$r = v\sqrt{(La/gA)}$$

The displacement curve of a simple harmonic motion is sinusoidal and so the surge tank water level moves sinusoidally in relation to time. Thus

$$y = r \sin \Omega t$$

$$y = v \sqrt{\left(\frac{La}{gA}\right)} \sin \left(\sqrt{\left(\frac{ga}{LA}\right)} \, t\right) = r \sin \left(\frac{2\pi}{T} \, t\right)$$

assuming that the origin is taken at the reservoir static water level.

9.2 Frictional analysis

To obtain a comparable result when friction is included the sudden complete shut-down must be considered as before but unfortunately the resulting differential equation becomes unintegrable.

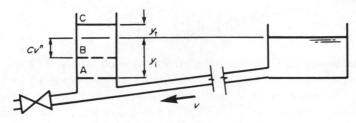

Fig. 9.6

In steady state the water level in the surge tank will be at A an amount y_i below the reservoir static water level (see Fig. 9.6) and

$$y_i = -\frac{4fLv_i^2}{2gd}$$

or

$$y_i = -Cv_i^n$$

where v_i is the steady state velocity, that is the initial velocity. The negative sign is introduced as y is measured positively upwards from the reservoir static water level.

When valve closure occurs

$$av = A \, dy/dt$$

as in equation (9.1) but the dynamic equation must be modified to allow for friction. If at time t the water level has risen to C due to the valve closure, and the velocity in the pipeline at that instant is v, the difference in level between the actual position of the water surface and the position it would be in if the flow were steady at velocity v (position B) is $y + Cv^n$. This means that the pressure applied to the pipe by the water in the surge tank is greater than that required for steady flow by an amount of $y + Cv^n$ and so the water in the pipeline must experience a deceleration given by

$$y + Cv^n = -\frac{L}{g}\frac{dv}{dt} \tag{9.4}$$

From equation (9.3)

$$\frac{dv}{dt} = \frac{A}{a}\frac{d^2y}{dt^2}$$

so

$$\frac{LA}{ga}\frac{d^2y}{dt^2} + C\left(\frac{A}{a}\right)^n\left(\frac{dy}{dt}\right)^n + y = 0$$

if $n = 2$ then

$$\frac{d^2y}{dt^2} + C\left(\frac{gA}{La}\right)\left(\frac{dy}{dt}\right)^2 + \frac{ga}{LA}y = 0 \tag{9.5}$$

(Note that by putting C equal to 0 the frictionless result is obtained.)

The expression for the head loss due to friction Cv^n where $n \approx 1\cdot83$ is valid for many commercial pipes. As mentioned elsewhere this expression is often inconvenient to use and as $C = 4\,fL/(2gd)$ if n is made equal to 2 and much information exists about the value of f if this is so, it is often better to use the form Cv^2 instead of Cv^n. When Cv^2 is used the value of C strictly is variable with v as opposed to the constant value that C takes if $n = 1\cdot83$. This variation of C can be dealt with by the use of finite difference techniques and presents no particular difficulties.

The result given in equation (9.5) is applicable to a rising surge but for a falling surge the friction changes sign. The expression giving the value of friction Cv^2 does not change sign although v reverses its direction and a modification of it is therefore desirable. If the expression is written $Cv|v|$ meaning C multiplied by the velocity multiplied by the positive value of the velocity (ignoring any negative sign) this difficulty is resolved so in future work the symbol $|v|$ will be used to indicate the absolute value of v.

In this way the two equations

$$v = \frac{A}{a}\frac{dy}{dt} \qquad \text{(See (9.1))}$$

and

$$y + Cv|v| = -\frac{L}{g}\frac{dv}{dt} \tag{9.6}$$

can be used to describe rising and falling surges.

The differential equation (9.5) cannot be integrated and finite difference computations have to be undertaken to estimate the magnitude of any surges. It is easier to integrate the two basic equations than to integrate the derived second degree equation. Before discussing the method for integrating the equations it is necessary to deal with other cases that can arise in conjunction with surge tank operation. These are:

(1) Partial sudden closure of a downstream valve, etc.
(2) Sudden opening of a downstream valve or turbine control.
(3) Partial opening of a downstream valve.
(4) Slow valve opening or closing.

(1) *Partial sudden closure of a downstream valve or turbine gate* (see Fig. 9.7)

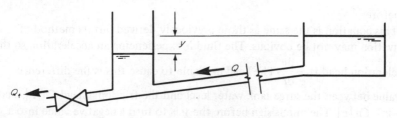

Fig. 9.7

The initial steady flow is denoted by Q, and the final flow by Q_f. Then

$$Q - Q_f = A \, dy/dt$$

so

$$dy/dt = (Q - Q_f)/A$$

The dynamic equation remains unchanged:

$$y + Cv|v| = -\frac{L}{g}\frac{dv}{dt}$$

(2) *Sudden valve opening or turbine start-up* (see Fig. 9.8)

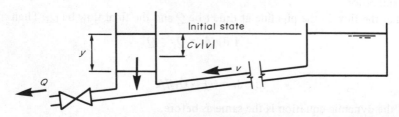

Fig. 9.8

Initially the water levels in the reservoir and surge tank are the same. When the turbine starts up the flow down the penstock suddenly takes the value Q.

At a time t then

$$Q = -A\frac{dy}{dt} + av$$

(The minus sign must be introduced because y is taking progressively larger negative values). In other words the flow Q is partly supplied by flow out of the surge tank and partly by flow from the pipeline. Initially there is no flow out of the pipe and all the flow comes from the surge tank. Finally, all the flow comes from the pipeline and none from the surge tank. The dynamic equation is

$$y + Cv|v| = -\frac{L}{g}\frac{dv}{dt}$$

as before.

This equation is the same as those previously derived but its method of derivation may not be obvious. The fluid is experiencing an acceleration so the acceleration head is $\dfrac{L}{g}\dfrac{dv}{dt}$. The head available to cause this is the difference in value between the surge tank water level and the frictional head $Cv|v|$, that is $-y - Cv|v|$. The minus sign before the y is to turn a negative value into a positive value. (Again the reader is reminded that y is measured positively in the upward direction.) This head causes an acceleration in the pipeline so

$$-y - Cv|v| = \frac{L}{g}\frac{dv}{dt}$$

$$\therefore \qquad y + Cv|v| = -\frac{L}{g}\frac{dv}{dt}$$

(3) Partial opening of a valve turbine

If the initial state of the system is total shut-down there is no difference between a sudden complete and sudden partial opening. If the initial state is that the system is running under part load then a further partial opening up needs separate consideration.

Let the flow in the pipeline at time t be Q and the final flow be Q_f. Then

$$-A\ dy/dt = Q_f - Q$$

so

$$dy/dt = (Q - Q_f)/A$$

and the dynamic equation is the same as before.

Note this result is the same as for a partial shut-down.

(4) *Slow valve opening or closing*

At any time t after a valve has begun to close the flow through it is a function of time

$$Q_t = Q(t)$$

If this value is then substituted for Q_f the equation becomes

$$dy/dt = (Q - Q(t))/A$$

The dynamic equation remains unaltered.

From the foregoing it can be seen that all cases can be described by the two equations

$$dy/dt = [Q - Q(t)]/A$$

and

$$y + Cv|v| = -\frac{L}{g}\frac{dv}{dt}$$

For a sudden closure

$$Q(t) = 0$$

For a partial sudden closure

$$Q(t) = Q_f$$

For a sudden opening

$$Q_i = 0 \quad \text{and} \quad Q(t) = Q_f$$

and so on.

9.3 Complex surge tanks

The simple surge tank is efficient in preventing the development of pressure transients in the pipeline but if suffers from a number of disadvantages.

(1) Frictional damping is small so oscillation can continue for significantly long periods.
(2) If it is to be near the pipe exit it may have to be unacceptably high.

There are a number of alternative designs of surge tanks in use today but lack of space forbids their description here. The use of these different types largely overcomes the various problems listed above.

9.4 Surge tank modelling

In many cases surge tank analyses can be of great difficulty and the results, when obtained, of doubtful value. For example, a pipeline of great length may require a number of surge tanks distributed along its length to protect it adequately and the interaction between such surge tanks may lead to resonance, the system being liable to instability as a consequence. Again, for engineering reasons it may be necessary to build a group of surge tanks instead of a single tank and, as before, instability in operation is a possibility. The analyses of such systems are very complex, and liable to mathematical instabilities which would invalidate any analysis. For these reasons it is necessary in the case of the more complex surge tanks to predict behaviour by the use of modelling techniques rather than by purely analytic methods.

A dimensional analysis of any surge tank system must lead to an expression which involves both the Froude and the Reynolds numbers, and as demonstrated in Chapter 3 it is not possible simultaneously to satisfy the Reynolds and Froude number criteria for modelling. If the Reynolds number criterion is ignored the Froude number criterion leads to a model in which velocities are proportional to the square root of the length scale. During periods in which the velocities in the prototype are low, the Froudian velocities in the model will be even lower. If it is to be possible to ignore the Reynolds number criterion it is necessary that the Reynolds number for both prototype and model should be large. In this case this does not occur and so Froudian modelling will be subject to scale effect and will not be suitable.

Other modelling criteria are possible, however, and for surge tanks the technique illustrated below is found to work well.

The two basic equations of surge tanks are

$$dy/dt = av/A \quad \text{(See (9.1))}$$

and

$$y + Cv^n = -\frac{L}{g}\frac{dv}{dt} \quad \text{(See (9.4))}$$

Writing these two equations for the model and the prototype and putting $R = A/a$

$$\left(\frac{dy}{dt}\right)_m = \left(\frac{av}{A}\right)_m = \frac{v_m}{R_m}$$

and

$$\left(\frac{dy}{dt}\right)_p = \left(\frac{av}{A}\right)_p = \frac{v_p}{R_p}$$

so

$$\frac{(dy/dt)_m}{(dy/dt)_p} = \frac{R_p}{R_m}\frac{v_m}{v_p}$$

Now considering a typical surge, the slope of a point on the graph of the surface displacement against time for both the model and prototype is defined by the ratio y/t. This is not to say that $dy/dt = y/t$, but given similarity between the model and prototype displacement versus time curves the ratio of $(dy/dt)_m$ to $(dy/dt)_p$ equals the ratio $(y/t)_m/(y/t)_p$.

Then

$$\frac{(y/t)_m}{(y/t)_p} = \frac{R_p}{R_m} \frac{v_m}{v_p}$$

(9.7)

Next considering the dynamic equation

$$y_m + C_m v_m^n = -\frac{L_m}{g}\left(\frac{dv}{dt}\right)_m$$

and

$$y_p + C_p v_p^N = -\frac{L_p}{g}\left(\frac{dv}{dt}\right)_p$$

Then

$$\frac{y_m + C_m v^n}{y_p + C_p v_p^N} = \frac{L_m}{L_p}\frac{(dv/dt)_m}{(dv/dt)_p}$$

Dividing through by $C_m v_m^n$ and $C_p v_p^N$

$$\frac{y_m/(C_m v_m^n) + 1}{y_p/(C_p v^N) + 1} = \frac{L_m(dv/dt)_m/C_m v_m^n}{L_p(dv/dt)_p/C_p v_p^N}$$

Now *if*

$$y_m/(C_m v_m^n) = y_p/(C_p v_p^N)$$

(9.8)

then

$$\frac{L_m(dv/dt)_m}{C_m v_m^n} = \frac{L_\rho(dv/dt)_p}{C_p v_p^N}$$

(9.9)

Before, dy/dt was said to be defined by y/t and similarly it may be said that dv/dt is defined by v/t

∴ equation (9.9) may be written

$$\frac{L_m v_m/t_m}{C_m v_m^n} = \frac{L_p v_p/t_p}{C_p v_p^N}$$

rearranging gives

(9.10)

$$\frac{t_p}{t_m} = \frac{L_p}{L_m}\frac{C_m}{C_p}\frac{v_m^{n-1}}{v_p^{N-1}}$$

and from equation (9.8)

$$\frac{y_p}{y_m} = \frac{C_p v_p^N}{C_m v_m^n} \tag{9.11}$$

From equation (9.7)

$$\frac{R_p}{R_m} = \frac{y_m v_p t_p}{y_p v_m t_m}$$

substituting for y_m/y_p and t_p/t_m from equations (9.11) and (9.10) gives

$$\frac{R_p}{R_m} = \left(\frac{C_m}{C_p}\right)^2 \frac{v_m^{2n-2}}{v_p^{2N-2}} \frac{L_p}{L_m}$$

If both the model and prototype pipes possess values of n that are close enough in value to be assumed equal then

$$\frac{R_p}{R_m} = \left(\frac{C_m}{C_p}\right)^2 \left(\frac{v_m}{v_p}\right)^{2n-2} \frac{L_p}{L_m}$$

and if $n = 2$ (as in the case of rough pipes)

$$\frac{R_p}{R_m} = \left(\frac{C_m}{C_p}\right)^2 \left(\frac{v_m}{v_p}\right)^2 \frac{L_p}{L_m} \tag{9.12}$$

If equation (9.12) is rewritten as follows

$$\left(\frac{v_m}{v_p}\right)^2 = \frac{R_p}{R_m} \left(\frac{C_p}{C_m}\right)^2 \frac{L_m}{L_p}$$

then as all terms on the right hand side are constant the left hand side must also be constant.

Thus if $n = N$, v_m/v_p is also a constant. If during a surge the model and prototype velocities change from v_{m1} to v_{m2} and v_{p1} to v_{p2} respectively then

$$\frac{v_{m1}}{v_{p1}} = \frac{v_{m2}}{v_{p2}} = \frac{R_p}{R_m} \left(\frac{C_p}{C_m}\right)^2 \frac{L_m}{L_p}$$

$$\therefore \qquad \frac{v_{m1}}{v_{m2}} = \frac{v_{p1}}{v_{p2}}$$

This result can be useful on occasions.

From these results it is possible to establish the dimensions of a model of the prototype which will satisfy the requirements that velocities should never be excessively small in the prototype so that laminar flow will not develop and also maintain a modelling scale for the surges that will give an easily measured model surge. These modelling criteria work very well and are far more satisfactory than those derived from considerations of dimensional analysis.

Worked examples

(1) A simple surge tank of 6 m diameter is connected to a pipeline of 1 m diameter at its downstream end. The pipeline is 3 km long and the value of the Darcy f for it is 0·005. When the flow through the pipeline is in steady state, the velocity in it is 3 m/s. Calculate the difference in level between the water surface in the upstream reservoir and that in the surge tank. Estimate the period of oscillation of the water surface in the surge tank when the flow in the pipe-line is in an unsteady state. Using a Δt interval of 10 seconds and a simple initial value method of finite difference integration, calculate the level of the water surface in the surge tank referred to the level in the reservoir as datum and the velocity in the pipeline at a time of 40 seconds after flow rejection.

The level difference between the surge tank and reservoir water surfaces is

$$\frac{4fLv^2}{2gd} = \frac{4 \times 0·005 \times 3000}{19·62} \times 9$$

$$= 27·52 \text{ m}$$

The period of oscillation

$$T = 2\pi \sqrt{\left(\frac{LA_s}{a_p g}\right)} = 2\pi \sqrt{\left[\frac{3000}{9·81}\left(\frac{6}{1}\right)^2\right]}$$

$$T = 659·3 \text{ s}$$

Equations applicable are

$$v = \frac{A}{a}\frac{dy}{dt}$$

and

$$y + Cv|v| = -\frac{L}{g}\frac{dv}{dt}$$

$$\Delta t = 10 \text{ (given)}; C = \frac{4fL}{2gd} = 3·0581$$

so

$$\Delta y = \frac{a}{A}v\Delta t = \tfrac{1}{36} \times 10v = 0·2778v$$

and

$$\Delta v = -\frac{g}{L}(y + Cv|v|)\Delta t = -0·0327(y + 3·058v|v|)$$

Initially

$$y_0 = -27 \cdot 52 \text{ m} \quad \text{and} \quad v_0 = 3 \text{ m/s}$$

The finite difference integration is best done in tabular form but as this may not be quite as clear a presentation, each stage of the calculation will be set out separately here

At $t = 0$; $y_0 = -27 \cdot 52$; $v_0 = 3$

$\Delta y = 0 \cdot 2778 \times 3 = 0 \cdot 8334 \text{ m}$; $\Delta v = 0$.

At $t = 10$; $y_{10} = -26 \cdot 69$; $v_{10} = 3$.

$\Delta y = 0 \cdot 2778 \times 3 = 0 \cdot 8334$; $\Delta v = -0 \cdot 0327(-26 \cdot 69 + 27 \cdot 523) = -0 \cdot 0273$

At $t = 20$; $y_{20} = -25 \cdot 86$; $v_{20} = 2 \cdot 972$

$y = 0 \cdot 2778 \times 2 \cdot 972 = 0 \cdot 826$; $\Delta v = -0 \cdot 0327(-25 \cdot 857 + 3 \cdot 058 \times 2 \cdot 972^2)$

At $t = 30$; $y_{30} = -25 \cdot 03$; $v_{30} = 2 \cdot 934$

$\Delta y = 0 \cdot 2778 \times 2 \cdot 934 = 0 \cdot 815$; $\Delta v = -0 \cdot 0327(-25 \cdot 03 + 3 \cdot 058 \times 2 \cdot 934^2)$

At $t = 40$; $y_{40} = -24 \cdot 2 \text{ m}$; $v_{40} = 2 \cdot 89 \text{ m/s}$.

Questions

(1) For what purposes are surge tanks used? Using the notation of Fig. 9.9, justify the equations

$$\frac{\mathrm{d}z}{\mathrm{d}t} = \frac{+av}{A} \quad \text{and} \quad z + Cv^2 = -\frac{L}{g}\frac{\mathrm{d}v}{\mathrm{d}t}$$

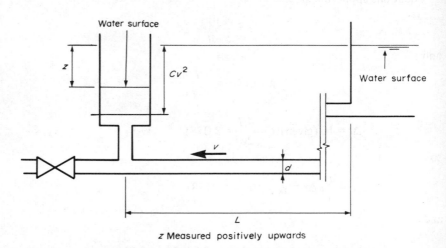

z Measured positively upwards

Fig. 9.9

for instantaneous closure of the valve. The table gives some particulars of a model and a full size surge tank system. Find (a) the correct initial velocity of flow for the model and (b) the ratio of the maximum surge in the model to that in the prototype.

	L	Pipe dia.	Surge tank dia.	C	Initial velocity
Model	40 ft	2·0"	3·5"	0·1	?
	[12 m]	[0·05 m]	[0·0875 m]	[0·0328]	
Prototype	640 ft	4·0 ft	10·0 ft	0·08	6 ft/s
	[192 m]	[1·2 m]	[3 m]	[0·0271]	[1·8 m/s]

Answers: 1·72 ft/s [0·53 m/s], $y_m/y_p = 0·103$ [0·105].

(2) What is the purpose of a surge tank? Show that when comparing a model and a prototype surge tank $R_p/R_m = (C_m/C_p)^2 L_p/L_m (V_m/V_p)^2$. Find a suitable diameter of model tank to satisfy the conditions set out below.

	L	Pipe dia.	Tank dia.	C	Initial velocity
Model	40 ft	2"	d	0·1	1·25
	[12 m]	[0·05 m]		[0·0328]	[0·375 m/s]
Prototype	600 ft	48"	120"	0·1	5·00
	[180 m]	[1·2 m]	[3·0 m]	[0·0328]	[1·5 m/s]

Answer: $d = 5·16"$ [0·13 m]

(3) In a simple surge chamber z is the height of the water level at any instant relative to reservoir water surface level. Develop the continuity and dynamic equations describing z for a sudden partial rejection of flow at the turbines. Hence derive a differential equation completely describing z in terms of t. Determine constants of the differential equation for a surge chamber 100 ft diameter, tunnel diameter 20 ft, tunnel length 4000 ft and $f = 0·005$, if the flow to the turbines is suddenly reduced from 2000 to 1000 cusecs.

Answer: $\dfrac{d^2z}{dt^2} + 0·0125 \dfrac{(dz)^2}{(dt)} + 0·003185 \dfrac{(dz)}{(dt)} + 0·000322z + 0·000203 = 0.$

(4) In a hydro-electric scheme the water velocity in the low-pressure penstock under steady full load conditions is 11 ft/s [3 m/s]. The low-pressure penstock is 3220 ft [981 m] long, has a diameter of 7 ft [2 m] and $f = 0·005$. It connects to the base of a simple surge tank 20 ft [6 m] diameter, and this connection is 70 ft [21 m] below reservoir static level. When starting the turbines the gates are suddenly completely opened to the steady full load flow. Estimate the minimum depth of water in the surge tank due to mass oscillation. Use the basic equations of mass oscillation to form a step-by-step integration.

Answer: Using initial value integration and 10 s steps: approximately 15 ft [7·9 m].

(5) An approximately horizontal tunnel 20 ft [6 m] diameter and 4 miles [7 km] long connects a reservoir to a surge tank which is cylindrical and 80 ft [24 m] dia. Find as a first approximation the periodic time of one complete oscillation of the water level in the tank by neglecting pipe friction. After a period of suddenly increased load, the outflow and inflow are momentarily equal to 1200 cusec [35 m^3/s] and the level in the surge tank is 16 ft [5 m] below reservoir level. The outflow remains constant and friction head lost in the tunnel may now be taken as $v^2/3$ ft [$1\cdot11v^2$] where v is the mean water velocity in the tunnel. Calculate the level in the surge tank after 60 seconds. A step-by-step method is suggested, taking three intervals of 20 seconds each.

Answer: 14·78 ft [4·66 m] below reservoir level
        Time of oscillation = 643·7 s [671·4 s] .

10 Rotodynamic Machines

The term 'rotodynamic machine' is used to describe machines which cause a change of total head of the fluid flowing through them by virtue of the dynamic effect they have upon the fluid. Machines exist which change the head of the working fluid without employing a dynamic effect, for example reciprocating pumps, and such machines will not be dealt with in this book.

Two classes of rotodynamic machines, turbine and pumps, are of particular interest to the engineer. The basic theory of both these classes of machines is the same.

10.1 Flow through rotating curved passages

Rotodynamic machines all make use of the effect that occurs when fluid passes through a rotating curved passage. In the case of turbines this effect consists of a transfer of energy from the fluid to the rotating passage and in the case of pumps it consists of the transfer of energy from the passage to the fluid.

Consider a flow through a rotating curved passage as illustrated in Fig. 10.1.

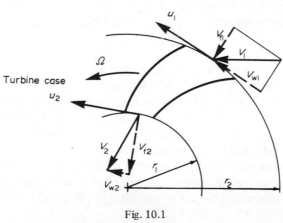

Fig. 10.1

329

Denote the peripheral velocity of the tips of the passage (at the inlet diameter) by u_1 and at the exit diameter by u_2. Let V_1 and V_2 denote the absolute velocity of the fluid at inlet and exit respectively and let V_{f1} and V_{w2} and V_{f2} and V_{w2} denote the radial and tangential components of the absolute velocity at inlet and exit. The subscripts f and w denote the terms flow and whirl respectively. The radial component of velocity V_f is called the velocity of flow because when it is multiplied by the peripheral area of the passage it gives the rate of flow in units of volume per second. The tangential component of the absolute velocity is called the velocity of whirl for obvious reasons.

The rate of change of angular momentum that a mass flow M of fluid experiences in passing through the curved passage is

$$M(V_{w1}r_1 - V_{w2}r_2)$$

Therefore

$$M(V_{w1}r_1 - V_{w2}r_2) = \text{torque applied to the passage}$$

The work done per second = torque × angular velocity

$$= M(V_{w1}r_1\Omega - V_{w2}r_2\Omega)$$

but

$$r_1\Omega = u_1 \quad \text{and} \quad r_2\Omega = u_2$$

So, work done per second is

$$M(V_{w1}u_1 - V_{w2}u_2)$$

If W is the number of units of weight flowing, $W = Mg$. Dividing through by W gives the work done per unit weight of fluid as

$$(V_{w1}u_1 - V_{w2}u_2)/g \tag{10.1}$$

This result applies whichever way the fluid is flowing, that is inwards or outwards. Being a momentum equation it is valid without further correction.

10.1.1 Runners and impellers In practice a single curved passage is not a practical device and a *runner* or *impeller* is used. (The term runner is used when talking about the rotating portion of a turbine and impeller is used to describe the rotating part of a centrifugal pump.) A typical runner is shown in Fig. 10.2. Here a number of curved passages have been formed by dividing up the volume between the two plates by fitting runner blades into it.

The impeller of a centrifugal pump is very similar except that the flow passes through it in the opposite direction. The geometry of the blading shown in Fig. 10.2 is not appropriate for use in a centrifugal pump and a suitable impeller is illustrated in Fig. 10.3.

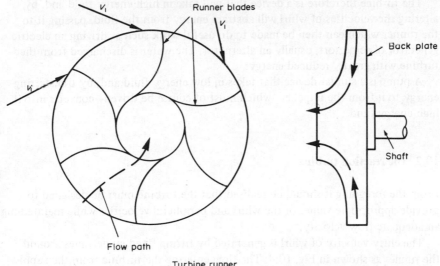

Runner blades

Back plate

Shaft

Flow path

Turbine runner

Fig. 10.2

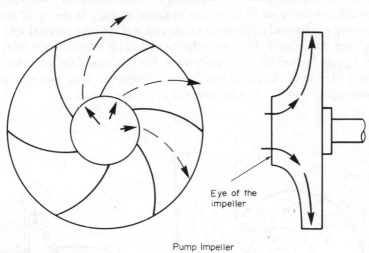

Eye of the
impeller

Pump impeller

Fig. 10.3

It should now be clear that turbines and pumps are devices that operate on
the equation (10.1)

$$\text{Work done/unit weight of fluid} = (V_{w1}u_1 - V_{w2}u_2)/g$$

and that the turbine or pump is merely a device that will generate optimum
values of V_{w1}, V_{w2}, u_1 and u_2, while passing an adequate volume of fluid.

The turbine therefore is a device that will take in high energy fluid and, by altering the velocities of whirl will abstract energy from the fluid, passing it to the runner which can then be made to do useful work such as driving an electric generator of some sort, usually an alternator. The water is discharged from the turbine with greatly reduced energy.

A pump is a similar device that takes in low energy fluid and, by transferring energy to it from the impeller—which must of course be driven—converts into it high energy fluid.

10.2 The reaction turbine

From the foregoing it should be realised that the turbine must be designed to provide appropriate values of the whirl and peripheral velocities while maintaining an adequate flow velocity.

The entry velocity of whirl is generated by fitting fixed guide vanes around the runner as shown in Fig. 10.4. The flow entering the turbine from the supply pipeline is compelled to travel in a circular path by entering a casing (the scroll casing). At the inner periphery of this casing the fluid enters the guide vane passages and is directed on to the runner blades at an angle to the radial direction so generating a tangential component of the absolute velocity—the velocity of whirl at inlet. The magnitude of this velocity of whirl is related to the absolute velocity V_1 and the setting of the guide vane. For this reason the guide vane setting is made adjustable, all the guide vanes rotate about their own axes and are linked together so that an adjustment of one blade adjusts all.

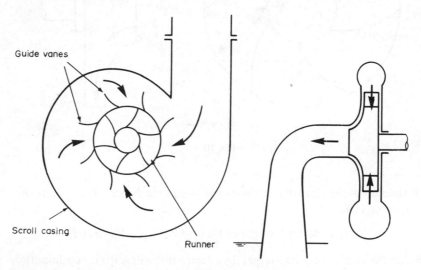

Guide vanes

Scroll casing

Runner

Fig. 10.4

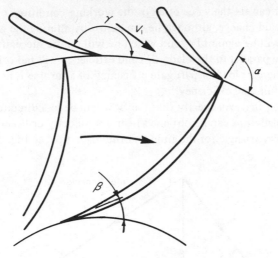

Fig. 10.5

We now examine the situation in a guide vane passage and a runner blade passage (see Fig. 10.5). Three angles must be involved in an analysis of the flow: α, β and γ. (All measured relative to the forward tangent.) γ is the angle made by the absolute inlet velocity V_1 to the forward tangent and is controlled by the guide vane setting. α and β are both fixed values and are the angles that the tangent to the blade tip makes with the forward peripheral tangent at inlet and exit to and from the blade respectively.

If a stream of water traverses an edge of a boundary in a direction parallel to the boundary no significant loss of energy results but if it traverses it in any other direction heavy losses will be caused due to boundary layer separations— see Fig. 10.6. It is therefore necessary to ensure that water leaving the guide vanes has a velocity *relative* to the runner blades that is parallel to the blade tips as otherwise energy will be lost and the efficiency of the turbine correspondingly reduced. It should be appreciated that this condition can be attained for only limited conditions.

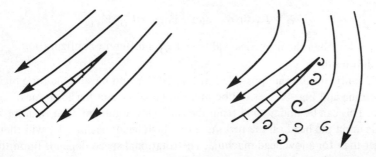

Fig. 10.6

The designer can fix the value of α for the working conditions of the turbine but if these should change, although he can adjust by altering the guide vane settings, there will be a point beyond which he will not be successful. The flow then will not be parallel to the blade tips and efficiency will fall off. This explains why the part load or part gate efficiency of a turbine is never as high as its designed full load efficiency.

The condition that will specify that the flow will be in a direction parallel to the runner blade tips can be obtained from the velocity or Euler triangles. Draw the velocity triangle for the fluid and the blade tip (see Fig. 10.7). V_{r1}

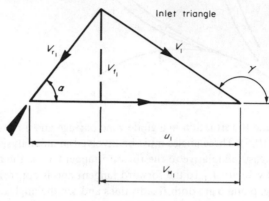

Fig. 10.7

denotes the velocity of the flow relative to the runner blade tip. It can be seen that for no energy loss at entry to the blade

$$\tan \alpha = \frac{V_{f1}}{u_1 - V_{w1}} \tag{10.2}$$

Now

$$V_{w1} = -V_1 \cos \gamma \quad (\cos \gamma \text{ is negative})$$

and

$$V_{f1} = V_1 \sin \gamma \quad \text{and} \quad \tan \gamma = V_{f1}/V_{w1}$$

If u_1 is less than the numerical value of V_{w1} a different configuration (Fig. 10.8) results.

Both configurations are possible but Fig. 10.8 would be suitable for a high head turbine and Fig. 10.7 would be more suitable for a rather lower head turbine. This can be understood from the following argument. The value of V_1 depends upon the head on the turbine. For a high head machine V_1 will therefore be larger than for a low head machine. The rotational speed depends upon the peripheral speed u and if the rotational speed is not to be too low the value of

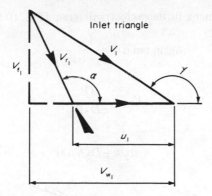

Inlet triangle

Fig. 10.8

u must be large. In Fig. 10.7 u_1/V_1 is large whereas in Fig. 10.8 u_1/V_1 is low. Therefore in the case of a low head turbine a large ratio u_1/V_1 is needed to give an adequate value of u_1 so the type of triangle illustrated in Fig. 10.7 is the more suitable. It should also be remembered that when there is a low head large flows are necessary to give adequate power. (Power = wQH x efficiency) so a large diameter of the runner is needed to avoid having an excessively large value of V_{f1}.

$$Q = \pi D_1 B_1 V_{f1} \qquad (10.3)$$

where B_1 is the axial length of the runner blade at inlet. As the rotational speed of the runner N is given by

$$N = \frac{u}{60\pi D} \text{ revs/min} \qquad (10.4)$$

and D must be large for a low head machine again a large value of u must be possible if N is not to be excessively low.

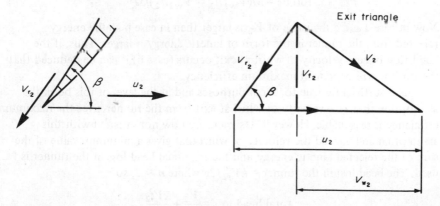

Exit triangle

Fig. 10.9

At exit from the runner a further velocity diagram (Fig. 10.9) can be drawn.

$$\text{Again } \tan \beta = \frac{V_{f2}}{u_2 - V_{w2}} \tag{10.5}$$

$$V_{f2} = \frac{Q}{\pi D_2 B_2} \tag{10.6}$$

and

$$u_2 = \pi D_2 N/60 \tag{10.7}$$

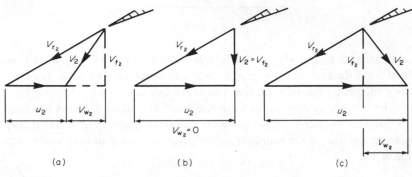

Fig. 10.10

B_2 being the axial length of the blade at exit. As for the entry triangle there are several forms of this triangle; some are shown in Fig. 10.10. The difference between figures a, b, and c depends on the value of u_2 which in turn depends upon the rotational speed N. If the turbine is overloaded and therefore running below the designed speed, Fig. 10.10a applies. V_{w2} is then negative and the torque applied to the wheel is increased because

$$\text{torque} = M(V_{w1}u_1 - V_{w2}u_2)/\Omega$$

Now in case a and c the value of V_2 is larger than in case b so the energy rejected from the runner in the form of kinetic energy is larger. Thus, if the condition of zero velocity of whirl at exit occurs (case b) it can be deduced that the turbine will operate at maximum efficiency.

This is sufficiently true for most purposes and at the level of this book the assumption that zero velocity of whirl at exit from the runner gives the maximum efficiency is reasonable. However designers are now not satisfied with this assumption and instead the velocity of whirl that gives a minimum value of the sum of the rejected kinetic energy and the frictional head loss in the runner is used. The head lost in the runner = $kV_{r2}^n/2g$ where $n \approx 2$, so

$$\text{total head loss} = \frac{V_2^2}{2g} + \frac{kV_{r2}^2}{2g}$$

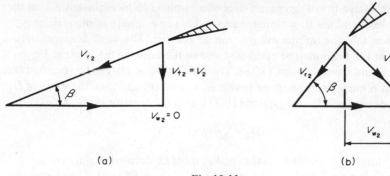

(a)   (b)

Fig. 10.11

As both V_2 and V_{r2} depend upon V_{w2} it will be realised that a minimum total head loss will be given when V_{w2} has a small positive value. (See Fig. 10.11a and b.) Assuming zero velocity of whirl at exit

$$\beta = \tan^{-1}(V_{f2}/u_2)$$

where

$$u_2 = \pi D_2 N/60 \quad \text{(from equation (10.7))}$$
$$\gamma = 180° - \tan^{-1}(V_{f1}/V_{w1})$$

The hydraulic or ideal efficiency E_h is given by

$$E_h = (\text{work done/unit weight})/H$$

∴ $$E_h = \frac{V_{w1}u_1}{gH}$$ (10.8)

assuming zero velocity of whirl at exit. This ideal efficiency cannot be achieved in practice because part of the energy given to the runner must be lost in supplying bearing friction and disc friction. This last is the friction between the runner and the fluid surrounding it. The overall efficiency E_0 is given by

$$E_0 = \text{shaft power/water power}$$

where water power = $wQH/550$ in horse power or $wQH/1000$ in kilowatts. The mechanical efficiency, E_m

$$E_m = \frac{\text{shaft horse power}}{wQV_{w1}u_1/550g} \text{ or } \frac{\text{shaft kW}}{wQV_{w1}u_1/1000g}$$

It follows that

$$E_0 = E_h \times E_m$$

At this stage the design parameters of a turbine can be evaluated. Given the supply head and the flow available and making an estimate of the efficiency, the power that the turbine will give can be assessed. The work done per unit weight of fluid can then be calculated and so the value of the product $V_{w1}u_1$ can be found from equation (10.8). The ratio of B_1 to D_1 can be specified (this decision is made on the basis of previous experience) and then the value of D_1 can be determined from equation (10.3) which gives, assuming $B_1 = CD_1$,

$$D_1 = \sqrt{(Q/\pi C V_{f1})}$$

To perform this calculation a value of V_{f1} must be decided. Values chosen usually lie between 8 ft/s and 14 ft/s [about 2 m/s and 5 m/s]. (The larger the value of V_{f1} chosen the smaller the resulting turbine, but the larger all the velocities within the turbine and hence the larger the frictional losses and the lower the efficiency.) Thus D_1 can be found. u_1 is then given by $u_1 = \pi N D_1 / 60$ and V_{w1} can be calculated.

From this, α, β and γ can be calculated if V_{w2} is assumed to be zero. From equation (10.2)

$$\alpha = \tan^{-1} \frac{V_{f1}}{u_1 - V_{w1}}$$

In many designs the velocity of flow is maintained constant so

$$B_2 D_2 V_{f2} = B_1 D_1 V_{f1}$$

and as V_{f1} then equals V_{f2}

$$B_2 D_2 = B_1 D_1$$

If D_2 is made too small B_2 will become too large so a satisfactory compromise must be made and D_2 can hence be determined.

The essential components of a turbine have been dealt with but two other parts of the machine have important functions; these are the scroll casing which leads water into the guide vanes and the draft tube which takes water from the runner and passes it to the tail race (the channel that carries off the water from the turbine.)

10.2.1 The scroll casing and the draft tube The scroll casing (Fig. 10.12a) conducts water into the guide vanes and the cross sectional area has to be reduced progressively around the periphery of the runner to maintain an approximately constant velocity of flow of the same order of magnitude as the velocity of entry to the guide vanes within it. It is wise to avoid sudden large velocity changes within any flow system as these can lead to energy losses even when flow is convergent. Details of methods of design of scroll casings are beyond the scope of this text.

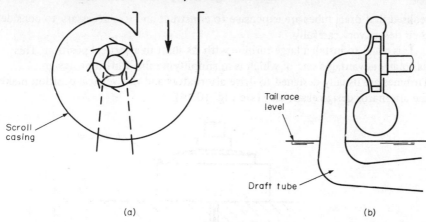

Scroll
casing

Tail race
level

Draft tube

(a)   (b)

Fig. 10.12

The draft tube (Fig. 10.12b) has two functions. First it applies a suction head
to the outlet of the runner approximately equal to the height of the runner above
the tail race level; without the draft tube, delivery from the runner would be
straight to atmosphere and the head on the turbine would consequently be reduced.
Secondly, by diverging the draft tube the rejected kinetic energy can be greatly
reduced so giving an even higher efficiency for the whole machine. If the angle of
divergence is large, boundary layer separations will occur and losses due to
turbulence will offset the recovery of rejected kinetic energy. Small angles of
divergence must therefore be used and this means that the draft tube must be
long if a significant reduction in rejected kinetic energy is to be achieved. Such
a long draft tube, if straight and vertical may involve raising the turbine level so
high as to risk gaseous cavitation or even vaporous cavitation with all the
attendant risks of damage. The draft tube is therefore often made in the shape
of an L. The bend needs careful design to avoid the occurrence of the boundary
layer separations. A method that can be used is to converge the duct in the plane
of the bend but diverge it in the perpendicular plane (see Fig. 10.13). It will be

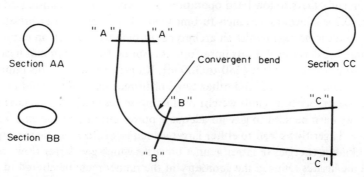

"A"   "A"

Section AA

Convergent bend   Section CC

"B"

"C"

Section BB

"B"   "C"

Fig. 10.13

realised that draft tubes are expensive to construct and it is necessary to consider their design very carefully.

It is usual to install a large turbine with its shaft in a vertical position. This has many advantages one of which is in simplifying the draft tube design. Turbines are usually designed to drive alternators and this vertical position makes the alternator design easier also (see Fig. 10.14).

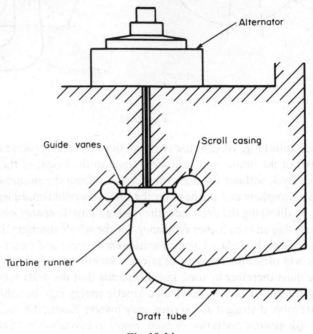

Fig. 10.14

10.2.2 Types of reaction turbine It was stated earlier that different types of turbine runner could be designed, some of which were best suited to high head operation and others to low head operation. The type of turbine discussed in Section 10.2 is known as a Francis turbine (sometimes a reaction or pressure turbine) and can be regarded as an archetypal turbine from which an entire class of turbines has developed. In its pure form it is suited for operation at heads ranging from 750 to 1000 ft [250 to 300 m]. If operated outside this limited range its efficiency falls off and other types of turbine can produce better results. At lower heads, because a unit weight of fluid possesses less potential energy, larger flows must be used to give an adequate power output. At the exit from the runner larger flows lead to either larger velocities of flow and hence larger rejected kinetic energies or larger runner breadths which give larger flow areas. Thus as the head is reduced the geometry of the runner must be altered to give progressively larger flow areas. (The alternative of progressively higher flow velocities with high rejected kinetic energies and high frictional losses is not acceptable.) Thus as heads reduce the runner shape must alter as shown in Fig. 10.15.

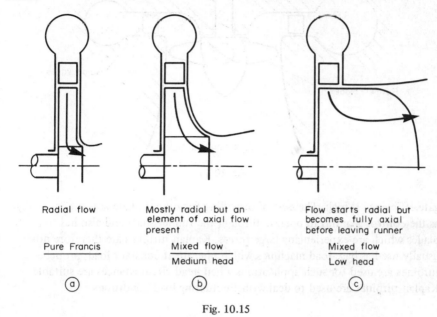

Radial flow	Mostly radial but an element of axial flow present	Flow starts radial but becomes fully axial before leaving runner
Pure Francis	Mixed flow	Mixed flow
	Medium head	Low head
(a)	(b)	(c)

Fig. 10.15

In the three turbines illustrated an evolution has occurred in which an element of axial flow has developed as the working head is reduced and the exit flow area has progressively increased. Runners in which flow occurs in the radial and axial direction are called mixed flow runners. It should be noted that as the head becomes lower the suitable runner progressively turns flow into the axial direction. In the pure Francis runner the change of direction occurs after the flow has left the runner but in a runner suitable for the lowest heads the change of direction occurs before the flow enters the runner. This suggests that as the head to which the turbine is suited is reduced its runner should produce a greater component of axial flow within itself. This leads to the idea of a purely axial flow runner for very low head operation. Turbines using such axial flow runners are called propellor turbines. (See Fig. 10.16.)

In the propeller turbine the flow is given its whirl by the guide vanes and then turned into the axial direction before entering the runner. The exit from the runner has the largest diameter possible, that is the diameter of the runner itself. In a standard propeller turbine the position of the guide vanes is adjustable but the runner blades are fixed. As a result this type of turbine has a high efficiency at full gate opening when operating under full load but this efficiency drops off rapidly at part gate openings. For this reason, this turbine is sometimes equipped with adjustable runner blades. It is then known as a Kaplan turbine and its part

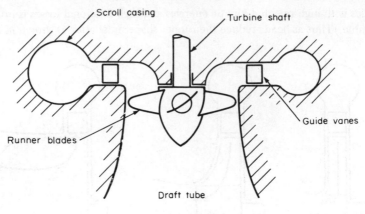

Fig. 10.16

gate efficiency is high. The cost of providing for runner blade adjustment is high as the control gear has to operate through the runner shaft and also has to control blades which are experiencing large forces. Kaplan turbines are therefore not usually used as base load machines which operate at constant load; propeller turbines are used for such applications when head circumstances are suitable. Kaplan turbines are used to deal with fluctuating load conditions.

10.3 Impulse turbines

Above the head range to which Francis turbines are best suited there is a naturally occurring head range for which reaction turbines are not well suited. This head range is from 1000 ft [330 m] to 2500 ft [800 m] and the turbines that work well in this range are known as impulse turbines. There are two main types of impulse turbines, Pelton wheels and Turgo wheels.

10.3.1 The Pelton wheel In the Pelton wheel (Fig. 10.17) the water under high head is passed through a nozzle in which its pressure head is converted into kinetic energy according to the standard nozzle equation

$$V_n = C_v \sqrt{(2gH_n)}$$

This jet then impinges upon a double blade (see Fig. 10.18) in which it is deflected through a large angle. This of course involves a change of momentum and a reaction upon the double blade (usually called a bucket) which, acting at a radius equal to that of the pitch circle of the buckets, provides the torque necessary to drive the wheel. As one bucket moves out of the line of the jet another moves in and experiences the same impulsive force as it moves through the arc in which it intercepts the jet. The nozzle is of a special type known as a

Fig. 10.17

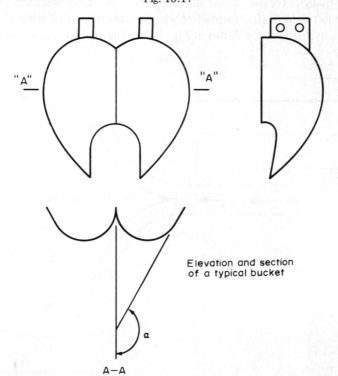

Elevation and section
of a typical bucket

A—A

Fig. 10.18

spear nozzle. The flow through the nozzle is controlled by moving the spear farther into or out of the orifice of the nozzle. The flow runs over the spear reforming into a smooth jet without turbulence thus giving control of flow area without energy loss. This is important from the viewpoint of part gate efficiency (see Fig. 10.19).

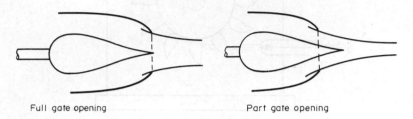

Full gate opening   Part gate opening

Fig. 10.19

Power developed

Consider a Pelton wheel bucket, for which the deflection angle γ is less than $90°$ (Such a bucket would never be made but it is slightly easier to imagine the velocity triangles for such a case and the signs of the trigonometrical functions of γ can be left to look after themselves when γ takes its normal value of $165°$). The velocity triangles are shown in Fig. 10.20. The value of V_{r2} is less than V_{r1}

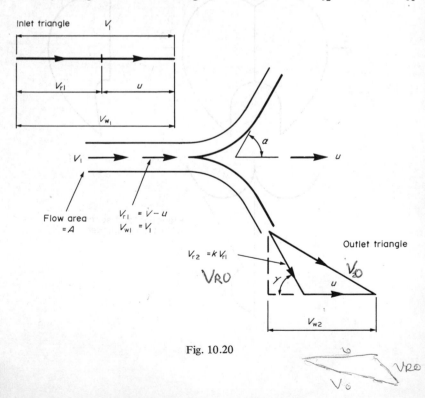

Fig. 10.20

due to friction losses incurred by flow over the blade surface. Usually $V_{r2} = kV_{r1}$ where k has a value of approximately 0.85.

The inlet triangle has degenerated into a straight line, and it can be seen that the following relationships must be true

$$V_{w1} = V_1$$

$$V_{r1} = V_1 - u$$

u is the same for both the inlet and outlet condition as inlet and outlet occurs at very nearly the same radius. From the outlet triangle

$$V_{w2} = u + kV_{r1} \cos \gamma$$

Therefore work done/unit weight of fluid is

$$(V_{w1}u_1 - V_{w2}u_2)/g = u[V_1 - (u + kV_{r1} \cos \gamma)]/g$$

Substituting for $V_{r1} = V_1 - u$, work done/unit weight is

$$u[V_1 - u - k(V_1 - u) \cos \gamma]/g = u(V_1 - u)(1 - k \cos \gamma)/g$$

The power generated is

$$wAV_1u(V_1 - u)(1 - k \cos \gamma)/550\,g \text{ hp}$$

$$= \rho A V_1^3(u/V_1)(1 - u/V_1)(1 - k \cos \gamma)/550 \text{ hp}$$

$$= \rho A V_1^3(u/V_1)(1 - u/V_1)(1 - k \cos \gamma)/1000 \text{ kW}$$

Efficiency of the wheel

$$\text{Input energy/unit weight} = V_1^2/2g$$

Therefore efficiency

$$E_h = 2(u/V_1)(1 - u/V_1)(1 - k \cos \gamma)$$

This is the hydraulic or ideal efficiency of the wheel. If the nozzle is included as part of the turbine the input pressure head upstream of the nozzle is H_n which equals $V_1^2/(C_v^2 2g)$. Therefore the ideal efficiency (nozzle included) is given by

$$E_h = 2C_v^2(u/V_1)(1 - u/V_1)(1 - k \cos \gamma) \tag{10.9}$$

These two expressions give parabolic relationships in which the variables are the two dimensionless groups E and u/V. The efficiency E is a maximum when u/V is 0.5. This can be easily checked by differentiation and equating to zero.

Thus, theoretically, the maximum efficiency and power should be obtained at a rotational speed such that the peripheral blade speed is one half the jet speed and this maximum efficiency can then be obtained from

$$E = 2 \times 0.98^2 \times 0.5 \times 0.5(1 - 0.85 \cos 165°)$$

In this expression usual values have been inserted for C_v, k and γ and E then is calculated as $87\frac{1}{2}\%$. In fact the maximum value of efficiency occurs at a value of u/V of 0·46 to 0·47 due to the distortion of the efficiency versus u/V curve by bearing and windage losses. The efficiency should be zero when $u/V = 0$ and again at $u/V = 1·0$ but due to windage and bearing losses the efficiency falls to zero at a value of u/V of about 0·8.

At the lower end of the Pelton wheel head range the power output tends to become inadequate because of a limitation on the maximum size of the nozzle. This limitation arises because if the bucket is to be able to deflect the jet through $165°$ its diameter must be at least three times that of the jet. If the nozzle diameter is made big enough to deliver a flow adequate to give large powers under low heads the buckets must be made correspondingly larger and because the wheel diameter cannot be greatly increased if an adequate rotational speed is to be maintained, the buckets become large by comparison with the wheel radius. This means that the assumption that the arc traced out by the bucket during the time that the jet is impinging upon it can be taken to be a straight line is not even approximately true. The maximum efficiency will thus fall off if the buckets are excessively large in relation to the pitch circle radius so the power of a machine to work under low heads cannot be adequately increased by increasing the nozzle diameter. If such a hydraulic situation is encountered it is possible to increase the flow on to the wheel by fitting another nozzle $120°$ away from the first nozzle. This is not a particularly good solution as there is a risk of water from one nozzle, after deflection by the buckets, interfering with the flow on to the wheel from the other nozzle. The best solution is to fit another wheel on the same shaft with its own nozzle. As many wheels may be fitted as necessary to give the required power (see Fig. 10.21).

The process of fitting additional wheels to the shaft to give the required power can lead to a clumsy design. When the flow parameters lie between those suitable for multi-Pelton wheels and Francis turbines another type of machine is available—the Turgo wheel.

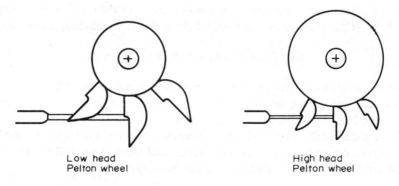

Low head
Pelton wheel

High head
Pelton wheel

Fig. 10.21

10.3.2 The Turgo wheel The Turgo wheel, like the Pelton wheel is an impulse turbine, the water being brought down to atmospheric pressure by being passed through a nozzle before entry to the wheel. The difference is that entry is made at an angle to the plane of the wheel as illustrated in Fig. 10.22. Flow through the blades is in the axial direction and there is little risk of interference between nozzles so there is no particular limitation on the number of nozzles.

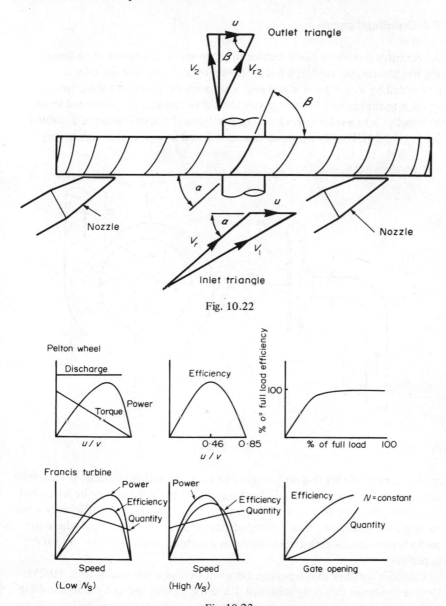

Fig. 10.22

Fig. 10.23

Velocity triangles are easy to draw and blade angles can therefore be calculated as for Francis turbines. It is cheaper and more efficient to use a Turgo wheel than to develop a multi-wheel, multi-jet Pelton wheel.

The turbine characteristics of the Pelton wheel and Francis turbine are illustrated in Fig. 10.23.

10.4 Centrifugal pumps

The centrifugal pump is like a turbine in many ways. It consists of an impeller, very like the turbine runner, which is driven by a motor. The impeller is surrounded by a casing and water passes through the pump in a direction opposite to that in which water passes through a turbine. It is delivered from the impeller into a volute which is very similar to the scroll casing of a turbine (see Fig. 10.24). Usually no guide vanes are fitted however, although in very

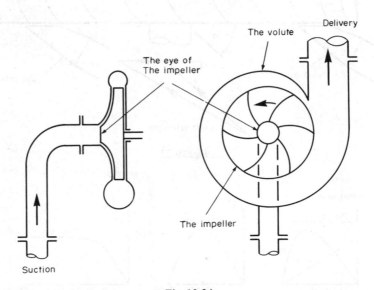

Fig. 10.24

large pumps a diffuser ring which acts like a set of fixed guide vanes is sometimes installed. In the largest of installations adjustable guide vanes may be fitted but this is rare, the type of installation has to be very large and high efficiency must be of great importance, as in a pumped storage scheme for example where the pump is periodically called upon to act as a turbine. Such machines are very expensive.

Consider the flow in the passage between two adjacent blades (Fig. 10.25). Denote the larger radius by subscript 1 and the smaller radius by 2 as was done in the section on turbines. Assuming that as the water travels up the suction pipe

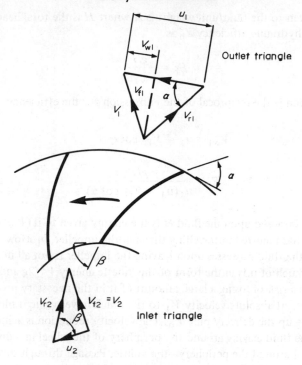

Outlet triangle

Inlet triangle

Fig. 10.25

it does not start to whirl before actually entering the impeller, then $V_{w2} = 0$. As for the turbine (equations (10.5), (10.6) and (10.7))

$$\tan \beta = \frac{V_{f2}}{u_2} = \frac{Q/(\pi B_2 D_2)}{\pi D_2 N/60}$$

$$\tan \beta = \frac{60Q}{\pi^2 B_2 D_2^2 N}$$

where N = rotational speed of the impeller in revs/min and B_2 = axial length of the impeller at entry. From equations (10.2) and (10.1),

$$\tan \alpha = \frac{V_{f1}}{u_1 - V_{w1}}$$

The work done per unit weight of fluid = $(V_{w1}u_1 - V_{w2}u_2)/g$ and as $V_{w2} = 0$, work done per unit weight is

$$V_{w1}u_1/g$$

The work given to the fluid/unit weight is H where H is the total head imposed on it, so the hydraulic efficiency E_h is

$$E_h = \frac{gH}{V_{w1}u_1}$$

This expression is the reciprocal of the expression for the efficiency of a turbine. Now

$$V_{w1} = u_1 - V_{f1} \cot \alpha$$

so

$$E_h = \frac{gH}{u_1(u_1 - V_{f1} \cot \alpha)}$$

The head imposed upon the fluid H is the energy given to it $(V_{w1}u_1/g)$ minus any losses it has incurred in travelling through the impeller, h_i. However the energy that the fluid possesses upon leaving the impeller is not all in pressure head form, much of it is in the form of the kinetic energy $V_1^2/2g$ and as this is large there is a risk of losing a large amount of it in the necessary process of reducing the exit absolute velocity V_1 to the much lower value it must have when moving up the delivery pipe V_p. The velocity reduction is achieved by collecting the fluid leaving around the periphery of the wheel in a divergent duct wrapped around the periphery—the volute. Passage through this divergent duct reduces the velocity but some loss must occur as it does in all cases of divergent flow. If the head loss in the volute is h_v

$$V_{w1}u_1/g = H + h_i + h_v + V_p^2/2g$$

The head loss in the impeller, h_i, must depend upon the velocity of the fluid relative to it, that is upon V_{r1}. Assume this loss is given by

$$h_i = k_1 V_{r1}^2/2g$$

Similarly the loss in the volute must depend upon V_1 so

$$h_v = k_v V_1^2/2g$$

The head rise across the pump is H and so

$$H = V_{w1}u_1/g - k_i V_{r1}^2/2g - k_v V_1^2/2g$$

including the $V_p^2/2g$ term in the volute loss.
From before

$$V_{w1} = u_1 - V_{f1} \cot \alpha$$

and

$$V_1^2 = V_{w1}^2 + V_{f1}^2$$

so
$$V_1^2 = (u_1 - V_{f1} \cot \alpha)^2 + V_{f1}^2$$

∴
$$V_1^2 = u_1^2 - 2u_1 V_{f1} \cot \alpha + V_{f1}^2 \csc^2 \alpha$$

also
$$V_{r1}^2 = V_{f1}^2 \csc^2 \alpha$$

∴
$$H = [2u_1^2 - 2u_1 V_{f1} \cot \alpha - k_i V_{f1}^2 \csc^2 \alpha - k_v(u_1^2 - 2u_1 V_{f1} \cot \alpha + V_{f1}^2 \csc^2 \alpha)]/2g$$

∴
$$H = [(2 - k_v)u_1^2 - 2u_1 V_{f1}(1 - k_v) \cot \alpha - V_{f1}^2 \csc^2 \alpha (k_i + k_v)]/2g.$$

Now
$$u_1 = \pi N D_1/60 \quad \text{and} \quad V_{f1} = Q/(\pi D_1 B_1)$$

so
$$H = AN^2 - BNQ - CQ^2 \tag{10.10}$$

where A, B and C are constants and B may be positive or negative depending upon the value of k_v.

This suggests that the characteristic curve of H versus Q for a pump should belong to the family of curves two members of which are illustrated in Fig. 10.26.

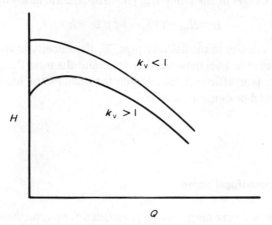

Fig. 10.26

Because the flow through the pump is directed outwards it is divergent and therefore more turbulent than the convergent, inwardly directed flow that occurs in a turbine. Hence pumps are less efficient than turbines, a figure of 75–80% being usual for a pump while 85–95% is usual for a turbine. The efficiency versus Q curve is as illustrated in Fig. 10.27. The curve has a shape like a distorted parabola, and this distortion is explained as follows. At a particular flow rate the inlet velocity triangle is as designed and maximum efficiency is obtained, at higher flow rates the losses in the volute and the runner increase rapidly and cause the rapid fall off in the efficiency.

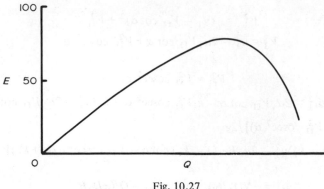

Fig. 10.27

10.4.1 Manometric head and efficiency

The head defined by equation (10.10) is the amount of energy given to a unit weight of fluid by the pump and is the sum of the kinetic energy plus the pressure head. However, this head is related to the head that would be measured by a manometer connected across the outlet and inlet flanges of the pump H_m (the manometric head) by the equation

$$H = H_m + (V_d^2 - V_s^2)/2g + h$$

where V_d is the velocity in the delivery pipe, V_s the velocity in the suction pipe and h the difference in level between the outlet and the inlet flanges. The manometric efficiency is an efficiency based upon the manometric head H_m rather than upon the total or dynamic head H

$$E_m = \frac{gH_m}{V_w u}$$

10.5 Types of centrifugal pump

Pumps, like turbines, may range from a pure radial flow type through all varieties of mixed flow to a pure axial flow type. As in the case of turbines, pure radial flow machines are suited to low flows and high heads and pure axial flow machines to large flow and low heads. However, pumps are very seldom as large as turbines and it is not usually economic to fit the parts required to give high efficiency.

To obtain high efficiency special care must be taken with the volute design as this is the part of the machine in which most of the losses occur. The function of the volute is to take the high velocity fluid leaving the impeller and by passing it through a gradually divergent passage to reduce its velocity to the value required in the delivery pipe. This process must be done with minimum energy loss if high efficiency is to be obtained and attempts to keep volute casings small and hence inexpensive are bound to give low efficiency machines. There are three basic volute types: (1) constant velocity volutes, (2) variable velocity volutes and (3) whirlpool volutes.

In the constant velocity volute the flow velocity in the volute is kept constant, the increasing flow around the impeller being compensated by the increasing area

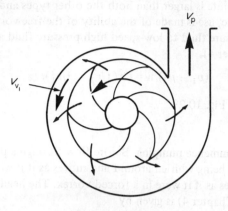

Fig. 10.28

of the volute (see Fig. 10.28). An energy loss is thus created which is approximately given by

$$h_L = (V_{w1} - V_v)^2/2g$$

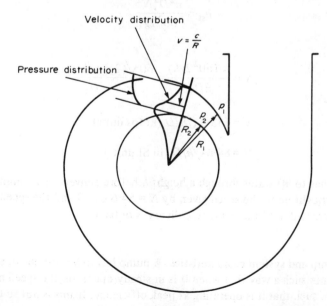

Fig. 10.29

This type of loss is often called a shock loss. If V_v is made equal to $0.5V_{w1}$, h_L is $\frac{1}{4}V_{w1}^2/2g$. In the variable velocity volute the flow area increases more rapidly than does that of a constant velocity volute, the flow velocity in the volute falls from V_{w1} at its smallest cross section to V_p—the velocity in the delivery pipe. This is more efficient but also more expensive than the constant velocity volute.

The whirlpool volute is larger than both the other types and is also more efficient. In this type, use is made of the ability of the free vortex to convert high-speed low-pressure fluid to low-speed high-pressure fluid according to the equation (see Chapter 4):

$$(p_1 - p_2)/w = C^2(1/r_2^2 - 1/r_1^2)/2g$$

This is illustrated in Fig. 10.29.

10.5.1 Speed to commence pumping Before flow through a pump starts the fluid in the pump is being whirled around and moves as it it were a solid. This means that it behaves as if it were in a forced vortex. The head developed in a forced vortex (see Chapter 4) is given by

$$h_{st} = u_1^2/2g$$

where u_1 has been used instead of V_1. But

$$u_1 = \pi D_1 N/60$$

$$\therefore \qquad h_{st} = \frac{\pi^2 D_1^2 N^2}{60^2 2g}$$

so

$$N = \frac{\sqrt{(60^2 2gh)}}{\pi D_1} = \frac{60\sqrt{(2g)}}{\pi}\frac{\sqrt{h}}{D_1}$$

$$\therefore \qquad N = 153.3\sqrt{h}/D \text{ (in fps units)}$$

$$N = 84.6\sqrt{h}/D \text{ (in SI units)}$$

If a pump has to lift water through a height h before delivery can commence it must be brought up to the speed given by $N = 84.6\sqrt{h}/D$ and the speed must then be increased until the required delivery is obtained.

10.5.2 Pump and system characteristics A pump has to be chosen for a particular application in such a way that when it is in steady operation the speed and delivery are such that it is operating at peak efficiency. If this is not so the pump will waste power and will be operating uneconomically.

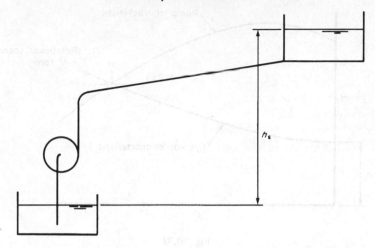

Fig. 10.30

It is first necessary to decide the operating point at which any pump will function when connected to any particular pipe system. Consider, as an example, the case of a rising main (see Fig. 10.30).

The head that the pump must supply, H, is given by

$$H = h_s + \frac{4fL V_p^2}{2gd}$$

or

$$H = h_s + \frac{fLQ^2}{3d^5}$$

that is

$$H = h_s + kQ^2 \quad \text{where} \quad k = fL/(3d^5)$$

This result can be plotted on to a graph of H versus Q. If the pump characteristic for H versus Q is also plotted on the graph a result such as that illustrated in Fig. 10.31 is obtained. The system will operate at point P.

Now it may be that the available pump cannot produce acceptable results; for example, the pump characteristic may at all places lie below the system characteristic or the pump characteristic may fall off so rapidly that an inadequate flow results. In such cases it may be necessary to combine pumps in series or parallel. Two similar pumps connected in series will give a combined pump characteristic of twice the height, as shown in Fig. 10.32. Two similar pumps in parallel will contribute equal flows for the same head so the Q values of the characteristic curve for a single pump should be doubled to give the combined characteristic for parallel operation. Parallel operation is more difficult

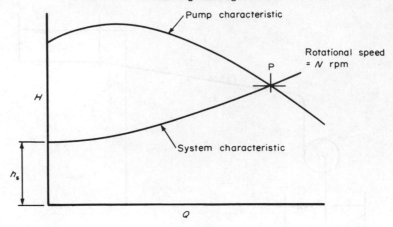

Fig. 10.31

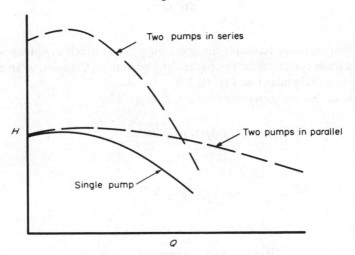

Fig. 10.32

to achieve successfully than is series operation. There is a danger that one pump may take the entire load while the other makes no contribution and, if not prevented by fitting reflex valves, it may happen that one pump attempts to drive the other as a turbine.

Instability may also occur when a pump with a characteristic head flow equation (10.10) in which B is negative is used on a pipe system which has a high value of h_s (see Fig. 10.33). There are two possible operating points in this case. It will not be possible to get the pump to deliver any flow at all unless either h_s can be temporarily reduced until flow has started and then increased again or unless the pump can be temporarily over-speeded. It is improbable that anyone would design a pump to operate in these circumstances except possibly when a

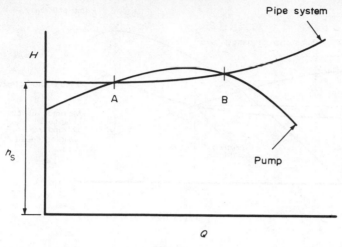

Fig. 10.33

pump is used to fill a reservoir. Initially Fig. 10.31 would apply apply but finally, if the rise in the reservoir surface required were sufficiently large Fig. 10.33 would apply. The operating point will move from P in Fig. 10.31 to B in Fig. 10.33. The pump will tend to operate at point B but a speed fluctuation or a momentary flow reduction could cause it to operate at point A. Immediately flow falls below that for point A, it will cease completely and it will not be possible to restart it until the level in the upstream reservoir has been lowered. It is undesirable to operate a pump in such conditions and this must be borne in mind when designing a rising main.

10.5.3 Selection of a pump for long term operation The pipe system characteristic is determined by the conditions which the system is designed to meet and there is only a small element of choice in this. This element comes in from the need to decide the diameters of the pipes in the rising main. If a large diameter pipe is chosen then the system characteristic will be flat and friction losses low. The capital cost will be larger than if a smaller diameter pipe had been chosen but the operating costs will be lower. An economic analysis will decide what diameter should be chosen. In this the annual cost of borrowing the capital plus the annual cost of operating the pumping station should be minimised. Having decided upon the diameter of the rising main it is necessary to select the pump. Pumps come in families, each member of the family being geometrically similar to the other members but of a different size. By plotting a pump characteristic on to the system characteristic it can be seen just how suitable or otherwise a particular pump is. For example in Fig. 10.34 it can be seen that the pump is not well suited to the system characteristic as it would operate at an efficiency E_1 which is far less than the maximum E_m that the pump can produce.

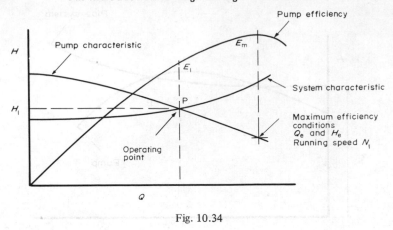

Fig. 10.34

The question arises, what can be done to make the pump more suited to the conditions under which it is to operate. If the pump is producing a satisfactory flow Q_1 it will not be an acceptable solution to merely run the pump faster so as to make the pump characteristic intersect the system characteristic just below the maximum of the efficiency curve. Such a solution will give a much larger flow than is required and will also require a considerably more powerful (and hence more expensive) motor to drive the pump. The best solution then is to pick a geometrically similar pump which will have the acceptable flow Q_1 but which will be of such a size and will operate at such a speed that it will be functioning at maximum efficiency.

Let the subscript r denote the parameters of the required pump. This pump is to run under maximum efficiency conditions. When this occurs the values of the dimensionless numbers quoted below (derived later in the text) must equal one another as both apply to maximum efficiency conditions.

$$\frac{Q_e}{N_1 d_1^3} = \frac{Q_r}{N_r d_r^3} \tag{10.11}$$

$$\frac{gH_e}{N_1^2 d_1^2} = \frac{gH_r}{N_r^2 d_r^2}$$

so as Q_r must equal Q_1 and H_r must equal H_1, Q_r and H_r are known. Q_e and H_e can be taken from the graph. So

$$\frac{N_r}{N_1} = \left(\frac{d_1}{d_r}\right)^3 \frac{Q_r}{Q_e}$$

and

$$\frac{N_r}{N_1} = \frac{d_1}{d_r} \sqrt{\left(\frac{H_r}{H_e}\right)}$$

$$\therefore \qquad \left(\frac{d_1}{d_r}\right)^3 = \left(\frac{N_r}{N_1}\right)^3 \left(\frac{H_e}{H_r}\right)^{3/2}$$

substituting for $\left(\dfrac{d_1}{d_r}\right)^3$

$$\frac{N_r}{N_1} = \left(\frac{N_r}{N_1}\right)^2 \left(\frac{H_e}{H_r}\right)^{3/2} \frac{Q_r}{Q_e}$$

$$\therefore \qquad \frac{N_r}{N_1} = \left(\frac{H_r}{H_e}\right)^{3/2} \frac{Q_e}{Q_r}$$

As everything here is known except N_r, it can be calculated. d_r can then be evaluated from equation (10.11) and the size and running speed of the pump which will give the required delivery in the specified conditions while working at maximum efficiency is known.

10.6 The dimensional analysis of rotodynamic machines

The analysis of rotodynamic machines is a simple example of the technique of dimensional analysis, the result only will therefore by quoted here.

$$f\left(\frac{Q}{ND^3}, \frac{gH}{N^2 D^2}, \frac{\rho N D^2}{\mu}, \frac{k}{D}, E\right) = 0 \qquad (10.12)$$

where k is the mean height of roughnesses on the runner.

A further group can be developed that could be used in place of one of its constituents.

$$\frac{Q}{ND^3} \times \frac{gH}{N^2 D^2} \times E \times \frac{\rho}{\rho} = \frac{wQHE}{\rho N^3 D^5}$$

but $wQHE = P$, the output power. The group is therefore

$$\frac{P}{\rho N^3 D^5}$$

The group $\rho N D^2 / \mu$ is a form of Reynolds number.

10.7 Unit speed, quantity and power

A turbine or pump designed for a particular application may be suitable for use in another application under different flow and head conditions. It is therefore helpful to be able to present the characteristics of any particular machine in a standardised form which can be scaled up to apply to any head conditions that may be required. This is done by reducing the characteristics to unit values. For example a certain pump delivers a quantity Q against a head H when running at a speed N. It is possible to scale down this quantity and speed to unit values, that is the quantity and speed which would be applicable to the pump when working against a head of unity (1 foot or 1 metre).

This scaling process can be very easily performed by using the dimensionless groups given above. As the results are to be obtained for the *same* machine, D does not change. So

$$\frac{Q}{ND^3} = \frac{Q_u}{N_u D^3}$$

$\therefore$
$$Q_u = QN_u/N$$

$$\frac{gH}{N^2 D^2} = \frac{gH_u}{N_u^2 D^2}$$

but
$$H_u = 1 \text{ by definition}$$

$\therefore$
$$N_u = N/\sqrt{H}$$

so
$$Q_u = Q/\sqrt{H}$$

also
$$\frac{P}{\rho N^3 D^5} = \frac{P_u}{\rho N_u^3 D^5}$$

$\therefore$
$$P_u = P(N_u/N)^3$$

so
$$P_u = P/H^{3/2}$$

Then
$$E_u = E$$

Results obtained by testing a machine can now be reduced to their unit values and these then plotted on graphs. The unit characteristics so obtained can be used to give characteristics applicable to any head required by multiplying the unit quantity by the head value raised to an index (either 0·5 or 1·5 according to which unit value is being manipulated). Unit characteristics can be used to compare the performance of different machines.

These unit values can be obtained by considering the velocity triangles of the machine when operating under full head with those applicable to the machine when operating under unit head. When the equivalent velocity triangles are similar to one another it is assumed that the efficiencies will be the same. This is not strictly true because the Reynolds numbers have not been made equal but this leads to a scale effect of relatively small magnitude.

It should be realised that considering the geometric similarity of the velocity triangles leads to exactly the same result as that given by the dimensional analysis. For example if the inlet velocity triangles are similar

$$V_f/u = V_{fu}/u_u$$

$$V_w/u = V_{wu}/u_u$$

$$\frac{gH}{V_w u} = \frac{gH_u}{V_{wu} u_u}$$

but

$$H_u = 1$$

and

$$V_{wu} = u_u V_w/u$$

so

$$\frac{gH}{V_w u} = \frac{g}{u_u^2 V_w/u}$$

∴

$$H/u^2 = 1/u_u^2$$

∴

$$N_u^2 D^2 = N^2 D^2/H$$

∴

$$N_u = N/\sqrt{H}$$

Similarly

$$V_f \propto Q/D^2$$

∴

$$\frac{Q}{uD^2} = \frac{Q_u}{u_u D^2}$$

$$u \propto ND$$

so

$$\frac{Q}{ND^3} = \frac{Q_u}{N_u D^3}$$

as before

$$Q_u = Q N_u/N = Q/\sqrt{H}$$

In effect the dimensional analysis has specified that for dynamical similarity to exist in a turbine operating under two different conditions the velocity triangles must be geometrically similar.

10.8 The specific speed

From what has been said so far in this chapter it should be clear that the type of turbine that should be used in any particular hydraulic situation is well defined and depends upon head. If a high head machine is operated under low head it will give an inadequate power output. However, if only a small power output is required a very small turbine of the high head type may be suitable. For example, the pure radial flow Francis turbine is said to be suited to operation between 750 and 1000 ft head yet a small model of such a turbine operating on heads less than 100 ft is to be found in almost every college fluid mechanics laboratory.

Therefore the head is not the only variable that determines the suitability of a turbine to any application. In fact there are three variables involved; these are head, rotational speed and either power or flow according to whether the machine is a turbine or a pump. It is necessary to develop a method of precisely defining the most suitable type of turbine for any hydraulic situation. A parameter called the specific speed can be developed which uniquely defines the type of turbine in terms of the head, speed and power or flow.

In the case of a turbine this specific speed is defined as the speed of operation of a model of the turbine which is so proportioned that it produces one horse-power (or kilowatt) when operating under unit head. It will be appreciated that if two geometrically similar turbines of different sizes are to be compared on this basis the model turbines that will produce unit power under unit head (the specific turbine) derived from either of the two geometrically similar turbines must be identically the same and will be operating at the same speed. Thus the value of the speed at which the specific turbine runs is a parameter which depends upon the geometry of the turbine and can be used to define the turbine type.

The same argument can be used to define a specific speed for pumps but as in this case the user is not usually interested in the water power generated by the pumps but by the flow delivered, the specific speed for pumps is defined as the speed of a model of the pump which is so proportioned that it delivers unit volume per second when generating unit head.

We now obtain expressions for the specific speed.

(1) Turbines

$$\frac{P}{N^3D^5} = \frac{P_s}{N_s^3 D_s^5}$$

For the specific turbine $P_s = 1$

$$\frac{gH}{N^2D^2} = \frac{gH_s}{N_s^2 D_s^2}$$

and $H_s = 1$. Eliminating D_s (the diameter of the specific turbine) from the above equation

$$D_s^2 = \frac{N^2}{N_s^2} \frac{D^2}{H}$$

$$D_s^5 = \frac{N^5}{N_s^5} \frac{D^5}{H^{5/2}}$$

$$\therefore \quad \frac{P}{N^3 D^5} = \frac{N_s^5 H^{5/2}}{N_s^3 N^5 D^5}$$

$$\therefore \quad N_s = N\sqrt{P}/H^{5/4}$$

(2) *Pumps*

$$\frac{Q}{ND^3} = \frac{Q_s}{N_s D_s^3}$$

and

$$Q_s = 1$$

$$\frac{gH}{N^2 D^2} = \frac{gH_s}{N_s^2 D_s^2}$$

and

$$H_s = 1$$

From the above equation

$$D_s^2 = \left(\frac{N}{N_s}\right)^2 \frac{D^2}{H}$$

$$D_s^3 = \left(\frac{N}{N_s}\right)^3 \frac{D^3}{H^{3/2}}$$

$$\therefore \quad \frac{Q}{ND^3} = \frac{N_s^3 H^{3/2}}{N_s N^3 D^3}$$

so

$$N_s = N\sqrt{Q}/H^{3/4}$$

Table 10.1.

Values of N_s for turbines

Turbine type	N_s range	
	P in hp, H in ft and N in revs/min	P in kW, H in m, and N in revs/min
Pelton wheel	2·5-6	9·5 – 2·3
Multiple wheel, multi-jet machines	6-10	23-38
Turgo	8-12	30-46
Radial flow Francis	12-20	46-76
Mixed flow Francis	20-100	76-380
Axial flow Francis	100-150	380-570
Kaplan and propellers	150-200	570-760

Note that 1 British unit = 3·82 metric units (P in kW)
    1 British unit = 4·45 metric units (P in metric hp)
    1 metric hp = 75 kgf m/s = 735 N m/s
    1 British hp = 76 kgf m/s = 746 N m/s

Table 10.2

Value of N_s for pumps

Pump type	N_s range	
	H in ft, Q in gallons/min	H in m, Q in m³/s
Pure radial centrifugal pump	500-2000	10-42
Mixed flow	3500-8000	74-170
Axial flow	7000-15 000	150-315

10.8.1 Machine efficiency The efficiencies of various types of turbine vary with the specific speed, that is, within the range of one particular type of turbine different runner geometries give different values for the maximum efficiency. As an example, consider a Pelton wheel. The specific speed can vary between 2·5 and 6 and the maximum efficiency will vary within this range. There is a similar relationship for other turbines (see Fig. 10.35).

A rotodynamic machine designed to work at a limiting value of its range of specific speeds will   have a lower efficiency than a machine designed to operate in the middle of the range. Design parameters have been pushed to their limits in such machines and somewhat excessive friction losses have been accepted in order to make the design possible, hence the low efficiency.

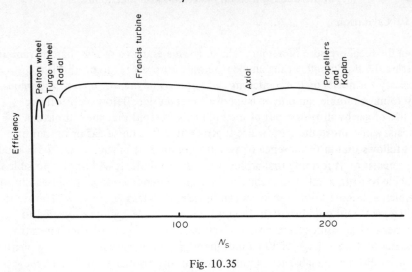

Fig. 10.35

10.9 Scaling of results from model tests

The full dimensional analysis of rotodynamic machines gives

$$f\left(\frac{Q}{ND^3}, \frac{gH}{N^2D^2}, \frac{H}{D}, \frac{k}{D}, \frac{\rho ND^2}{\mu}, E\right) = 0$$

The results obtained from a test on a model of the machine under investigation can be presented as a series of graphs of

$$E \text{ versus } Q/(ND^3) \quad \text{and} \quad gH/(N^2D^2) \text{ versus } Q/(ND^3).$$

The modelling of the roughness number k/D is extremely difficult as scaling down of prototype roughness to model roughness requires a smoother finish on the model than on the prototype which is already as smooth as it can conveniently be made. However, as the prototype is already hydraulically smooth in most cases, it is not necessary to model for roughness.

To model for Reynolds number is, as always, extremely difficult because of the requirement that for equality of the Reynolds numbers the model speed must be much greater than that of the prototype. In fact, in rotodynamic machines the effect of failing to achieve Reynolds number equality is not very important as the flow is so turbulent that the Reynolds number effect can almost be ignored. This leaves the H/D term and failure to model for this leads to a scale effect which can be corrected by the application of the following law due to L. F. Moody.

$$\frac{1 - E_p}{1 - E_m} = \left(\frac{D_m}{D_p}\right)^{0.25}\left(\frac{H_p}{H_m}\right)^{0.04}$$

(Subscript p denotes prototype and subscript m denotes model.)

10.10 Cavitation

When the local pressure of a liquid falls to a value equal to or less than its vapour pressure the liquid boils at the ambient temperature. For it to do this it is necessary for nuclei of size 5 microns and larger to be present on which bubbles may form. A bubble can only exist above a certain size. Below a critical size the loss of gas by diffusion out of the bubble is so rapid that the bubble disappears, while above the critical size it grows by diffusion of gas or vapour into it. It follows that in the absence of a particle or nucleus of such a size that vapour may deposit on it forming first a film and then a bubble, it will not be possible for a bubble to form at all. This is illustrated by the phenomenon of super-ebullition in which a fluid without any nuclei can be heated to temperatures well above the boiling point without ebullition. Liquid free from nuclei can be subjected to large negative pressures (tensions) without evolution of bubbles. Such negative pressures may be as large as 300 atmospheres. In ordinary water there are plenty of nuclei. Even in the absence of particles from the normal environment there is always a constant supply of fine particles falling from the upper atmosphere which come from the burning up of micro meteorites in the outer reaches of the atmosphere.

Under any circumstances in which liquid pressure is reduced the formation of gas filled bubbles (gaseous cavitation) or vapour filled bubbles (vaporous cavitation) is possible. Almost all man-made hydraulic devices can generate cavitation in one or more of the component valves, sluices, turbines, pumps, propellers, etc. It is probably true to say that the final cause of failure of a domestic tap is the pitting of the valve seat caused by cavitation. The phenomenon is thus of common occurrence.

When a vapour filled bubble forms, it will start to grow but as it is carried along in the moving fluid it will sooner or later reach a zone in which the pressure is above the vapour pressure. The vapour in the bubble then very rapidly diffuses back into the main body of the liquid and the bubble collapses. At the instant of collapse the opposite faces of the bubble impinge upon one another at a very high velocity so generating a localised but very intense pressure wave. Pressure intensities as high as 45 tonf/in^2 [approximately 700 MN/m^2] have been detected. This wave propagates as a spherical wave and its intensity falls off very rapidly. However, if the point of implosion of a bubble is very close to a boundary the boundary experiences an intense impulsive impact and may be eroded by it. As many thousands of bubbles are imploding every second in such a region the boundary is being subjected to a highly erosive effect which over a period of months may damage it. In pumps and turbines the damage done by cavitation is so great that care must be taken to prevent it occurring. The effect commonly occurs near low pressure areas of the runner (or impeller in the case of centrifugal pumps) and the damage may involve complete erosion of much of the rotating parts.

Gas in solution will come out of solution when the pressure is rapidly reduced to below a specific pressure, the *gas release pressure*. In normal water at ordinary temperatures this is about 8 ft (2·45 m) head absolute. This means that interspersed with the vapour filled bubbles caused by vaporous cavitation there will be larger gas filled bubbles generated when the pressure first fell below the gas release pressure. When a pressure greater than vapour pressure but less than gas release pressure is encountered the vapour filled bubbles will collapse but the gas filled ones will not and these will then act as miniature surge tanks and will damp out the spherical pressure waves generated by the collapse of the vapour filled bubbles.

Thus gaseous cavitation protects somewhat against the worst effects of vaporous cavitation but if pressures do not reach vapour pressure and fall below gas release pressure then pure gaseous cavitation occurs and this has a similar effect to vaporous cavitation. The implosion of gas filled bubbles is thought to be an altogether slower process than the collapse of vapour filled bubbles but is believed to be mildly damaging. Thus gas cavitation should be seen as a phenomenon to be avoided if possible but if it does occur it mitigates the more extreme damage of vaporous cavitation.

Vaporous cavitation is accompanied by noise and vibration. A cavitating turbine or pump sounds as if the water passing through it is carrying small particles of stone or gravel in suspension and these are impinging upon the inner boundaries of the machine. If air is added to the water in a carefully controlled manner the noise and vibration can be stopped. This added air stops the cavitation by a similar process to that by which gaseous cavitation mitigates vaporous cavitation.

The damage cavitation causes is not its only effect. The bubbles of gas and vapour generated cause the gross volume of fluid flowing to be increased and this in turn causes higher velocities and hence large frictional losses. These in turn cause decreased pressures and so the number and volume of bubbles present increases still further. When this effect occurs the characteristic curves of a turbine or pump are most adversely affected. This is illustrated in Fig. 10.36. The efficiency curve is, of course, similarly affected.

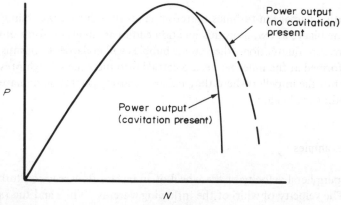

Fig. 10.36

Cavitation is thus a phenonenon to be avoided at almost any cost. If it cannot be avoided chrome steel should be used in places which are liable to cavitation attack as chrome steels are much more resistant to cavitation than other materials. It seems certain that an element of chemical attack occurs and some believe that this is due to the removal of protective oxide films by the mechanical effects of cavitation which then leaves the metal unprotected against chemical attack. Concrete is particularly prone to cavitation, and may need to be protected.

To decide if cavitation is likely to occur it is necessary to know whether zones of low pressure are likely to occur. In calculations of head drops in turbines it seems simple to estimate pressure heads at various points throughout the machine. In fact the pressure that is calculated is a mean pressure and in a rotodynamic machine such as a turbine or pump, local pressures at points quite near to the point for which the mean pressure was calculated may be very much lower than the mean pressure. Bubbles generated in such a zone may then cause damage in zones of higher pressure. It is not possible to calculate accurately the difference in pressure between zones in which local pressures are less than the mean pressure, yet it is precisely this difference in pressure that matters. Places in pumps and turbines in which such local low pressures develop are usually on the face of the blades at the inner radius (see Fig. 10.37).

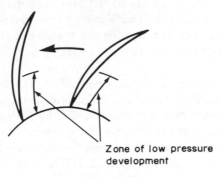

Zone of low pressure
development

Fig. 10.37

Cavitation damage in turbines therefore tends to occur on the trailing face of the turbine blades at exit as in this location eddies are likely to form in which, due to pressure fluctuations, the vapour bubbles will collapse. In pumps the bubbles formed at the inner radius are carried into the zone of high pressure at the outlet of the impeller where they collapse causing cavitation damage to the outlet end of the blade.

Worked examples

(1) The peripheral velocity of the wheel of an inward flow reaction turbine is 20 m/s. The velocity of whirl of the inflowing water is 17 m/s and the radial velocity of flow is 2 m/s. If the flow is 0·7 m^3/s and the hydraulic efficiency

is 80%, find the head on the wheel, the power generated by the turbine and the angles of the vanes. Assume radial discharge.

$$\text{Hydraulic efficiency} = \frac{V_w u}{gH} \quad \text{(from equation 10.8)}$$

$$\therefore \qquad 0 \cdot 8 = \frac{17 \times 20}{9 \cdot 81 \times H}$$

$$\therefore \qquad H = 43 \cdot 3 \text{ m}$$

$$\text{Power generated} = w \frac{V_w u}{g} \times \frac{Q}{1000} \text{ kW}$$

$$= \frac{9810 \times 34 \cdot 658 \times 0 \cdot 7}{1000}$$

$$= 238 \text{ kW}$$

$$\text{Angle of guide vane} = \arctan (V_{f1}/V_{w1})$$

$$= \arctan (2/17)$$

$$= 180° - 6 \cdot 71° = 173 \cdot 29°$$

$$\text{Entry angle of runner} = \arctan \left(\frac{V_{f1}}{u_1 - V_{w1}} \right)$$

$$= \arctan \left(\frac{2}{20 - 17} \right)$$

$$- 33 \cdot 7°$$

(2) A Pelton wheel is required to develop 4500 kW at 400 revs/min., operating under an available head of 360 m. There are two equal jets and the bucket angle is 170°. Calculate the bucket pitch circle diameter, the cross sectional area of each jet and the hydraulic efficiency of the turbine making the following assumptions.

(a) The overall efficiency E_o is 85% when the water is discharged from the wheel in a direction parallel to the axis of rotation (u/V may then be taken as 0·46).
(b) The loss in the nozzles is 3% of the available head.
(c) k for the buckets may be taken as 0·85.

Hydraulic efficiency (from equation 10.9) is

$$2C_v^2 \frac{u}{V} \left(1 - \frac{u}{V}\right)(1 - k \cos \gamma)$$

Nozzle loss $= 0.03 = 1/C_v^2 - 1$

$\therefore$ $\quad C_v = 0.985$

$\therefore$ $\quad E_h = 2 \times 0.985^2 \times 0.46 \times 0.54 \times (1 - 0.85 \cos 170°)$

$\quad = 88.54\%$

$$\text{Power} = E_0 \times \frac{wQH}{1000}$$

$$= \frac{0.85 \times 9810 \times Q \times H}{1000}$$

$\therefore$ $\quad QH = \dfrac{4500\,000}{0.85 \times 9810}$

$\quad = 539.67$

but

$$H = 360$$

so

$$Q = 1.499 \text{ m}^3/\text{s}$$

but for the nozzles

$$Q/2 = C_v \times a_n \sqrt{(2gH)} \text{ per nozzle}$$

where C_c is assumed to be unity

$\therefore$ $\quad a_n = \dfrac{0.145}{2 \times 0.985 \times \sqrt{2 \times 9.81 \times 360}}$

$\quad = 9.054 \times 10^{-3} \text{ m}^2$

$\therefore$ $\quad$ Area of each nozzle $= 90.54 \text{ cm}^2$

Velocity of buckets $= 0.46 \times 0.985 \sqrt{(2g \times 360)}$

$\quad = 38.08 \text{ m/s}$

Now

$$\pi ND/60 = \text{bucket velocity} = 38.08$$

$\therefore$ $\quad D = \dfrac{38.08 \times 60}{\pi \times 400}$

$$D = 1.82 \text{ m}$$

(3) A Francis turbine of specific speed 210 is to develop 30 MW at 180 revs/min. An experimental model is to be made to operate at a flow of 0·6 m³/s and head of 4·5 m. Assume an efficiency E of 88% and estimate the speed, power and scale ratio for the model. Using the same efficiency estimate the flow through the turbine.

$$\text{Power generated by model} = \frac{wQH}{1000}E$$

$$= \frac{9810 \times 0·6 \times 4·5}{1000} \times 0·88 = 23·309 \text{ kW}$$

$$N_{sm} = N_m \sqrt{P_m}/H_m^{5/4}$$

$$N_{sp} = N_p \sqrt{P_p}/H_p^{5/4}$$

$$\text{Now } N_{sp} = 210 = 180\sqrt{(30\ 000)}/H_p^{5/4}$$

$$H_p = (180\sqrt{(30\ 000)}/210)^{0·8}$$

$$= 54·61 \text{ m}$$

but as

$$N_{sm} = N_{sp}$$

$$N_m \sqrt{P_m}/H_m^{5/4} = 210$$

$\therefore$

$$N_m = 210 \times 4·5^{1·25}/\sqrt{(23·309)}$$

$$= 285 \text{ revs/min}$$

Scale ratio

$$\frac{gH_m}{N_m^2 D_m^2} = \frac{gH_p}{N_p^2 D_p^2}$$

so

$$\frac{D_p}{D_m} = \frac{N_m}{N_p}\sqrt{\Bigg/\left(\frac{H_p}{H_m}\right)} = \frac{285·08}{180}\sqrt{\Bigg/\left(\frac{54·61}{4·5}\right)}$$

$\therefore$

$$D_p/D_m = 5·517$$

$$Q_p = \frac{P_p}{wH_pE} = \frac{30\ 000 \times 1000}{9810 \times 54·61 \times 0·88}$$

$$= 63·6 \text{ m}^3/\text{s}$$

$$P_m = P_p(N_m/N_p)^3(D_m/D_p)^5 = 23·3 \text{ kW}$$

(4) In an inward flow turbine developing 200 kW under a head of 30 m the guide and runner vane outlet angles are 160° and 25° respectively, the inlet diameter is 1·5 times the outlet diameter, the ratio of the outlet area from the runner to that from the guides is 1·333, the outlet pressure is atmospheric and the outlet discharge is radial. Assuming that the loss in the guides is 10% of the velocity head generated at their outlet and the loss in the runner vanes is 20% of the outlet relative velocity head determine the outlet area of the guide vanes, the pressure head at the inlet to the wheel and the flow in m³/s.

$$P = 200 = \frac{wQHE}{1000}$$

$$= 9810Q \times 30 \times E/1000$$

$$= 294\cdot3\ QE$$

so

$$QE = 200/294\cdot3$$

$$= 0\cdot6796$$

$$V_{w1} = V_1 \cos 20° = 0\cdot9397 V_1$$

$$V_{f1} = V_1 \sin 20° = 0\cdot3421 V_1$$

$$0\cdot1\ V_1^2/2g + 0\cdot2\ V_{r2}^2/2g + V_{w1}u_1/g + V_2^2/2g = 30$$

$$u_2 = \tfrac{2}{3}u_1$$

$$V_2 = V_{f2} = u_2 \tan 25° = 0\cdot4663 u_2$$

so

$$V_2 = 0\cdot4336 \times 2u_1/3$$

$$= 0\cdot3109 u_1$$

$$V_{r2} = u_2/\cos 25° = (\tfrac{2}{3}/\cos 25°)u_1$$

$$= 0\cdot7356\mu_1$$

$$\frac{V_{f2}}{V_{f1}} = \frac{A_1}{A_2} = \frac{1}{1\cdot333} = \frac{3}{4}$$

so

$$V_{f2} = 0\cdot75 V_{f1}$$

but

$$V_2 = V_{f2} = 0\cdot3109 u_1 = 0\cdot75 V_{f1}$$

$$u_1 = \frac{0\cdot75}{0\cdot3109} V_{f1} = \frac{0\cdot75}{0\cdot3109} \times 0\cdot3420 V_1$$

$$u_1 = 0\cdot8252 V_1; \quad V_2 = 0\cdot2565 V_1$$

Questions

(1) Explain very briefly in what way a hydraulic reaction turbine differs fundamentally from an impulse turbine. Write down an expression for the work done
per unit weight of water passing through an inward-flow reaction turbine runner
and define the hydraulic efficiency of the turbine. An inward-flow reaction turbine
has a guide vane angle γ, its runner vanes are radial at inlet, it has a constant velocity of flow, and no tangential whirl at exit from the runner. If $180° - \gamma$ is
small, determine approximately the proportion of the total head which is
kinetic at the entrance to the runner.
Answer: 50%.

(2) In an inward-flow reaction turbine the guide vane angle is γ, the runner vanes
are radial at inlet and the velocity of flow is constant. Show that the maximum
theoretical hydraulic efficiency of the turbine is given by $E = 1/(1 + \frac{1}{2} \tan^2 \gamma)$.
Show also that for a Pelton wheel the hydraulic efficiency is a theoretical
maximum when the speed ratio is $\frac{1}{2}$. Neglect friction losses.

(3) The peripheral velocity of the wheel of an inward-flow reaction turbine is
70 ft/s [20 m/s]. The velocity of whirl of the inflowing water is 55 ft/s [17 m/s]
and the radial velocity of flow is 7 ft/s [2 m/s]. If the flow is 24 ft^3/s [0·7 m^3/s]
and the hydraulic efficiency is 80%, find the head on the wheel, the hp [kW]
of the turbine and by drawing to scale or otherwise, from the velocity triangle
find the angles of the vanes. Assume discharge radial.
Answers: 326 hp [238 kW], 149·5 ft head [43·3 m], $\alpha = 25°$ [33·7°],
         $\gamma = 172·8°$ [173·3°].

(4) A Pelton wheel has a mean bucket speed of 40 ft/s [12 m/s] and is supplied
with water at a rate of 150 galls/s [700 kg/s] under a head of 100 ft [30 m].
If the buckets deflect the jet through an angle of 160° (when stationary) find
the hp [kW] and the efficiency of the wheel. Assume $C_v = 0·98$ and k = 1.
Answers: 254 hp [192 kW], 93·11% [93·14%].

(5) Derive an expression giving the efficiency of a reaction turbine in terms of
the velocity of whirl of the water, the peripheral velocity of the runner and the
head under which the turbine operates. The velocity of the whirl at the inlet edge
of an inward-flow reaction turbine runner is $0·7\sqrt{(2gH)}$ and the velocity of flow
at this edge is $0·12\sqrt{(2gH)}$ where H is the head. The velocity of whirl at the
outlet edge of the runner is zero. The hydraulic efficiency is 84%. Determine
(a) the peripheral speed of the inlet edge of the runner and the relative velocity
of the water at this point in terms of H, and (b) the direction of this relative
velocity.
Answers: (a) $0·6\sqrt{(2gH)}$, $0·156\sqrt{(2gH)}$, (b) 50°.

(6) For an inward-flow water turbine of specific speed 25 operating under a head
H ft the outer peripheral speed of the wheel is $0·65\sqrt{(2gH)}$ and the breadth of the
outer periphery is $0·125$ of the diameter of the wheel.

Find the velocity of whirl and velocity of flow at the entrance of the wheel, both in terms of $\sqrt{(2gH)}$ and hence find the guide blade angle and the vane angle at entrance to the wheel. Assume the overall efficiency to be $82\frac{1}{2}\%$ and the hydraulic efficiency to be 87% when the discharge from the wheel is radial.

Answers: $0.67\sqrt{(2gH)}$, $0.2135\sqrt{(2gH)}$, $95.4°$.

(7) A Pelton wheel of 5 ft [1.5 m] dia. is operated by two jets each of 3.75 in. [9.5 cm] dia. The total head at the nozzles is 900 ft [275 m]. The nozzle velocity coefficient is 0.98, the bucket angle of deflection is 165° and the turbine develops 3000 hp [2250 kW] when the wheel is running at the theoretically best speed. Show how this speed is determined. Calculate the efficiency of the system and the velocity coefficient for the buckets.

Answers: 0.71 [0.727], 81% [81.73%].

(8) Establish the formula for the specific speed of a water turbine and apply the formula to find the specific speed of a single-jet Pelton wheel in which the diameter of the wheel is 15 times the diameter of the jet, given that the mean speed of the cups is $0.46\sqrt{(2gH)}$ and the velocity of the jet $0.985\sqrt{(2gH)}$ when the overall efficiency is 85%. H is the available head to the turbine.

Answer: 3.65 [13.9].

(9) The water available for a Pelton wheel is 150 ft³/sec [4.3 m³/s] and the total head from reservoir level to nozzles is 900 ft [275 m]. The turbine has two runners with two jets per runner. All four jets have the same diameter. The pipeline is 10 000 ft [3000 m] long. The efficiency of power transmission through the pipeline and nozzle is 91% and the efficiency of each runner is 90%. The velocity coefficient for each nozzle is 0.975 and the coefficient for the pipeline is 0.0045. Determine (a) the hp [kW] developed by the turbine, (b) the diameter of the jets and (c) the diameter of the pipeline.

Answers: 12 600 hp [9.5 MW], 5.47″ [138 mm], 4.84 ft [1.48 m].

(10) A Pelton wheel driven by two similar jets transmits 5000 hp [3750 kW] to the shaft when running at 375 revs/min. The head from reservoir level to nozzles is 670 ft [200 m] and the efficiency of power transmission through the pipe and nozzles is 90%. The centrelines of the jets are tangential to a 4.8 ft [1.5 m] dia. circle. The relative velocity decreases by 10% as the water traverses the bucket surfaces which are so shaped that they would, if stationary, deflect the jet through an angle of 165°. Neglecting windage losses find (a) the efficiency of the runner, and (b) the diameter of each jet.

Answers: 93.3% [93.5%], 6″ [156 mm].

(11) The output power of a Pelton wheel is 8000 hp [6 mw]. If the coefficient of velocity of the nozzle is 0.97, the relative velocity of the water to the bucket at exit is 0.95 times the relative velocity at inlet, the bucket speed is 0.46 times the jet speed and the deflecting angle of the buckets is 165°, calculate the hydraulic efficiency of the turbine. If the flow through the nozzle is reduced by

10% by closing a throttle valve, the jet needle position being unaltered and the wheel speed remaining constant, calculate the new output power. Assume that the mechanical efficiency of the turbine remains a constant fraction of the hydraulic efficiency.

Answers: 89·6% [89·6%]; 5870 hp [4·40 MW].

(12) In the spiral casing of a reaction turbine the head is 48 ft above atmospheric and the water speed is 16 ft/s. The shaft is vertical and the discharge is 400 cusec. The top of the draft tube and level of the tail race are 3 ft and 10 ft respectively below the horizontal central plane of the spiral casing. At the draft tube inlet the water has a velocity of 16 ft/s without whirl, and it leaves the tube at 8 ft/s. Assuming that the overall efficiency is 80% and the hydraulic efficiency is 84% and the loss of head in the draft tube is 1·6 ft, find (a) the friction loss of head in the turbine proper, (b) the head at the top of the draft tube and (c) the value of the specific speed. $N_s = N\sqrt{P}/H^{5/4}$ if the speed of rotation of the turbine is 250 revs/min.

Answers: (a) 950 ft, (b) 8·38 ft below atmospheric, (c) 70·42.

Centrifugal pumps

(13) A centrifugal fan has to deliver 150 ft³/s [4·25 m³/s] of air when running at 750 revs/min. The diameter of the impeller at inlet is 21″ [0·53 m] and at outlet is 30″ [0·76 m]. It may be assumed that the air enters the fan radially. The vanes are set backward at outlet at 70° to the tangent and the width at outlet is 4″ [0·1 m]. The volute casing gives 30% recovery of the outlet velocity head. The losses in the impeller may be taken as equivalent to 25% of the outlet velocity head. The specific volume of the air is 12·8 ft³/lb [0·082 m³/N] and blade thickness effects may be neglected. Determine the efficiency, the power required and the total head at discharge.

Answers: 42·05%, 5·193 hp, 1·54″ water [41·23%, 3·68 kW, 36·4 mm water].

(14) Develop an expression for the specific speed of a centrifugal pump. Making reasonable approximations show that the ratio B/D increases as specific speed increases. B = impeller breadth at outlet, D = impeller diameter. A centrifugal pump delivers 5000 galls/min [0·38 m³/s] of water against a head of 160 ft [49 m] with a speed of 750 revs/min. Calculate the head and discharge at 400 revs/min and the power required to drive it if the efficiency is 80%.

Answers: Head = 45·5 ft, discharge = 2667 galls/min, power = 46·0 hp [Head = 13·94 m, discharge = 0·203 m³/s, power = 34·7 kW].

(15) Draw a section through a two-stage centrifugal pump having a pair of balanced opposed impellers. The impeller of a centrifugal pump is 13″ [0·33 m] dia. and 3/4″ [0·019 m] wide at the outlet. The blade angle at outlet is 35° to the tangent at the periphery and the water at inlet has radial flow. The

suction lift, including pipe and valve losses, is 12 ft [3·7 m] and the impeller speed is 1600 revs/min. The delivery pipe losses are estimated to be 30 ft [9·1 m] and the static lift from the pump centre is 130 ft [40 m]. The suction and delivery pipes are both 5″ [0·125 m] dia. If the manometric efficiency is 76% and the overall efficiency is 72%, calculate the discharge in cusec [m³/s] and the hp [kW] to be supplied.

Answer: 1·56, 42·25 hp [0·0414 m³/s, 29·8 kW].

(16) Explain what is meant by the manometric head of a centrifugal pump. Prove that in general for such a pump running at speed N and giving a discharge Q the manometric head can be expressed in the form: $H = AN^2 + BNQ + CQ^2$, where A, B and C are constants. A centrifugal pump impeller is 10″ [0·25 m] external dia. and the water passage is 1·25″ [0·03 m] wide at exit. The circumference is reduced by 12% on account of the thickness of the vanes. The impeller vanes are inclined at 140° to the forward tangent at exit. If the manometric efficiency is 83% when the pump runs at 1000 revs/min and delivers 630 galls/min [0·05 m³/s], calculate the fraction of the kinetic energy of discharge from the impeller which is subsequently recovered in the casing, assuming no loss of head in the impeller. If none of the exit kinetic energy were so recovered what would be the manometric efficiency?

Answers: 59·5%, 58% [59%, 58·8%].

Further Reading

The following list contains books that present the whole subject of fluid mechanics in a way that I feel is particularly useful. They should be studied to extend and amplify the material that has been presented in this book.

General

Engineering Fluid Mechanics by C. Jaeger. Blackie, London (1961).
Mechanics of Fluids by B. S. Massey. Van Nostrand, London (2nd ed. 1971).
Fluid Mechanics by R. F. Pao. Wiley, New York (1961).
A Textbook of Fluid Mechanics by J. R. D. Francis. Arnold, London (3rd ed. 1969).
Fluid Mechanics by L. Prandtl.
Fluid Mechanics with Engineering Applications by R. L. Daugherty and J. B. Franzini. McGraw-Hill, New York (1965) (6th ed. of Daugherty and Ingersoll, A. C.; rev. by Daugherty and Franzini.
Elementary Fluid Mechanics by J. K. Vennard. Wiley, New York (4th ed. 1961).
Fluid Mechanics for Hydraulic Engineers by Hunter Rouse. Dover, New York (1961).

Waterhammer Analysis

Hydraulic Transients by G. R. Rich. Dover, New York (2nd. ed. 1963).
Waterhammer Analysis by J. Parmakian. Dover, New York (1963).
Engineering Fluid Mechanics by C. Jaeger. Blackie, London (1961).
Analysis of Surge by J. Pickford. Macmillan, London (1969).

Hydrodynamics

Theoretical Hydrodynamics by L. M. Milne Thomson. Macmillan, London (5th ed. 1968).
Hydrodynamics by Horace Lamb. C.U.P., Cambridge (6th ed. 1962).

379

Viscous Flow Theory, vols I and II by Shih I. Pai. Van Nostrand, Princeton (1956).
Applied Hydrodynamics by H. R. Vallentine. Butterworths, London (1959).

Boundary Layer Theory

Mechanics of Fluids by B. S. Massey. Van Nostrand, London (2nd ed. 1971).
Fluid Mechanics by R. F. Pao. Wiley, New York (1961).
Essentials of Fluid Mechanics by L. Prandtl. Blackie, London (1952) (translation
of *Führer durch die Strömungslehre* vieweg und Sohn. Brunswick (1949)).

Open Channel Hydraulics

Open Channel Hydraulics by Ven Te Chow. McGraw-Hill, New York (1959).
Open Channel Flow by F. M. Henderson. Collier-Macmillan, London (1966).

Unsteady Flow Theory

Mathematical Methods for Digital Computers by Ralston and Wilf. Wiley, New
York.
Engineering Fluid Mechanics by C. Jaeger. Blackie, London (1961).
Analysis of Surge by J. Pickford. Macmillan, London (1969).
Open Channel Flow by F. M. Henderson, Collier-Macmillan, London (1966).

Index